大飞机出版工程

总主编 顾诵芬

民机驾驶舱人机工效
综合仿真理论与方法研究

Theory and Methodology of Integranted Simulation of
Ergonomics for Civil Flight Deck

陈迎春 主 编

上海交通大学出版社
SHANGHAI JIAO TONG UNIVERSITY PRESS

内容提要

本书汇集了国内在航空人机工效领域走在前沿的北京航空航天大学、西北工业大学、上海交通大学、复旦大学、中国民航大学等院校,以及中国商用飞机有限公司负责承担的国家重点基础研究计划(973 计划)项目的研究成果。

论文集收集并筛选了各个课题在前两年研究过程中在国内外期刊、会议上发表的部分有代表性的研究论文,涵盖了理论与方法、建模仿真、实验研究 3 个部分:理论与方法部分以人机工效基础理论、分析和评价方法研究为主;建模仿真部分以人机工效研究对象的建模和仿真方法研究为主;实验研究部分主要包括前期研究过程中的基础实验研究,研究结果可为仿真建模提供依据。

图书在版编目(CIP)数据

民机驾驶舱人机工效综合仿真理论与方法研究/陈迎春
主编. 一上海:上海交通大学出版社,2013
(大飞机出版工程)
ISBN 978 - 7 - 313 - 09717 - 0

Ⅰ.①民… Ⅱ.①陈… Ⅲ.①民用飞机-座舱-
工效学-仿真-研究 Ⅳ.①V223

中国版本图书馆 CIP 数据核字(2013)第 099930 号

民机驾驶舱人机工效综合仿真理论与方法研究

主　　编:陈迎春
出版发行:上海交通大学出版社　　　　　　　地　　址:上海市番禺路 951 号
邮政编码:200030　　　　　　　　　　　　　电　　话:021 - 64071208
出 版 人:韩建民
印　　制:浙江云广印业有限公司　　　　　　经　　销:全国新华书店
开　　本:787mm×1092mm　1/16　　　　　印　　张:25
字　　数:490 千字
版　　次:2013 年 12 月第 1 版　　　　　　　印　　次:2013 年 12 月第 1 次印刷
书　　号:ISBN 978 - 7 - 313 - 09717 - 0/V
定　　价:98.00 元

大飞机出版工程

丛书编委会

总主编

顾诵芬（中国航空工业集团公司科技委副主任、两院院士）

副总主编

金壮龙（中国商用飞机有限责任公司董事长）

马德秀（上海交通大学党委书记、教授）

编　委(按姓氏笔画排序)

王礼恒（中国航天科技集团公司科技委主任、院士）

王宗光（上海交通大学原党委书记、教授）

刘　洪（上海交通大学航空航天学院教授）

许金泉（上海交通大学船舶海洋与建筑工程学院工程力学系主任、教授）

杨育中（中国航空工业集团公司原副总经理、研究员）

吴光辉（中国商用飞机有限责任公司副总经理、总设计师、研究员）

汪　海（上海交通大学航空航天学院副院长、研究员）

沈元康（中国民用航空局原副局长、研究员）

陈　刚（上海交通大学副校长、教授）

陈迎春（中国商用飞机有限责任公司常务副总设计师、研究员）

林忠钦（上海交通大学常务副校长、院士）

金兴明（上海市经济与信息化委副主任、研究员）

金德琨（中国航空工业集团公司科技委委员、研究员）

崔德刚（中国航空工业集团公司科技委委员、研究员）

敬忠良（上海交通大学航空航天学院常务副院长、教授）

傅　山（上海交通大学航空航天学院研究员）

总　序

国务院在 2007 年 2 月底批准了大型飞机研制重大科技专项正式立项,得到全国上下各方面的关注。"大型飞机"工程项目作为创新型国家的标志工程重新燃起我们国家和人民共同承载着"航空报国梦"的巨大热情。对于所有从事航空事业的工作者,这是历史赋予的使命和挑战。

1903 年 12 月 17 日,美国莱特兄弟制作的世界第一架有动力、可操纵、重于空气的载人飞行器试飞成功,标志着人类飞行的梦想变成了现实。飞机作为 20 世纪最重大的科技成果之一,是人类科技创新能力与工业化生产形式相结合的产物,也是现代科学技术的集大成者。军事和民生对飞机的需求促进了飞机迅速而不间断的发展,应用和体现了当代科学技术的最新成果;而航空领域的持续探索和不断创新,为诸多学科的发展和相关技术的突破提供了强劲动力。航空工业已经成为知识密集、技术密集、高附加值、低消耗的产业。

从大型飞机工程项目开始论证到确定为《国家中长期科学和技术发展规划纲要》的十六个重大专项之一,直至立项通过,不仅使全国上下重视起我国自主航空事业,而且使我们的人民、政府理解了我国航空事业半个世纪发展的艰辛和成绩。大型飞机重大专项正式立项和启动使我们的民用航空进入新纪元。经过 50 多年的风雨历程,当今中国的航空工业已经步入了科学、理性的发展轨道。大型客机项目其产业链长、辐射面宽、对国家综合实力带动性强,在国民经济发展和科学技术进步中发挥着重要作用,我国的航空工业迎来了新的发展机遇。

大型飞机的研制承载着中国几代航空人的梦想,在 2016 年造出与波音 B737 和空客 A320 改进型一样先进的"国产大飞机"已经成为每个航空人心中奋斗的目标。然而,大型飞机覆盖了机械、电子、材料、冶金、仪器仪表、化工等几乎所有工业门类,集成了数学、空气动力学、材料学、人机工程学、自动控制学等多种学科,是一个复杂的科技创新系统。为了迎接新形势下理论、技术和工程等方面的严峻挑战,迫切需要引入、借鉴国外的优秀出版物和数据资料,总结、巩固我们的经验和成果,编著一套以"大飞机"为主题的丛书,借以推动服务"大型飞机"作为推动服务整个航空科学的切入点,同时对于促进我国航空事业的发展和加快航空紧缺人才的培养,具有十分重要的现实意义和深远的历史意义。

2008 年 5 月,中国商用飞机有限公司成立之初,上海交通大学出版社就开始酝酿"大飞机出版工程",这是一项非常适合"大飞机"研制工作时宜的事业。新中国第一位飞机设计宗师——徐舜寿同志在领导我们研制中国第一架喷气式歼击教练机——歼教 1 时,亲自撰写了《飞机性能捷算法》,及时编译了第一部《英汉航空工程名词字典》,翻译出版了《飞机构造学》、《飞机强度学》,从理论上保证了我们飞机研制工作。我本人作为航空事业发展 50 年的见证人,欣然接受了上海交通大学出版社的邀请担任该丛书的主编,希望为我国的"大型飞机"研制发展出一份力。出版社同时也邀请了王礼恒院士、金德琨研究员、吴光辉总设计师、陈迎春副总设计师等航空领域专家撰写专著、精选书目,承担翻译、审校等工作,以确保这套"大飞机"丛书具有高品质和重大的社会价值,为我国的大飞机研制以及学科发展提供参考和智力支持。

编著这套丛书,一是总结整理 50 多年来航空科学技术的重要成果及宝贵经验;二是优化航空专业技术教材体系,为飞机设计技术人员培养提供一套系统、全面的教科书,满足人才培养对教材的迫切需求;三是为大飞机研制提供有力的技术保障;四是将许多专家、教授、学者广博的学识见解和丰富的实践经验总结继承下来,旨在从系统性、完整性和实用性角度出发,把丰富的实践经验进一步理论化、科学化,形成具有我国特色的"大飞机"理论与实践相结合的知识体系。

"大飞机"丛书主要涵盖了总体气动、航空发动机、结构强度、航电、制造等专业方向,知识领域覆盖我国国产大飞机的关键技术。图书类别分为译著、专著、教材、工具书等几个模块；其内容既包括领域内专家们最先进的理论方法和技术成果，也包括来自飞机设计第一线的理论和实践成果。如：2009 年出版的荷兰原福克飞机公司总师撰写的 *Aerodynamic Design of Transport Aircraft*（《运输类飞机的空气动力设计》），由美国堪萨斯大学 2008 年出版的 *Aircraft Propulsion*（《飞机推进》）等国外最新科技的结晶；国内《民用飞机总体设计》等总体阐述之作和《涡量动力学》、《民用飞机气动设计》等专业细分的著作；也有《民机设计 1000 问》、《英汉航空双向词典》等工具类图书。

　　该套图书得到国家出版基金资助，体现了国家对"大型飞机项目"以及"大飞机出版工程"这套丛书的高度重视。这套丛书承担着记载与弘扬科技成就、积累和传播科技知识的使命，凝结了国内外航空领域专业人士的智慧和成果，具有较强的系统性、完整性、实用性和技术前瞻性，既可作为实际工作指导用书，亦可作为相关专业人员的学习参考用书。期望这套丛书能够有益于航空领域里人才的培养，有益于航空工业的发展，有益于大飞机的成功研制。同时，希望能为大飞机工程吸引更多的读者来关心航空、支持航空和热爱航空，并投身于中国航空事业做出一点贡献。

2009 年 12 月 15 日

序

  大型飞机是《国家中长期科学和技术发展规划纲要(2006—2020 年)》中规划的重大专项,是建设创新性国家和新时期改革开放的标志性工程。其中中国商用飞机有限责任公司承担了大型客机项目并统筹干线客机和支线客机的发展。

  飞机驾驶舱是飞行员与飞机交互的唯一界面,在飞行中起着至关重要的作用。因此在飞机研制过程中,驾驶舱设计占有举足轻重的地位,其中,对驾驶舱与飞行员之间交互行为规律的人机工效学研究受到了国内外航空界的普遍重视。科技部对中国商用飞机有限责任公司承担的大型客机研发寄予厚望,于2009 年在 973 计划中立了"民机驾驶舱人机工效综合仿真理论与方法研究"项目。目的是通过开展基础研究,进行相关积累,以提高我国民机驾驶舱人机界面设计的原始创新能力。

  "民机驾驶舱人机工效综合仿真理论与方法研究"项目是 973 计划中首个关于人机工效研究的重大基础项目。项目以中国商用飞机有限责任公司为依托单位,由上海飞机设计研究院 C919 大型客机常务副总设计师陈迎春担任首席科学家,集中了上海飞机设计研究院、北京航空航天大学、西北工业大学、上海交通大学、复旦大学以及中国民航大学等院校的研究队伍开展研究工作。研究的主要内容是通过综合人-机-环全方位信息,在仿真的环境和任务过程中进行飞行员建模、飞行驾驶舱环境建模、界面仿真、驾驶舱人机工效评估等研究,明确飞行员与驾驶舱环境之间的相互作用机理,达到服务民用航空领域,指导型号设计的目的。项目组在首席科学家带领下,聘请民航总局、航空公司、航空院校的专家成立了项目专家组,并多次向科技部 973 项目责任专家进行汇报,保证了项目研究方向的正确性。项目组主要成员在项目专家的帮助下赴南方航空公司、东方航空公司等相关单位进行了大量的需求调研,并与飞行员进行座谈交流,明确了项目研究的目标和要求。项目参研单位在收集国内外相关研究进展的基础上把民机驾驶舱人-机-环境综合作用机理作为项目研究目标,并以大型客机为研究对象进行了大量实验、验证,部分研究成果在型号研究中已得到应用,实现了项目立项的宗旨。项目执行过程中组织管理有效,研究团队精诚合作。通过项目研究还培养了一批优秀的青年科技人员,并已初步形成了一支具有国际水平的研究队伍。

本论文集收集了该项目组在前两年研究中发表在国内外期刊、会议上的部分代表性研究论文。从论文发表内容的分布情况来看,前两年的研究工作主要集中于理论与方法的研究以及仿真平台的构建,并开展了部分实验研究。这与项目组前期研究计划相吻合。从发表的论文来看,具有较高的学术水平,体现了973计划的理论高度。这表明在前期有足够的积累后,项目组有望在后续研究中集中开展仿真及实验验证工作,达到解决人机工效综合评估方法体系的重大基础问题,为民机安全、高效、舒适的驾驶舱设计提供有力支持的目的,并在此基础上形成具有我国创新特点的驾驶舱人机工效综合仿真理论与方法。

民机驾驶舱人机工效以飞行员-驾驶舱-环境作为研究的基本对象,是一个交叉性和工程性背景很强的研究领域,其中需要开展研究的内容很多。项目研究已经取得的成果为项目的后续研究和民机驾驶舱人机工效的未来发展奠定了良好基础。论文集的出版,将促进人机工效学学科理论方法的交流,创新性研究成果的推广,对该学科的理论发展和设计实践均有重要的推动作用和现实意义。

王浚

2012 年 10 月 22 日

目　录

第一篇　理论与方法

第二篇　建模仿真

第三篇　实验研究

第一篇　理论与方法

基于感知控制论的系统分析与设计方法

栾义春,薛红军,宋笔锋

西北工业大学航空学院,西安 710072

摘要 传统的系统分析与设计方法把操作者看做输入、输出机器,采用自上而下的方式进行,分析的目的是描述操作者输入、输出和两者的关联规则。基于感知控制论的系统分析方法把操作者看过一个感知控制系统,采用自上而下和自下而上的方式进行,分析的目的是确定操作者控制的变量,实现控制所采取的手段,把系统功能分析、功能设计、功能分配、人机界面设计以及人员培训等集成到一起。最后的实例分析论证了这种方法使各过程的关联更加紧密,逻辑性更强。

关键词 感知控制论,系统分析,层级目标分析,人机界面

A Method of System Analysis and Design Based on Perceptual Control Theory

Luan Yichun, Xue Hongjun, Song Bifeng

College of Aeronautics, Northwestern Polytechnical University, Xi'an 710072

Abstract Conventional system analysis and design views the operator as an input-output device. The goal is to describe operator inputs, outputs and the rules that relate them. Decomposition of function and task traces top-down. A method of system analysis and design based on Perceptual Control Theory views the operator as a perceptual control system, the goal of the analysis is to determine the variables that the operator is to keep under control and the means the operator must have to affect this control. Decomposition of function and task traces top-down and bottom-top. This new method integrates the analysis, allocation, design of interface and training process. The instance proved that the combination of those processes was better and more logical.

Keywords Perceptual Control Theory, system analysis, hierarchical goal analysis, man-machine interface

1 引言

随着计算机等智能技术的发展,系统的复杂化和任务难度的增加,单靠某个人完成一项复杂的任务是不现实的,更多的是人与人、人与机器的协作。因此,关于人与人间的协同、人与机的匹配的研究被重新重视,并具备了新的特点。找寻人与人、人与机共同的工作机制,以此作为系统分析和设计的基础,才能使设计出的协作方案和系统工作绩效更高。

感知控制论(Perceptual Control Theory, PCT)汲取了信息理论、控制理论和系统论的精华,把人的有目的的认知简化为具备输入、输出以及处理的信息加工过程[1]。感知作为输入,操作行为作为输出,知觉、注意、记忆、思维与决策等作为加工过程。外界信息作用于人的感觉器官,通过肢体等行为对外界环境做出反应,同时接收到外界的反馈信息,这一切以目的为导向。感知控制论也可作为机器的工作机制,两者都包括输入、处理与输出,人与人之间、人与机之间的协作通过对共同所处的外界环境的作用来实现。因此,基于感知控制论的系统分析和设计是一种新型的方法,它分析得更全面,设计出的协作方案和系统工作绩效更高。

2 系统分析与设计过程

系统分析与设计过程一般包括建立目标、功能分析、功能分配、界面设计、系统综合与评价等[2, 3]。系统功能分析关系到系统功能设计与分配,影响系统详细设计、系统软件和操作界面的开发。传统的系统分析方法把操作者看作输入、输出机器,采用自上而下的方式进行,分析的目的是描述操作者输入、输出和两者的关联规则。功能分析、功能分配与界面设计是分开进行的,往往是一项任务完成后,才能进入下一项任务,出现问题时,重新展开分析。

传统的系统分析方法围绕目标进行,即在确定系统目标后,寻求实现该目标的手段,此时可能有多个可行方案。可行方案越多,选择的余地越大,从而需要在一定的限制条件下选出较好的方案,往往采用树状具有关联的功能设计展开法,其核心是从"为什么"到"怎么做"。每一等级中上位是目标,下位是功能。各等级的目标和相应的功能之间,从目标找功能时是"怎么办",从功能找目标则是"为什么"。

3 感知控制论

感知控制论的核心是所有行为都是感知的控制。人通过采取行为来减少目前状态感知和目标期望(参考)的差异,如果目前状态的感知没有趋向于期望的目标则会改变行动,这个循环称为基本控制单元(Elementary Control Unit, ECU)。如

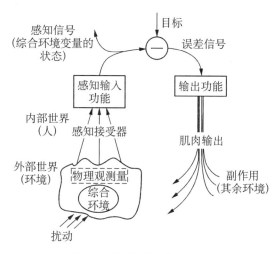

图 1 感知控制论核心理念

图 1 所示,"感知输入功能"的形式决定或定义了综合环境变量,是外界环境的物理观测量的函数,也是系统的内部变量。感知信号与目标进行比较,差异被认定为误差信号。输出功能将这些误差信号产生分类输出,对环境产生不同的物理作用,在控制有效的情况下,将感知信号的值趋向目标值进行了改变,减少误差。基于目标的层级分析,PCT 成为多层 ECU 构成的层级结构。其组织方式是:n 水平的 ECU 的输出信号影响几个 $n-1$ 水平的 ECU 的目标输入,而 $n-1$ 水平的 ECU 加工的感知信号则服务于 n 水平的 ECU 的感知输入功能的输入,最终最低水平的输出引发肌肉行为。

随着感知控制论的完善和发展,结合认知心理学中记忆和学习机制,Powers, Hendy 提出了具备记忆和学习机制的感知控制论[4, 5]。在进行感知信息获取时,有两种信息:一是与目标期望进行比较的信号;二是改变内部记忆的信息。进行记忆和学习时,没有行为输出,不会对外界环境造成改变,但改变了人对外界的认知和记忆,如图 2 所示。

人通过一些变量参数对外界环境施加他们的影响,如同人通过目标来确定不同控制循环的参考点,机器也有系统设计者事先设定的不同的参考值或程序设定的目标。因此感知控制模型能够同时应用于人和机器,被看做是人和具有智能的机器相通的模型,人与人、人

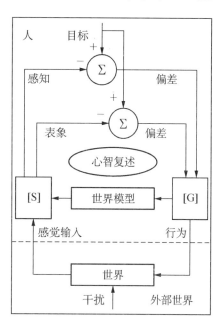

图 2 具备记忆和学习机制的
感知控制论模型

与机的协作是这些智能体对他们共同所处的外界环境的感知、理解和作用。图3中人甲与人乙都对他们共同所处的环境做出一些行为,由于行为的多样性他们的控制循环可能存在不耦合。Powers指出甲与乙的耦合程度依赖于甲与乙试图对同一环境变量做出的行为,以及这些变量经过线性转化形成感知的深度[6]。耦合程度很低的时稳定性特别重要,目标、感知输入和行为输出的转换必须达到很好的契合。因此为了实现智能体协作的总体稳定性,控制循环必须分层级,并基于共同的目标和相同的智能模型,为每个智能体定义角色和功能分配,是多智能体的协作最重要的。

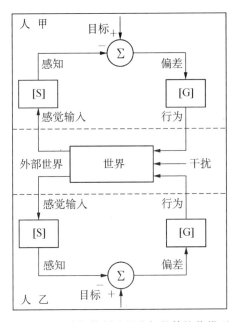

图3　基于感知控制论的多智能体协作模型

4　基于感知控制论的系统分析与设计方法

4.1　基于感知控制论的系统分析与设计方法

　　Hendy等人建议按照目标层级、网络、次序或其他连接方式连接为逻辑网络来分析系统功能,然后进行建模或预测行为[5]。基于感知控制论的系统分析方法把操作者看过一个感知控制系统展开分析,操作者可能是人,也可能是机器,目的是确定操作者控制的变量,以及实现控制所采取的手段。基于感知控制论的系统分析方法把系统分析、功能设计、功能分配、人机界面设计以及人员培训等集成到一起,关联紧密,逻辑性强。基于感知控制论的系统分析方法实现包括三个步骤:

　　(1)自上而下的目标层级分析,采用目标-手段分析法,从最抽象到与现实世界关联的变量参数。设计每个目标中人或机的功能和任务,定义每个目标对应的操作变量参数。

（2）自下而上的任务分析,尽量充分考虑系统运行期间存在的所有情况,针对每种情况,设计完备的操作序列,与第一步中的目标对应。

（3）设计操作序列表,以图示的方式描述系统对每种态势的处理和衔接,智能体之间的通信和联系。

基于感知控制论的系统分析与设计方法具有三个特点:

（1）功能分析、设计与分配紧密结合。基于感知控制论的系统分析方法以目标为主线,目标采用手段-目的分析法自上而下展开层级分析,从最抽象到最具体与现实世界相关联的变量参数的目标层级结构,表征了系统生命周期中被激活的控制循环层级,任何不被循环支持的目标对外界环境没有影响力,也没有行为输出。同样,所有受影响的系统变量都与某个目标有关联。所有的控制循环包含一个或多个影响变量,一个目标必须至少对应一个变量。如果任一水平的反馈中断的话,就没有闭环控制,如果循环不可控的话,也就没有偏差引导。因此,所有的目标必须被分配给人或机,即对人与机进行角色分配和控制权限的划分,这也是基于感知控制论的系统分析方法的独到之处。

（2）最低层目标对应的变量是与现实世界相关联的可直接操作的变量,是人或机通过操作行为控制和改变的变量,亦即交互界面上的操作变量。人与机之间的功能分配不是 0 或 1 的关系,存在多种交互水平,如机为人提供的各种辅助方式,召唤、建议、否决等[7, 8]。在进行人机界面设计时,一是获取操作变量,二是设计这些变量改变的方式。通过基于感知控制论的系统分析与设计方法,直接得到人机界面操作变量,从而为人机界面设计提供良好的基础。

（3）在感知控制论中,行为一般是控制感知的"副产品",而不是有意识的以某种特殊方式行动的结果。基于感知控制论系统功能/任务分析的结果不是描述一组操作者为了实现任务目标必须采取的特定行为,而是操作者在所有可能的外部条件影响下采取的有效行为的范围。依据所有的行为都是感知控制行为这一理论,推出固定不变的结果通过变化的行为得到,固定的行为却得到变化的结果。这种新型的功能分析方法与传统任务分析不同之处在于任务目标的达成,不再是依靠一组固定的行为,而是存在多种方式实现同一任务目标。

4.2　实例分析

对无人机系统任务规划与控制站展开分析。无人机系统参与作战时,通常包括无人机,具有辅助决策性能的任务规划与控制站,操作员和指挥员,无人机又分为长机和僚机等。这些具有智能和自主能力的作战单元之间的协作基于感知控制论的系统分析法实现。

（1）从分析专家讨论议定的作战场景和任务脚本开始,综合各作战单元性能等建立功能库,然后从这些任务或功能中剖析出第一、第二层目标,甚至第三、第四层目标,直到与现实世界相关联的最低一层目标,而与一般任务体系结构相关的功能库在对特定任务场景的剖析和分解过程中重新加以评定,如表1所示,包括每项功能操作的参数或变量。

表 1　自上而下层级功能/任务分析

一级目标	二级目标	三级目标	参数
3 战术任务决策和管理	3.1 任务规划	3.1.1 装载任务 ……	任务信息 ……
	3.2 任务决策与管理	3.2.1 任务执行监控 ……	态势信息 ……
	3.3 搜索和探测区域	3.3.1 规划搜索区域 ……	经纬度/相对方位 ……
	3.4 目标的搜索和定位	3.4.1 机载传感器参数调整 ……	传感器参数 ……

表1所示为自上而下层级功能/任务分析,表2所示为自下而上功能/任务分析。

表 2　自下而上功能/任务分析

序号	任务描述	执行等级	对应自上而下任务
14	空空数据链测试	4	1.2.1
15	测试完好消息	4	5.2.13
16	无人机状态确认	0	5.1.1.5
17	下达任务开始执行指令	0	5.1.1.1
18	接收到任务开始指令	0	5.1.1.1

(2) 场景的每时每刻,都有指挥员、操作员、无人机在执行最底层的任务目标。每个时间点只有一项功能或任务发生,但同一项功能或任务却可以在场景中不同时间发生,且具有一个不同的、独一无二的任务号。除了最底层的任务,还包括任务执行等级、执行者、与自上而下的任务对应的序号。任务执行等级包括0手动,1召唤,2准许,3否决,4自动。建议与否决是介于自动与手动之间过渡的操作方式,在自动的可靠性无法完全被验证之前,采用建议与否决的方式,即减轻了操作人员的负担,又能掌握无人机的执行能力,当无人机执行能力下降,或者说任务执行的可靠度降低时,操作人员能够超控或进行调整[8]。某些任务是随机的,可能发生,也可能不发生,比如说系统报警、威胁报警等。

(3) 操作序列图基于任务活动时间,使用图标来图示化动作、检查、数据处理(比如传输,接收,存储)、时间延迟、决策等,展示系统整个信息流和操作员功能,图形化地解释任务想定进行中操作员任务的逻辑互联和信息流。如图4所示,这种视觉表征很适合层级目标分析和网络模型的构建,特别用在高度复杂系统,多个人员一起完成时间紧急的信息感知、处理、决策、执行等功能和任务。

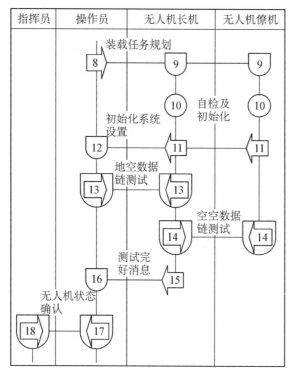

图 4　操作序列图

5　结论

　　基于感知控制论的系统分析与设计方法将功能分析、功能分配与人机界面设计集成到一起进行。与传统的系统分析设计方法相比,这种方法支持任何时间段的功能或任务的自上而下和自下而上的扩展分析;所有的目标必须被分配给人或机,即对人与机进行角色分配和控制权限的划分,充分考虑了人与人的协同、人与机的协作;通过分析,直接获取操作变量,为人机界面设计提供了强有力的支持;更易于实现多智能体交互的系统的分析和设计。

致谢

　　在薛红军老师的帮助下完成了论文的撰写工作,在此表示感谢;除了国家重点基础研究发展计划的资助,论文的工作内容还基于与某研究所的合作项目,在此也一并表示感谢。感谢国家重点基础研究发展计划(2010,CB734101)的资助。

参考文献

[1]　Powers W T. Feedback:beyond behaviorism, Science [J]. 179(4071),351 - 356.

[2] MIL-HDBK-46855A, Human Engineering Program Process and Procedures [M], Washington, DC, USA: Department of Defense.

[3] IEEE-1220. Application and Management of the Systems Engineering Process. Piscataway [M], NJ, USA: Institution of Electrical and Electronic Engineers.

[4] Powers W T. Learning and Evolution. In W. T. Powers, Ed. Living Control Systems II, Gravel Switch, KY: The Control Systems Group. 1992, 233 – 240.

[5] Hendy K C, Beevis D, Lichacz F, et al. Analyzing the cognitive system from a perceptual control theory point of view. In, Cognitive systems engineering in military aviation environments: Avoiding cogminutia fragmentosa: A report produced under the auspices of the Technical Cooperation Programme Technical Panel HUM TP – 7 Human Factors in Aircraft Environments. Human Systems IAC, Wright-Patterson AFB, OH. 2002.

[6] Powers W T. C T psychology and social organizations. In, G. Williams(Ed.) Living Control Systems II-Selected Papers of William T. Powers, Gravel Switch, KY: USA: The Control Systems Group. 1992, pp 91 – 127.

[7] Parasuraman R, Sheridan T B, Wickens C D. A model for types and levels of human interaction with automation [J]. IEEE Transactions on Systems. Man & Cybernetics, 2000, 30(3):286 – 297.

[8] 栾义春,薛红军,宋笔锋,等. 无人作战飞机系统的辅助决策等级实验研究[J]. 计算机工程与应用,2009,45,5:222 – 223.

[9] Luan Y C, Xue H J, Song B F, et al. Experiment on levels of decision aiding in UCAV system [J]. Computer Engineering and Applications, 2009,45(5):222 – 224.

飞行员失误机理及失误修复研究

薛红军[1],张晓燕[1],陈迎春[2],周　琳[2]

1. 西北工业大学航空学院,西安 710072
2. 中国商用飞机有限公司,上海 200232

摘要　通过对现有航空事故的分析,从飞行员自身认知和操纵能力的限制出发,提出飞行员失误的三种机理:速率极限、粗犷认知和认知锁死,并设计实验验证了这三种失误机理。对这三种失误机理的深层次原因进行分析,发现人的反应极限、认知惰性以及认知资源有限性是引起三种失误的根本原因。最后从驾驶舱设计和飞行员训练两个方面提出修复三种失误的建议,在驾驶舱界面设计时,信息的显示时间不能低于 500 ms,以便给予飞行员足够的时间进行信息认知和反应;在飞行员必须同时执行多项任务时,各项信息的出现应设计合理的梯度化,以便飞行员合理的分配注意资源;在飞行员训练方面建议加强飞行员工作责任心的培养和章程及规章的学习,避免粗犷认知引起的失误。

关键词　飞行员失误,失误修复,速率极限,粗犷认知,认知锁死,驾驶舱设计,飞行员训练

Pilot Error and Error Recovery

Xue Hongjun[1] , Zhang Xiaoyan[1] , Chen Yingchun[2] , Zhou Lin[2]

1 School of Aeronautics，Northwestern Polytechnical University，Xi'an 710072

2 Commercial Aircraft Corporation of China，Ltd Shanghai 200232

Abstract　Three pilot error mechanisms have been presented from the analysis of aviation accidents and the limit of pilot performance. The three mechanisms are speed limited，learned carelessness and cognitive locked which is all validated through experiments. The essential causes for the three errors are the capability limit, human cognitive inertia and the limited cognitive resource. The recovery suggestions have already been presented from the cockpit design and pilot training. For the cockpit HCI design, the persistent time for the information should be

500 ms at least to give pilot enough time to react; if the pilot has to execute multitasks at the same time, the information should be designed in ladder to avoid unreasonable attention resource allocation; for the pilot training, the responsibility of the job should be enhanced and the study of rules and procedures should be stricter to avoid carelessness learned.

Keywords　pilot error, error recovery, speed limited, learned carelessness, cognitive locked, cockpit design, pilot training

1　前言

近年来的航空事故统计结果如图 1 所示,由图可以看出机组原因占据了整个飞行事故原因的将近 70%。航空事故调查结论认为认知和环境因素是引起事故的主要原因,其中环境因素主要是粗糙的界面设计[1],难以理解的自动化操作[2]和恶劣的天气条件[3];认知因素主要和飞行员相关,情景意识较差[4],不遵守操作程序[5],以及机组之间的协调沟通[6]等三方面的因素。国外学者建立了一些飞行员的认知模型[7],如 ACT‑R,AIR‑MIDAS,A‑SA 等来模拟飞行员的决策失误,模型能够对飞行员在进近着陆阶段的滑行道选择错误进行仿真,得到了理想的结果。尽管这些团队建立了飞行员的认知能力模型,但是仿真团队致力于提高对人的仿真能力[8],并没有将失误的机理与实际飞行事故很好地联系起来,对航空安全的指导意义并不明显。本文通过对目前飞行事故的调查研究,提出导致飞行员失误的三种机理,即速率极限、粗犷认知和认知锁死,通过实验对这三种机理进行验证,并从飞机驾驶舱界面设计、操作程序设计以及飞行员训练等方面提出了对三种失误的修复建议。

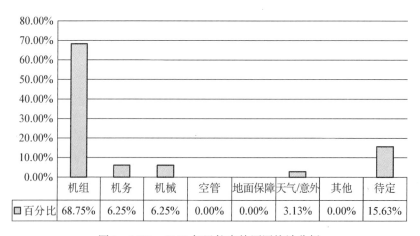

图 1　1999～2008 年民航事故原因统计分析

2　飞行员失误机理

飞行员失误是指飞行员的一种作为或者不作为,导致偏离机组的意愿或者偏离情景的要求,如法规、规章和标准操作程序[9]。影响飞行员失误的主要因素有情景意识、疲劳、工作负荷、界面设计水平、组织文化以及外界环境等。现从飞行员自身能力的局限性出发探讨飞行员失误的三种机理,即速率极限、粗犷认知以及认知锁死。

1)速率极限

速率极限认为人的能力有固有的局限性,在一定的速度限制下不可能做出正确的反应。尤其是在恶劣和应急的操纵条件下,飞机处于特殊的状态下,很难留给飞行员足够的时间进行认知和反应,从而引发失误。

2)粗犷认知

飞行员在执行一项任务时,如果第一次没有按照规程做而没有引起事故或事故征候,那么后续就会有简化操作程序的倾向,最终导致失误。

3)认知锁死

认知锁死认为飞行员的认知资源是有限的[10],在同时执行多项任务时,即使给予足够的时间,飞行员依然会注意资源分配不当,造成失误。

下面通过实验对引起飞行员失误的三种机理进行验证,探讨了飞行员进行操作时的速率极限、飞行员简化操作程序的倾向性以及飞行员认知资源的限制等。

3　失误机理验证实验及分析

3.1　速率极限实验

1)被试

在校学生 12 人,其中女生 3 人,右手为优势手。认知水平限定为在保证正确的前提下越快越好,且实验时被试精神状态正常,无疲劳等异常情况。

2)实验设计

本实验为单变量实验,变量为刺激的出现时间,依次为 800,500,400,350,300,275,250,200 ms,每个变量水平要求被试实验 20 次。实验时,屏幕上会出现红色和绿色的告警灯,表明不同的告警等级,被试需根据出现告警灯的不同颜色点击不同的按键,若反应错误或没有反应,计算机会给出警告音,记录被试的反应时和正确率。要求被试在对实验达到熟练的程度之后再进行正式实验。

3)实验结果与分析

图 2 所示为速率极限实验结果。

观察实验结果发现,被试的失误率随着呈现时间的缩短呈近似线性的升高,告警信息的呈现时间为 200 ms 时,被试的失误率达到 90%,在 216 次失误中,无反应失误占 195 次,占总失误次数的 90.3%,在出现时间增加到 800 ms 时,被试的失误

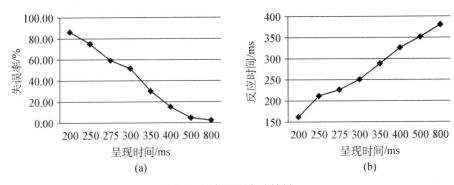

图 2　速率极限实验结果

(a) 时间与失误率的关系；(b) 时间与反应时的关系

率降低为 2.9%，其中无反应失误只有 1 次。被试对信息进行判断，然后做出反应是一个完整的认知操纵过程，当出现时间不能满足过程时间时，人会出现无反应错误，或者下意识的反应错误。

表 1 所示为不同失误类型失误次数与呈现时间的关系。图 3 所示为不同失误类型与呈现时间的关系。

表 1　不同失误类型失误次数与呈现时间的关系

失误类型＼呈现时间/ms	200	250	275	300	350	400	500	800
n_{non}/n	195/216	148/180	107/143	94/126	61/73	27/38	6/12	1/7
n_{err}/n	21/216	32/180	36/143	32/126	12/73	11/38	6/12	6/7

其中 n_{non} 为无反应失误次数；n_{err} 为有反应，但是反应错误的次数；n 为总的失误次数。

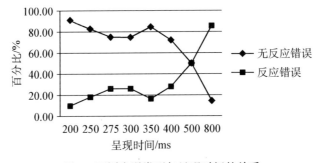

图 3　不同失误类型与呈现时间的关系

另外，实验结果表明被试的反应时随着呈现时间的缩短呈近似线性的减小，在刺激的呈现时间为 200 ms 时，人的正确反应时间只有 163.5 ms，而人类学家的研究表明，人的反应速度极限为 0.2 s。说明在时间压力下，会激发人的潜能，使人能够超越能力限制进行反应。但是这种超越能力限制的状态不会持续很长时间，当人发现反应的频率跟不上刺激呈现的频率时,会有挫败感,倾向于放弃反应。

速率极限实验是在要求被试对操作程序非常熟悉的情况下进行的。结果表明,人存在着能力局限性,不管是认知能力或操纵能力,在留给被试的反应时间小于400 ms时,失误率将达到30%以上,而在反应时间高于800 ms时,被试的失误率逐渐趋于0。

3.2 粗犷认知实验

1) 被试

同上。

2) 实验设计

在飞行员进行告警信息判断的同时,要求被试检查告警信息旁边的飞行高度表,高度表处于异常范围时按压不同的按键。在实验设计时,先将高度表设置为正常的高度范围,之间穿插着异常的高度表范围,记录被试的反应时和正确率,整个实验进行 20 次。要求被试在对实验达到熟练的程度之后再进行正式实验。

3) 实验结果与分析

在粗犷认知的实验设计时,主试有意在实验前 5 次将飞行高度表设置在正常范围内,观察被试有没有忽略检查高度表的倾向。实验结果表明(见表2),实验开始时被试的反应时较长,说明他们在按照程序要求做高度表检查,但是在发现连续几次高度表都正常之后,被试会忽略对高度表的检查,简化操作流程,反应时缩短,错误率上升。在 240 次实验中,失误的情况几乎都出现在实验设置的拐点处,即由高度表指示正常转向异常的高度表指示时。说明人产生了认知惯性,并且发现不坚持高度表检查并没有引起预先告知的事故或事故征候,就会倾向于忽略对高度表的检查,这就验证了粗犷认知的存在,这也是飞行员对操作程序的有意简化引发失误的机理。

表 2   粗犷认知实验结果

反应时间/ms	失误率/%
744	6.25

3.3 认知锁死实验

1) 被试

同上。

2) 实验设计

在飞行员进行告警信息判断的同时,要求被试注意屏幕右下角的是否出现其他提示信息,若出现其他提示信息按压相应的按键,记录被试的反应时和正确率,整个实验进行 20 次。要求被试在对实验达到熟练的程度之后再进行正式实验。

3）实验结果与分析

在认知锁死实验设计时，有意地让被试同时执行两项任务，判断不同的告警信息级别，并随时注意观察屏幕右下角是否出现异常，做出相应的反应。被试的反应时要比粗犷认知的反应时短，但是错误率确明显升高。在对被试的失误分析发现，被试的失误出现在对屏幕右下角异常的反应上；另外，在对被试的访谈中得知，他们倾向于将注意力集中在对告警信息等级的判断上，导致忽略了对屏幕右下角异常的判断。说明人的认知资源是有限的，在同时执行多项任务时，往往会造成注意力分配不当，忽略重要信息而导致失误，并且这种失误的几率比粗犷认知的几率要高，且不容易修复。表 3 所示为认知锁死实验结果。

表 3　认知锁死实验结果

反应时间/ms	正确率/%
496.0	12.9

4　结论

三个实验的结果验证了飞行员的三种失误机理，即速率极限、粗犷认知和认知锁死。文中对三种失误机理的原因进行了详细的分析。飞行员出现失误就需要有相应修复失误的措施，才能达到提高航空安全性的目的。本文从驾驶舱界面设计和飞行员训练两方面对三种失误的修复提出了一些建议：

（1）速率极限表明人的能力限制是人产生失误的根源之一，因此在进行驾驶舱界面设计时，要考虑到人的能力限制，信息的呈现时间不能低于 500 ms；在出现故障时尽量采用不只出现一种故障警告措施，防止飞行员忽略警告信息。

（2）粗犷认知表明人有思维惯性和操作惰性，因此在进行飞行员训练时，要着重培养飞行员的工作责任心；加强章程和操作规则的学习；在驾驶舱界面设计时，若飞行员没有按照操作程序操作，应该有合理的警告提醒。

（3）认知锁死表明人的认知资源是有限的，因此在驾驶舱界面设计时应注意使刺激呈现形式梯度化，尽量不在同一时间要求飞行员执行多项任务。若必须同时执行多项任务，应该有打断飞行员认知锁死的提醒装置。同时从飞行员训练的角度来讲，应该强化飞行员在同时执行多项任务时的能力训练，因为在飞行过程中不可避免地会出现需要飞行员同时执行多项任务的情况，尤其是在恶劣和应急条件下。

尽管本文从飞行员能力局限性的角度提出了飞行员的三种失误机理，并提出了相应的修复措施，但是影响航空安全的因素十分复杂，严重的事故或者灾难通常是环境、人、驾驶舱设计等多项因素同时作用导致的，因此后续将继续研究恶劣和应急条件下影响飞行员失误的综合因素，致力于提高航空安全。

参考文献

［1］ Degani A，Shafto M，Kirlik A. Modes in human-machine systems：Review，classification and application ［J］. The International Journal of Aviation Psychology，1999，9：125 - 138.

［2］ Olson W A，Sarter N B. Automation management strategies：Pilot preferences and operational experiences ［J］. The International Journal of Aviation Psychology，2000，10，327 - 341.

［3］ Wiegman D A，Goh J.（2001）. Pilots' decisions to continue visual flight rules（VFR）flight into adverse weather：Effects of distance traveled and flight experience（Tech. Rep. No. ARL - 01 - 11/FAA - 01 - 3）［R］. Savoy：University of Illinois，Aviation Research Laboratory.

［4］ Endsley M R，Smolensky M W. Situation awareness in air traffic control：The big picture In M. W ［M］. Smolensky & E. S. Stein（Eds.），Human factors in air traffic control. San Diego，CA：Academic，1998.

［5］ Bisantz A M，Pritchett A R. Measuring judgement in complex，dynamics environments：A lens model analysis of collision detection behavior ［J］. Human Factors，2003，45：266 - 280.

［6］ Fousee H C，Helmreich R L. Group interaction and flight crew performance. In E. L ［M］. Wiener & D C Nagel（Eds.），Human Factors in Aviation. San Diego，CA：Academic. 1988.

［7］ Foyle D C，Hooey B L，Human performance modeling in aviation ［M］. CRC Press，2007.

［8］ Michael D Byrne，Alex Kirlik. Using computational cognitive modeling to diagnose possible sources of aviation error ［J］. The International Journal of Aviation Psychology，2005，15（2），135 - 155.

［9］ Reason J，Human error ［M］. Cambridge：Cambridge University Press. 1990.

［10］ Eduardos，Dan M. Human factors in aviation ［M］（second edition）. Academic Press，2010.

白光 LED 颜色质量评价方法研究

程雯婷,孙耀杰,童立青,林燕丹

复旦大学电光源研究所,先进照明技术教育部工程研究中心,上海 200433

摘要 随着对光源颜色质量评价研究的深入及新光源 LED 的出现,现行的显色指数(Color Rendering Index, CRI)的可适用性受到了质疑,而存在取代可能的新指数 CQS (Color Quality Scale)可靠性则有待验证。因此需要基于视觉实验重新评价光源颜色质量。本文测试了包括 LED 在内的 5 种光源光谱,对显色指数 CRI 和 CQS 进行计算,并进行了颜色还原性和偏好度的视觉实验,对其颜色质量进行评价。实验发现:①显色指数 CRI 无法反映光源显色性的实际情况,更不能用以评价人们的偏好度;②在评价光源颜色质量的准确性上,CQS 与 CRI 相比未见明显优势,若要取代 CRI,仍有待进一步的完善;③光源的颜色质量是一项综合性的评价,需考虑显色性、偏好度和应用需求等多种因素;④LED 在提高某些颜色饱和度,从而增加人们偏好度方面有一定优势,在实际很多场合中可加以利用。

关键词 颜色质量评价,显色性,显色指数(CRI),LED,CQS

The Method of Color Quality Evaluation of White Light LED

Cheng Wenting, Sun Yaojie, Tong Liqing, Lin Yandan

Institute for Electric Light Sources; Engineering Research Center of Advanced Lighting

Technology, Ministry of Education, Fudan University, Shanghai 200433

Abstract The failure of the current CIE Color Rendering Index (CRI) for modernlight sources (especially LED) has been demonstrated, and the reliability of the a new index of CQS(Color Quality Scale) is yet to be verified. These call for the research on evaluation of color quality based on visual experiments. In this article, spectra of 5 sample light sources were measured, and their CRI and CQS values were calculated, and then visual experiments were conducted to test both color fidelity and color preference. The results show that: ① CRI can properly evaluate neither color fidelity nor preference of the light sources；② Comparing with CRI，CQS doesn't show big improvement on evaluating the color fidelity and color preference；③ Color quality of the light sources should be evaluated with consideration of many factors，including color fidelity，preference，and application fields，etc；④ In some practical applications，white LED can play their advantages to increase chroma saturation for some colors.

Keywords color quality evaluation，color rendering，CRI，LED

1 显色指数与光源颜色质量评价

如今人们对室内照明质量的要求越来越高,照明评价也从过去以照度为主要指标,向基于舒适性的综合评价发展。其中光源的颜色质量(color quality)是决定室内照明效果的重要指标之一。

1.1 显色指数(CRI)存在的缺陷

在过去数十年,国际照明学会(CIE)制定的显色指数(CRI)是最常用的评价颜色质量的指数。但随着新光源的出现和研究的深入,其存在的问题也日益凸显[1]。根据 CIE 对于显色性(color rendering)的定义,它是指与参考照明体下的色表相比较,一个光源对于物体的色表产生的效果[2]。用于评价的指数即显色指数 CRI (Color Rendering Index)。

美国[3, 4]、匈牙利[5—8]、法国[9—11]等研究者均发现,CRI 在用于评价光源,尤其是 LED 的显色性时,存在许多问题,如色空间不均匀、颜色样品数量少且饱和度过低等。不仅如此,用光源的显色性来评价颜色质量本身存在着问题。根据显色性的定义,其评价的是光源对于物体在参考照明体下的颜色还原性(color fidelity),所以不管物体的色表往什么方向偏离,显色指数都会降低。但实际运用中,如果光源使颜色饱和度增加,可以提高视觉清晰度和视亮度,所以单纯用显色指数来评判光源颜色质量是不全面的[12]。基于以上考虑,CIE 在 2007 年的技术报告[13]中明确提出:目前的显色指数 CRI 不能有效反映包括白光 LED 在内的白光照明光源的显色性优劣。

1.2 新的光源颜色质量评价体系的形成

为了修正 CRI 的一些问题,美国 NIST 的 Wendy Davis 和 Yoshi Ohno[14, 15]开发了一套新的评价系统 CQS(Color Quality Scale)。但该系统主要基于数学计算开发,其可靠性仍有待更多的视觉实验验证。例如 Nicola Pousset 等人[16]将其颜色偏好度实验结果与各光源的 CQS 值进行比较,结果发现 CQS 值在反映真实颜色质量上不尽如人意。

要全面地评价光源的颜色质量,需要考虑很多因素,如颜色还原性,色调的区分度(hue discrimination),颜色鲜艳度(vividness),人们对颜色的偏好度(preference),视觉清晰度(visual clarity),视觉舒适度等。到目前为止国际上未形成统一的光源颜色质量评价体系,但有一点可以肯定:决定光源颜色质量的是人的视觉感受。因此,我们需要基于大量的视觉实验来确定该评价体系。

目前的研究主要采用的方法可归纳为:保持桌面照度或视线方向亮度相同;并保证各待测光源具有接近的色温或色坐标。被试进行的视觉任务主要有:考察颜色还原性(即显色)的色差评价;考察偏好度、鲜艳度、协调度等的主观评分;考察颜色区分度的辨色实验。但目前大多研究均为视觉实验结果与显色指数 CRI 的比较,并已获得明确结论,对于 CQS 体系的验证仅含颜色偏好度,不含颜色还原性等其他方面的研究,因此还有待完善。此外,大部分视觉实验针对欧洲人群,由于东西方人群的颜色偏好差异,其结果未必适用于亚洲人。因此,本实验对光源的 CRI 和 CQS 均进行了计算,并基于视觉实验考量了颜色还原性和颜色偏好度两方面特性,以期在 CQS 系统的可靠性验证方面作出完善,并为颜色质量评价的视觉实验中补充亚洲人样本。

2　实验方法

实验选取了 5 种光源,其中 2 种卤素灯、2 种 LED 和 1 种紧凑型荧光灯(CFL),均为市场上的高端产品,其中卤素灯作为相同色温下的参考光源。光源基本信息见表 1,按照色温将 5 种光源分为 6,500 K 组与 5,000 K 组两组进行,分开进行视觉实验。首先采集了各光源的相对光谱功率分布,见图 1、图 2,光谱采集仪器为 STC3000 光谱仪,然后基于光谱计算了 CRI 和 CQS 两类指数。

表 1　实验用光源的相关信息

光源编号	标称相关色温 /K	实测相关色温 /K	显色指数 /Ra	光源类型	简称
1	6,500	6,558	90	卤素灯	Halogen
2	6,500	6,438	78	荧光粉 LED	LED
3	6,500	6,080	84	紧凑型荧光灯	CFL
4	5,300	5,200	92	卤素灯	Halogen
5	5,000	5,010	71	荧光粉 LED	LED

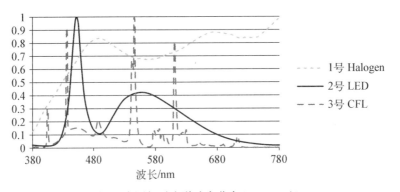

图 1　光源相对光谱功率分布(6,500 K 组)

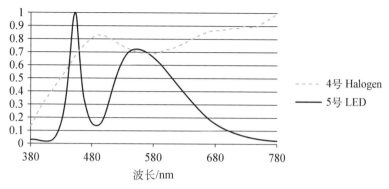

图 2　光源相对光谱功率分布（5,000 K 组）

视觉实验在一暗室内进行，分成两个独立的隔间，如图 3 所示。隔间顶部安装有光源，不产生眩光，均匀照射在桌面上，两隔间桌面平均照度均为 500 lx。

图 3　视觉实验场景

实验采用被试者内设计（within-subjects design）的方法，选取 10 名被试，其中 5 男 5 女，年龄为 20～24 岁，均具有正常的视力或矫正视力，且经过筛选，无色盲等其他眼疾。利用抵消实验条件的设计（reversal experimental condition design），每位被试均参与所有实验条件下的三组实验任务。第一项为色差的评分。实验采用的 MCC 色卡是 CIE 规定的用以比较待测光源与参考光源色差的标准方法。实验时，在两个隔间桌面上分别放置 MCC 色卡，隔间一边放置待测光源（CFL 或 LED），另一边放置相同色温的卤素灯作为参考光源，令被试观察两套色卡中的 24 对颜色样品，并对每一对的色差进行打分。打分的规则是，首先将色差感受归入 VS（very small，非常小），S（small，较小），M（medium，中等），L（large，大），VL（very large，很大）5 个大类（见表 2），然后再进行 5 个等级的细分，最后以诸如 VS4，L2 的形式表示。为便于统计数据，统计过时将得分转换为末尾行 1—25 的数字。

第二、三项实验均为偏好度的评价。隔间内放置了仿真花朵和水果，便于被试观察并作出判断。其中第二项为对各个光源的单独评分，包括从 0（非常不喜欢）到 100（非常喜欢）的评分，和从 1（非常不喜欢）到 5（非常喜欢）的评分。第三项为两两比较，令被试选出更喜欢的光源编号。

表 2　色差评定标准及得分转换

VS					S					M					L					VL				
1	2	3	4	5	1	2	3	4	5	1	2	3	4	5	1	2	3	4	5	1	2	3	4	5
1	2	3	4	5	6	7	8	9	10	11	12	13	14	15	16	17	18	19	20	21	22	23	24	25

3　实验结果与分析

3.1　CRI 和 CQS 计算结果

表 3 显示了各个光源的 CRI 和 CQS 计算值。对于显色指数 CRI,除了计算了一般显色指数(Ra),还计算了 4 个高饱和度特殊显色指数的平均值 R(9~12),另外还单独列出了 9 号样品的特殊显色指数 R9,因为 R9 是高饱和度红色样品,而目前 LED 往往在红色区域的显色性较差。由表可见,由于 9 号样品在 5 号光源下产生过大的色差,R9 呈现了负值。这也正证明了 CRI 指数存在问题。CQS 的计算值里包括了 Qa,Qp,Qf 3 种。CRI 中存在的一个问题是把有益于视亮度增加的饱和度增加也作为色差,引起 CRI 值的降低,因此 Qa 计算中将由饱和度增加引起的色差去除在外,而 Qf(fidelity)则完全依照显色性的颜色还原性这一定义,未剔除此类色差,而 Qp(preference)则从偏好度角度考虑,鉴于增加饱和度可以提高偏好度,所以饱和度增加引起的色差不但不减小分数,反而额外增加分数。因此从结果可以发现,三个 CQS 值不尽相同,1,2,3,4 号光源从整体上看都使颜色样品的饱和度在不同程度上有所增加。

表 3　测试光源的 CRI 和 CQS 计算值

光源编号		1	2	3	4	5
		Halogen	LED	CFL	Halogen	LED
CRI	Ra:	90	78	84	92	71
	R(9~12):	77	45	61	82	27
	R9:	57	12	50	67	−30
CQS	Qa	92	72	83	95	72
	Qf	90	71	82	93	72
	Qp	95	76	87	98	72

3.2　视觉实验结果与分析

24 个颜色样品在不同光源(2 号 6,500 K LED,3 号 6,500 K CFL,5 号 5,000 K LED)与对应的参考光源(同色温的卤素灯)下的色差得分结果如图 4 所示。方差分析(ANOVA)显示,对其中 8 个样品,三种光源在 0.05 的显著水平下有显著差异(见表 4)。通过进一步的两两比较得到各光源在一些颜色样品下的还原性优劣(见图 5):①对于样品 4,11,16,2 号(6,500 K LED)优于 3 号(6,500 K CFL);②对于样品 7，18，3 号(6,500 K CFL)优于 2 号(6,500 K LED)；③对于样品 1，2，4，16，21，2 号(6,500 K LED)优于 5 号(5,000 K LED)；④对于样品 1，2，7，18，3 号(6,500 K CFL)优于 5 号(5,000 K LED)。由此可见，对于不同颜色样品，三种光源(6,500 K LED；6,500 K CFL；5,000 K LED)在显色性能上各有利弊，显色指数高的CFL，未显示出特别优势。

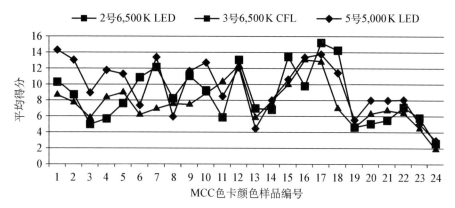

图 4　24 个 MCC 色卡的颜色样品在 3 种光源于其对应的参考光源下的色差得分平均值

表 4　ANOVA 和配对比较结果(本表仅显示了 ANOVA 得到显著差异 P < 0.05 的颜色样品)

样品编号	ANOVA 组内差异		LSD 配对比较		
			2~3 号	2~5 号	3~5 号
	F	Sig.	Sig.	Sig.	Sig.
1	5.780	**0.028**	0.591	**0.037**	**0.030**
2	6.104	**0.009**	0.597	**0.018**	**0.018**
4	7.616	**0.004**	**0.021**	**0.006**	0.135
7	5.944	**0.010**	**0.041**	0.278	**0.005**
11	8.168	**0.003**	**0.000**	0.065	0.222
16	3.878	**0.040**	**0.02**	**0.046**	0.947
18	9.958	**0.001**	**0.000**	0.240	**0.017**
21	3.549	**0.050**	0.077	**0.043**	0.382

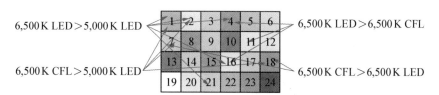

图 5　各测试光源对各 MCC 色卡上颜色样品的颜色还原性优劣比较

偏好度的主观评分中，0～100 的评分与 1～5 的评分基本一致，但方差分析(ANOVA)显示 6,500 K 组和 5,000 K 组之间的各光源都不存在显著差异(假设检

验结果分别为 $p = 0.069$ 和 $p = 0.572$）。

偏好度的比较实验中，根据每位被试评分计算，6,500 K 组的光源偏好度顺序为：CFL（3 号）＞LED（2 号）＞Halogen（1 号）；而 5,000 K 组中，Halogen（4 号）与 LED（5 号）相同。为了检验被试评分的一致性，根据实验结果进行了肯德尔和谐系数 W（Kendall's coeficient of concordance）检验，6,500 K 组中 $W = 0.31$，$P < 0.05$，说明被试对 6,500 K 组的评分具有统计学意义上的一致性。

3.3　视觉实验与 CRI，CQS 计算值的比较

关于颜色还原性，基于显色指数 CRI 和美国 NIST 提出的 CQS 指数的计算，6,500 K 和 5,000 K 的 LED 样品比 6,500 K 荧光灯样品的显色性（颜色还原性）差，但基于颜色样品色差评价的视觉实验，该荧光灯样品较 LED 未见明显优势，对于不同颜色的样品，它们的还原性各有利弊。

人们对各样品的偏好度更是与 CRI 和 CQS 大相径庭。其中，对饱和度因素进行补偿的 Qp 的计算值与 Qa 有所区别，但是仍然与实际的偏好度不符合。具有很高的 CRI 和 CQS 值的 6,500 K 卤素灯反而比同色温的荧光灯、LED 偏好度低。可能的原因是，荧光灯和 LED 的光谱具有一些峰值，提高了一些颜色的饱和度，使颜色更鲜艳，物体更清晰。

4　结论

（1）显色指数 CRI 无法反映光源显色性（即对被照物体的颜色还原性）的实际情况，更不能用以评价人们的偏好度。

（2）针对本次测试的样品，CQS 指数与显色指数 CRI 计算值接近，在评价光源颜色质量的准确性上，与 CRI 相比未见优势。若 CQS 要取代 CRI 成为新的标准评价体系，仍有待进一步的完善，提高其普适性。

（3）光源的颜色质量是一项综合性的评价，需考虑显色性、偏好度等多种因素。

（4）LED 在提高某些颜色饱和度从而增加人们偏好度方面有一定优势，在实际应用中可加以利用，如在商业照明中，提供白光普通照明的同时，在照明蓝色、黄绿色商品时，可增加其颜色饱和度，以提高商品吸引力。

参考文献

［1］　LED Measurement Series：Color Rendering Index and LED ［M］. DOE publications, PNNL - SA - 56891 January 2008.

［2］　Commission Internationale de l'Eclairage. International Lighting Vocabulary ［M］, item 845 - 02 - 59, CIE 17.4 - 1987.

［3］　Mark S. Rea, Jean P. Freyssinier-Nova, Color rendering：a tale of two metrics ［J］. Color research and application 2008, 33, 3：192 - 202.

［4］　Narendran N, Deng L. Color rendering properties of LED light sources. the Proceedings of

SPIE [M]. Volume 4776 Solid State Lighting II; Seattle, WA, 2002. 61 – 67.

[5] F Szabó, P Bodrogi, J Schanda, A colour harmony rendering index based on predictions of colour harmony impression [J]. Lighting Res. Technol. 2009,41:165 – 182.

[6] Ferenc Szabó, János Schanda, Peter Bodrogi, Emil Radkov, A Comparative study of new solid state light sources [C]. CIE 26th session, Beijing, 2007.

[7] N Sandor, J Scanda, Visual colour rendering based on colour difference evaluations [J]. Lighting Res, Technol. 2006,38,3:225 – 239.

[8] T. Tarczali, P. Bodrogi, J. Schanda. Colour Rendering Properties of LED sources [C]. 2nd CIE Expert Symposium on LED Measurement, Gaithersburg, USA, 2001.

[9] Elodie Mahler et al. Testing LED lighting for colour discrimination and colour rendering [J]. Color research and application, 2009,34,1:8 – 17.

[10] Vienot Francoise et al. Color appearance under LED illumination: the visual judgement of observers [J]. Journal of light and visual environment 2008,32(2):208 – 213.

[11] Vienot Francoise et al. Grading LED illumination: from coloour rendering indices to specific light quality indices [C], CIE 26th session, Beijing, 2007.

[12] Wendy Davis, Yoshi Ohno, Towards an improved color rendering metric [C], Proc. of SPIE Vol. 5941 59411G.

[13] CIE Technical Report 177:2007, Color rendering of white LED light sources [R], ISBN 978 3 901 906 57 2.

[14] Wendy Davis, Yoshi Ohno, Color quality scale [J], Opt. Eng. , Vol. 49,033602(2010); doi:10. 1117/1. 3360335.

[15] Wendy Davis, Yoshi Ohno, Towards an improved color rendering metric [C]. Proc. of SPIE Vol. 5941 59411G.

[16] Nicolas Pousset et al. Visual experiment on LED lighting quality with color quality scale colored samples [C]. CIE 2010: Lighting Quality and Energy Efficiency, 14 – 17 March 2010, Vienna, Austria.

基于视觉监控的操作手势
轨迹分布模式研究

刘振华，傅　山

上海交通大学航空航天学院，上海 200240

摘要　基于改进的层次自组织神经网络方法学习飞行员操作手势的轨迹分布模式。引进 Wilcoxon 秩和检验技术结合编辑距离来判断"内部网"的匹配程度，同时采用交叉验证技术，使验证集获得误差最小以自适应取得判断异常的阈值。最后利用学习到的轨迹分布模式检测操作过程中局部可能的异常，检测整个运动轨迹所表示的事件是否为异常事件，以及预测手势的将来行为轨迹。实验验证了改进方案的有效性。

关键词　飞行员手势，层次自组织神经网络，异常检测，行为预测，Wilcoxon 秩，交叉验证

Operation gesture trajectory distribution patterns research based on Visual Surveillance

Liu Zhenhua，Fu Shan

School of Aeronautics and Astronautics，Shanghai JiaoTong University，Shanghai 200240

Abstract　The trajectory distribution patterns of pilot operation gesture are learned based on an improved self-organizing neural network model. Wilcoxon rank sum test and the edit distance technology are introduced to determine the matches in the "internal net" neighborhood. And the cross-validation technique is used to get the threshold for abnormal detection. Using the learned patterns，the authors consider both local and global anomaly detection as well as object behavior prediction. Experimental results demonstrate the effectiveness of this approach.

Keywords　pilot gesture, hierarchical self-organizing neural network, anomaly detection, behavior prediction, wilcoxon rank, cross-validation

1　概述

飞行员的手势操作行为是影响飞行的最直接因素,也是飞机状态变化的直接控制输入。因此研究飞行员的手势操作特性,对于评估飞行员的操作技能,或从人机交互角度改进飞机驾驶舱的设计,有着极大的价值。本文借助视觉监控技术来研究飞行员操作手势的轨迹分布模式。

对于场景中的运动目标,轨迹分析是事件解释的基本问题[1]。本质上说,基于运动轨迹的目标行为分析是一个多维序列数据的分类问题,即将表征目标运动特征的轨迹序列与预先标定的代表典型行为模式的参考序列进行匹配的问题[2]。因此要理解飞行员操作事件,首先要学习手势轨迹分布,再基于轨迹分布进行分类和预测。近年来,轨迹模式分析引起了过内外学者的广泛关注,并提出了多种算法[2]。文献[3]通过由 Leaky 神经元连接的两层竞争网络来建立运动轨迹的统计模型:第一个学习网络建立流失量的分布模式,第二个用输出神经元对应轨迹,建立轨迹的分布模式。Sumpter 等[4]对文献[3]中的第二层竞争网络巧妙地引入了反馈机制,更好地实现了对轨迹分布预测的功能。但文献[3],[4]的网络权数庞大,训练速度慢。胡卫明等[1]去掉了文献[3,4]中的第二层竞争网络,利用神经网络的侧向连接,提出了层次自组织神经网络,通过运动轨迹到"内部网"的拓扑映射完成轨迹模式的学习,在"十"字路口的交通流轨迹上取得了较好的效果。本文参照此方法学习手势轨迹分布模式,但该算法中衡量相邻"内部网"只用到了轨迹样本分布的空间信息,没有考虑到轨迹序列的有序性,对手势分布模式的学习不理想,因此本文对原方法进行改进,采用 Wilcoxon 秩和检验结合编辑距离改善"内部网"的匹配度衡量方法,实验表明,本方法在手势轨迹分布模式的学习具有更好的效果。

2　轨迹编码和预处理

2.1　轨迹编码

运动轨迹就是运动目标在图像坐标中关于时间变化的质心点序列[3, 4]。可使用流失量序列来描述目标的运动轨迹:$f_i = (x_i, y_i, \delta_x, \delta_y)$($f_i$ 为向量),其中(x_i, y_i)为第 i 次采样时的质心点位置,(δ_{xi}, δ_{yi})为运动速度,($x_{i+1} - x_i$, $y_{i+1} - y_i$)可表示速度的大小和方向。则目标轨迹 Q 表示为 n 个流失量组成的序列:$Q = \{f_1, f_2, \cdots, f_i, \cdots f_{n-1}, f_n\}$。

速度组元 δ_{xi}, δ_{yi} 关于原点对称,流失量对旋转和平移变换敏感,因此本文将速度组元表示成(λ_i, θ_i),其中 λ_i 表示速率,θ_i 表示方向角度,值域为[0, 360]。于是新的流失量表示为

$$f_i = (x_i, y_i, \lambda_i, \theta_i).$$

为平衡计算流失量相似度时各分量的贡献,速度组元相对于位置组元依比例缩

放,使得每个组元都在[0,1]范围内。

2.2 预处理

采用时间上和空间上相结合的处理技术[5],对轨迹矢量进行重采样,以滤波改善由传感器或运动不稳定性噪声带来的影响。其中合适的空间采样阈值 D 和时间采样阈值 T 在能够降低噪声、减少数据量的同时保持较完整的轨迹信息,其数值在参考文献[5]中根据实验效果选取。

3 轨迹分布模式的层次自组织映射方法及其改进

3.1 自组织特征映射及层次自组织神经网络

Kohonen 自组织特征映射[6]是由输入层和输出层组成的两层神经网络,输入层中的每一个神经元通过权与输出层相连。在输出层中,竞争是这样进行的:对于"赢"的那个神经元 c,在其周围 N_c 区域内的神经元在不同程度上得到兴奋,而在 N_c 以外的神经元都被抑制。网络的学习过程就是网络的连接权数根据训练样本进行自调整、自适应、自组织的过程,逐渐把相似的输入样本映射到相近的输出节点上,实现从输入到输出的非线性降维,并保持拓扑有序关系。

文献[1]对自组织特征映射进行了巧妙的改进,用一个神经元集合代替自组织特征映射的输出节点,称之为"内部网",该神经元集合就是轨迹的流失量序列,给网

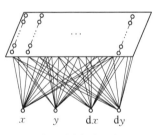

图 1　层次自组织
神经网络

络赋予了层次性,如图 1 所示。训练过程中,竞争获胜的"内部网"内的神经元及其邻域内的神经元得到不同程度的兴奋,而邻域外的神经元被抑制,实现原相近的流失量映射到相近的神经元;同样,"内部网"之间也采用这种学习方法,实现相近的轨迹序列映射到相近的输出层上。学习过程中通过神经元之间的协作移动,使得输出神经元的分布与流失量的分布相一致,并且使得输出神经元"线"的分布与训练样本轨迹的分布相一致。

3.2 层次自组织神经网络算法

3.2.1 神经元邻域及内部网邻域

本文采用文献[1]定义的邻域算法是:对于竞争获胜的神经元 c,邻域用"层"的概念来表示,"层"随着远离竞胜神经元 c 而逐级递增,不同时间 t 时的邻域 $N_c(t)$ 由相对于 c 的第 $k(t)$ 层以内的神经元组成。$k(t)$ 反应邻域的大小,随时间递减。使用"层函数"$L(k)$ 表示层 k 中的神经元与神经元 c 的激活值:$L(k) = 1/Z^{k-1}(Z \geqslant 2)$。类似,与输入轨迹有一定匹配关系的"内部网"构成"内部网"邻域,匹配最大的作为该邻域的中心。

3.2.2 竞争获胜算法及其改进

竞争获胜是确保把原多维空间样本准确映射到降维的神经元,并保持相似拓扑

关系不变的关键。文献[1]中采用欧式距离描述样本与神经元的匹配关系,并使用线上竞争获胜的神经元个数描述样本轨迹与"内部网"的匹配程度,线上获胜神经元越多,它与输入轨迹的匹配程度越高。

对于流失量的匹配程度,这种算法是合适的,但对于轨迹的匹配程度,由于线上竞争获胜的个数只能大概描述两个样本集的统计相似度,而轨迹序列与集合不一样的地方在于它的有序性,准确评价两条轨迹模式的匹配度需要能够度量:

(1) 组成轨迹的流失量集具有相似的分布,即轨迹穿过多维空间相近的点集;

(2) 同时轨迹穿过这些点集的顺序也应该近似相同,否则对应不同的轨迹。

原文衡量匹配程度的方法并没有同时考虑到这两点。对此,本文结合Wilcoxon 秩和检验技术和编辑距离进行改进。

Wilcoxon 秩和检验[7]用来决定两个对立样本是否来自相同或相等的总体,适用于总体分布未知的情况。如果原假设中两个独立样本来自相同的总体,那么秩将均匀分布在两个样本中,以此能衡量各个样本集的相似程度。

本文中取流失量的长度构成样本序列,如比较序列 T_X, T_Y 的匹配程度:

$$T_X:\{X_1, X_2, \cdots, X_M\} \quad (X_i = |f_{xi}|)$$

$$T_Y:\{Y_1, Y_2, \cdots, Y_N\} \quad (Y_i = |f_{yi}|)$$

将 T_X, T_Y 混合成一个样本,并从小到大排序,序号即对应 T_X, T_Y 的秩,设 T_X 的秩和为 W_X,对于观察数大于等于 8 的样本,可用标准正态分布近似检验两个序列的分布情况:

$$Z = \frac{W_X - u \pm 0.5}{\sigma} \sim N(0, 1)$$

其中:
$$u = \frac{M(M+N+1)}{2}$$

$$\sigma = \frac{MN(M+N+1)}{12}$$

选取符合要求的显著性水平(如 0.005),就可以判断 T_X 与 T_Y 是否相匹配。本文只把 z 的绝对值作为匹配程度的数值。

在上面的混合样本序列中,相近的数据对应相近的序号,构造关于 T_X 与 T_Y 经过混合序列样本的序号串,如 T_X 的 X_1 在混合序列中序号为 3,X_2 序号为 1,则 T_X 的数据在混合序列中经过的序号 O_X 为:$O_X:3 \rightarrow 1 \rightarrow \cdots \rightarrow M+N$;同样得到 O_Y 串。

若 T_X 与 T_Y 轨迹相近,则代表 T_X 与 T_Y 途径流失量顺序的 O_X 与 O_Y 串也应该近似相同,或者说,O_X 与 O_Y 串的距离较短。本文参照文献[8]采用编辑距离衡量整数串的距离。

设 T_X 与 T_Y 在混合样本中的序号串 O_X 与 O_Y 的编辑距离记为 $O(T_X, T_Y)$,结合衡量两个样本集匹配程度的 $|Z|$,给出以下模型作为衡量轨迹 T_X 与 T_Y 的匹配程度:

$$d(T_X, T_Y) = |Z|^{-1} \cdot e^{-O(T_X, T_Y)}$$

由此,轨迹的序列特性也得到量化。

3.3 参数训练

参照文献[1]训练步骤。

4 轨迹异常检测及行为预测

4.1 异常检测

轨迹模式训练完成后,采用下面的方法进行异常检测:对于运动轨迹的流失量集合 Q_0 中的每一个流失量 $f_i = (x_i, y_i, \lambda_i, \theta_i)$,求出与之最匹配的神经元,若与该神经元权值的欧式距离大于某一阈值 q,则认为此点异常,若一条轨迹上连续三个或以上点发生异常,则把该轨迹对应的事件作为异常事件。

对于阈值 q,文献[1]选取与各样本最匹配的神经元的距离的最大值,将该值的一半作为阈值,这种方法虽然统计了样本集与神经元的匹配情况,但是对于噪声点的随机干扰无法去除。本文采用交叉验证的算法,自适应求取阈值。交叉验证作为一种启发式技术,通过验证集上的误差率最小来得到自适应的阈值[7]。通过以下条件:

(1) 阈值 q 使得每个测试轨迹在轨迹模式中匹配被误判为异常的个数不超过轨迹长度的 5%;

(2) q 尽量小,以使模型对于异常检测比较灵敏。

每次选取总样本的 $1/10$ 作为测试集,循环验证以得到合适的 q 值。

4.2 行为预测

对于产生的部分轨迹,采用上节定义的竞争获胜算法求取与之最匹配的"内部网",该"内部网"则是目标未来最可能的运动轨迹。

5 实验及结果分析

本方法已在 Windows XP 系统上用 Visual C++2008 编程实现。由于进近着陆阶段是飞机操作很重要的阶段,也是航空事故高发阶段,因此本文选取飞机下降阶段,在高仿真度驾驶舱模拟器中研究飞行员操作手势轨迹。

5.1 手势轨迹提取及预处理

手势的位置由手心在图像中的二维坐标组成。提取到的数据如图 2 所示,椭圆包围检测到的双手,手心由点标出。本文仅研究右手轨迹。

图 2   手势检测与跟踪

通过反复跟踪得到轨迹样本数据如图 3 所示,图 4 表示采样过后的轨迹,箭头表示运动方向,箭头长度表示该处速率大小(本文取 $T=2$,$D=3$)。

图 3   轨迹图   图 4   重采样后的流失量轨迹图

5.2   手势轨迹分布模式学习

图 5 所示是样本轨迹学习后输出的神经元"线"的分布情况,图中共有 16 条"线"(驾驶舱中的操作目标主要涉及方向舵、油门杆、自动驾驶仪,因此主要轨迹应该覆盖这三个目标,不会有太复杂的操作轨线)。每条线上取 15 个神经元。从图中可以看出,学习后输出神经元"线"分布与样本轨迹的分布相一致,学习效果比较理想。

图 6 所示是采用文献[4]中算法在相同的网络稳定化判据下对轨迹样本训练的效果。不难看出,原来算法在手势存在回抖的地方检测出的方向比较凌乱,而图 5 中没有这种现象。其原因在于原文的竞争获胜机制只是考虑到了轨迹流失量的分

图 5   模型学习结果   图 6   改进前的算法训练效果

布情况,没有考虑轨迹流失量的分布秩序,失去了轨迹的序列特性,因此效果不理想。本文提出的改进方案收到了更好的效果。

图 7 所示是在学习过程当中,初始阶段、50 次迭代、100 次迭代后,输出神经元与"线"的分布情况。图中可以看出,随着迭代次数的增加,输出层中"线"的分布逐步与样本轨迹的分布相一致,网络收敛效果良好。

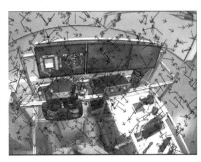
随机初始权值

第50次迭代

第100次迭代

图 7　算法学习过程

5.3　异常检测和行为预测

图 8 所示是一个异常检测的例子,当手离开方向舵向自动驾驶仪移动的过程中,先伸向驾驶舱前面的显示器,再迂回到自动驾驶仪。对于下降中的操作而言,这种动作是不应该存在的,当手势超过一定阈值时,异常点就被检测出来了,由图中的

图 8　操作手势异常检测

十字表示。当异常点数目超过一定数目，该轨迹所代表的事件即为异常事件。

图 9 所示是一个预测的例子，手离开驾驶杆，模型计算出将要出现的最可能轨迹是伸向方向舵。从图中看出，该路线跟人的预测很符合。

图 9　操作手势预测实例

6　结束语

层次自组织神经网络，通过"内部网"表示输出神经元之间的连接关系来实现运动轨迹与"线"的映射，学习出"线"的分布，并利用"线"的分布进行异常检测和行为预测。衡量轨迹样本之间的匹配程度是成功的关键，本文对原算法中的衡量方法进行改进，将轨迹流失量的秩序信息考虑进去，并运用在下降过程中的飞行员手势操作轨迹分析上，取得了较好的效果。未来的工作是进一步挖掘飞行员操作手势的规律，如结合眼动分析仪研究手眼配合，挖掘驾驶舱中手在执行任务时各操作的时延特性等。

参考文献

［1］ 胡卫明，谢单，谭铁牛，等. 轨迹分布模式学习的层次自组织神经网络方法［J］. 计算机学报，2003,26(4):417－426.

［2］ Hu W M, Tan T N, et al. A survey on visual surveillance of object motion and behaviors ［J］. IEEE Transaction on Systems, Man and Cybernetics — Part C: Application and Reviews, 2004,34(3):334－352.

［3］ Johnson N, Hogg D. Learning the distribution of activities in video ［J］. IEEE Transactions on Pattern Analysis and Machine Intelligence, 2000,22(8):844－851.

［4］ Sumpter N, Bulpitt A. Learning spatio-temporal pattern s for predicting object behavior ［J］. Image and Vision Computing, 2000,18(9):697－704.

［5］ Owens J, Hunter A, Application of the Self-Organizing Map to Trajectory Classification ［C］, in Proc. IEEE Int. Workshop Visual Surveillance, 2000,77－83.

［6］ Richard O D, Peter E H, David G S. 模式分类［M］. 李宏东，姚天翔，等，译. 北京：机械工业出版社,2003.

［7］ 吴赣昌. 概率论与数理统计［M］. 北京：中国人民大学出版社,2009:230－231.

［8］ 袁先平，仲红，黄宏升，等. 一种字符串近似匹配的安全查询协议［J］. 计算机工程,2011,37(20):142－144.

基于信息处理模型的机组行为
与事故关系的研究

王苹丽[1]，董大勇[2]，傅　山[2]

1. 上海交通大学电子信息与电气工程学院，上海 200240
2. 上海交通大学航空航天学院，上海 200240

摘要　人的行为失误是飞行事故的最主要原因，分析飞行事故必须以机组行为的研究为基础。本文基于人的信息处理模型，建立人信息处理过程中信息获取、信息分析和信息执行三个阶段与事故发生率之间的关系模型。结果表明，信息分析和获取过程对事故的发生影响较大，信息执行过程对事故的发生影响较小。

关键词　信息处理模型，飞行事故率，机组行为

Investigation on the Relationship between the Crew Behavious and the Accidents using Information Processing Model

Wang Pingli[1], Dong Dayong[2], Fu Shan[2]

1. School of Electronics, Information and Electrics, Shanghai Jiao Tong University, Shanghai 200240
2. School of Aeronautics and Astronautics, Shanghai Jiao Tong University, Shanghai 200240

Abstract　The human error is the main reason for aircraft accidents, so the analysis of the aircraft accidents is based on the research of human performance. In the article a relation model of information sensory, information analysis, information response and accidents rates has been established based on the information processing model. The result indicates that the stage of information analysis and information sensory have great effect on the accidents, however, the stage of information response has less influence on the accidents.

Keywords　Information processing model, Rate of aircraft accidents, Human performance

1 引言

　　70%左右的飞行事故是由于人的因素造成的[1]。飞行机组是飞行任务的核心

主体,研究机组行为是分析飞行事故的一个重要方法。目前,已经有很多机组行为研究方法,包括工效学、认知学、心理学等。人的信息处理模型是一种认知学的机组行为研究方法,它认为人的活动是一个连续的信息处理过程,即信息获取、分析和执行。人的信息处理模型已经广泛应用于机组行为的研究。本文根据信息处理模型,研究飞行员在执行任务时信息处理过程与飞行事故之间的关系。

2　信息处理模型

飞行员执行任务的过程也是一个不断地信息处理的过程。从信息处理模型假设激励到响应之间需要一系列的脑力操作过程,包括信息获取、信息分析和信息执行[2],如图 1 所示。

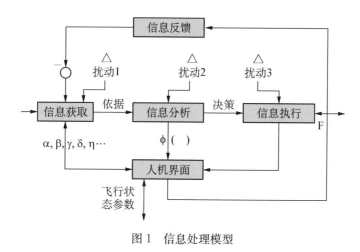

图 1　信息处理模型

信息获取:人的感官(眼、耳等及其他感觉器官)感知到外界物理刺激,并理解其含义,例如,飞行员理解文字的含义;根据环境和飞机特征理解飞机的状态等。它包括两个方面的内容:①信号检测问题,即飞行员能检测到的最小的激励信号;②选择问题,即检测到的信号与当前任务的映射关系。信息获取的能力受到注意力资源的限制。

信息分析:这一阶段主要指飞行员根据获取的信息进行决策和响应的选择。飞行员的判断或决策对飞行安全起着关键性的作用。飞行员根据获取的信息和已有经验理解分析当前的环境状态,设计可行的解决方案;然后,比较预测各个方案的结果,做出决策,选择最终的执行方案和执行步骤。这一阶段主要受脑力资源的分配、飞行员的精神状态如紧张程度等因素的影响。

信息执行:根据决策执行操作。这一阶段受触觉反馈、用力力度、响应时间等因素的影响。

3　机组信息处理过程与飞行事故的关系

上述三个阶段,任何一个阶段机组信息处理行为出现偏差,均可能对飞机造成不良的影响,甚至引发飞行事故。本文将根据不同任务的飞行事故统计数据,以及任务的复杂程度,建立飞行事故率(p)、任务复杂度因子(w)和信息处理模型中信息获取、分析、执行过程对事故率的影响因子(分别为 k_s, k_p, k_r)之间的关系模型,讨论信息处理过程中各个阶段的行为偏差对飞行事故的影响。

3.1 事故率统计数据

美国全国运输安全委员会(National Transportation Safety Board,NTSB)2005年通用航空事故统计表明[3]:2005 年与人为因素相关的事故总数为 1,372 起,其中与飞机控制任务相关的事故数为 990,与计划制订任务相关的事故数为 489,与飞行设备的应用任务相关的事故数为 14,与通信相关的事故数为 69。如表 1 所示。

表 1　不同阶段任务的事故率

	飞机控制	计划制订	飞行设备的使用	通信
事故数	990	489	148	69
事故率(p_i)/%	72.20	35.60	10.8	5.03

3.2 任务复杂度的分析

任务的复杂度由工作量进行衡量。本文根据四种任务的特点和功能,通过两个方向的工作量的比较联合确定任务复杂度:①相同任务在信息处理模型不同阶段的工作量的比较;②不同任务在信息处理模型相同阶段的工作量的比较。

"飞机控制"指在起飞、巡航以及着陆的整个飞行过程中,飞行员调节飞机的高度控制飞机在操纵面内[4]。"计划制订"指根据环境情况进行任务安排或决策。与信息处理模型中的决策不同,这里指宏观意义上的整体的决策。"飞行设备的使用"包括使用方法,设备的识别与选择,操作设备所需的力度等方面。"通信"包括人机通信,机组成员间的通信,以及机组与其他人员(如交通管制员,地勤人员等)的通信。

根据不同任务的特性,按照信息处理模型,将任务在不同阶段的工作量量化为不同的等级。表 2(a)为任务间的工作量等级表,等级越高工作量越大;表 2(b)为同一任务在信息处理不同阶段的工作量等级表,等级越高工作量越大;表 2(c)为两种比较的的综合结果。

表 2　任务量等级表
(a)

	信息获取	信息分析	信息执行
飞机控制	L3	L3	L3
计划制定	L2	L4	L1
设备使用	L1	L1	L4
通信	L4	L2	L2

(b)			
	信息获取	信息分析	信息执行
飞机控制	L3	L2	L1
计划制定	L2	L3	L1
设备使用	L2	L1	L3
通信	L3	L1	L2

(c)			
	信息获取	信息分析	信息执行
飞机控制	(L3, L3)	(L3, L2)	(L3, L1)
计划制定	(L2, L2)	(L4, L3)	(L1, L1)
设备使用	(L1, L2)	(L1, L2)	(L4, L3)
通信	(L4, L3)	(L1, L1)	(L2, L2)

表 2(c)得到不同任务不同阶段的工作量,共 7 个等级。假设工作量等级组合中只要一个分量高于另一个组合的同一分量,那么该组的工作量等级高,工作量大,因此工作量等级($q_1 < q_2 < \cdots < q_7$)由低到高依次为(L1, L1),(L1, L2),(L2, L2),(L3, L1),(L3, L2),(L3, L3),(L4, L3)。取 $q_1 = 1$,$q_2 = 2$,$q_3 = 3$,$q_4 = 4$,$q_5 = 5$,$q_6 = 6$,$q_7 = 7$ 等级工作量依次为 1,2,3,4,5,6,7,归一化得到各个等级工作量的权重因子(w_1,w_2,$\cdots$,w_7),其中:

$$w_i = \frac{q_i}{q_1 + q_2 + q_3 + q_4 + q_5 + q_6 + q_7}$$

因此得到 7 个等级工作量的权重因子以及各个任务不同信息处理阶段的工作量权重因子,分别如表 3 和表 4 所示。

表 3 7 个等级工作量的权重因子

w_1	w_2	w_3	w_4	w_5	w_6	w_7
0.035,7	0.071,4	0.107	0.143	0.179	0.214	0.25

表 4 各个任务不同信息处理阶段的工作量权重因子

	信息获取	信息分析	信息执行
飞机控制	w_6	w_5	w_4
计划制定	w_3	w_7	w_1
设备使用	w_2	w_2	w_7
通信	w_7	w_1	w_3

3.3　飞行事故率与飞行员信息处理过程的关系

假设飞行事故与飞行员的信息处理过程之间存在线性关系,同时也受到任务复杂程度的影响。因此建立飞行事故率与信息处理过程的关系模型:

$$w_{i1} \times k_{\mathrm{s}} + w_{i2} \times k_{\mathrm{p}} + w_{i3} \times k_{\mathrm{r}} = p_i$$

式中:w_{i1},w_{i2},$w_{i3} \in \{w_1,w_2,w_3,w_4,w_5,w_6,w_7\}$,表示第 i 个任务在信息处理过程三个阶段的工作量权重因子。因此得到

$$\begin{cases} w_{11} \times k_{\mathrm{s}} + w_{12} \times k_{\mathrm{p}} + w_{13} \times k_{\mathrm{r}} = p_1 \\ w_{21} \times k_{\mathrm{s}} + w_{22} \times k_{\mathrm{p}} + w_{23} \times k_{\mathrm{r}} = p_2 \\ w_{31} \times k_{\mathrm{s}} + w_{32} \times k_{\mathrm{p}} + w_{33} \times k_{\mathrm{r}} = p_3 \\ w_{41} \times k_{\mathrm{s}} + w_{42} \times k_{\mathrm{p}} + w_{43} \times k_{\mathrm{r}} = p_4 \end{cases} \Leftrightarrow \begin{cases} w_4 \times k_{\mathrm{s}} + w_6 \times k_{\mathrm{p}} + w_5 \times k_{\mathrm{r}} = p_1 \\ w_2 \times k_{\mathrm{s}} + w_7 \times k_{\mathrm{p}} + w_1 \times k_{\mathrm{r}} = p_2 \\ w_2 \times k_{\mathrm{s}} + w_3 \times k_{\mathrm{p}} + w_7 \times k_{\mathrm{r}} = p_3 \\ w_7 \times k_{\mathrm{s}} + w_2 \times k_{\mathrm{p}} + w_2 \times k_{\mathrm{r}} = p_4 \end{cases}$$

求解得到:$k_{\mathrm{s}} = 0.518,1$,$k_{\mathrm{p}} = 1.825,1$,$k_{\mathrm{r}} = 0.052,7$。

把 k_{s},k_{p},k_{r} 归一化得:$k_{\mathrm{s}} = 0.216$,$k_{\mathrm{p}} = 0.762$,$k_{\mathrm{r}} = 0.022$。

根据分析结果可以得到,信息处理过程中的分析阶段对事故的影响最大。Jensen 和 Benel 也曾指出,50% 以上的飞行事故是由于飞行员决断错误引起的[5]。这主要由于该阶段飞行员工作量较大,不仅需要建立正确的场景意识,判断有用的信息,还需要以往经验的指导,比较所有可能的操作结果,做出最有效的决策。信息获取对飞行事故的影响也较大,因为人们信息检测的能力受其对信号的敏感度和反应偏差的制约,受到其知识水平和对任务的理解程度的限制。目前,驾驶舱自动化和集成化程度越来越高,飞行员往往难以掌握其复杂的设计原理,当出现故障时,飞行员信息过载,分析负荷急剧增加,这也是该阶段发生事故的重要原因。执行阶段的对事故的作用较小。这一阶段主要指用肌肉运动完成某一操作或一系列的操作,自动化技术的引入,减少并简化了飞行员对飞机的操作,因此也减少了执行引发的飞行事故。但是,飞行员仍可能产生无意识的操作错误。

4　讨论与展望

计算结果表明,信息分析过程对飞行事故的影响远大于信息获取和执行阶段。驾驶舱新技术的引入和发展,为飞行员提供了丰富的信息获取的途径和方法,同时也简化了飞行员的操作,但能够帮助飞行员进行准确实时的信息分析的自动化设备仍然匮乏或不完备,有的设备甚至会增加机组的负担。因此,在新的驾驶舱设计中,应该改善和增强辅助机组信息分析的自动化设备,提高机组的信息分析能力。

虽然 Jensen 和 Benel 指出 50% 以上的飞行事故是由于飞行员决断错误引起的,而本文计算得到 $k_{\mathrm{s}} = 0.762$,大于 50%,因为,飞行事故的发生通常是由几个因素共同作用造成的,而不是由单一因素引起的,当事故由几个因素共同作用产生时,它会被重复地运用于相关原因事故率的统计中,例如文中引用的任务事故率总和并

非 100%。本文 k_s，k_p，k_r 的计算过程也存在类似的重叠问题，造成计算的结果大于 50%。因此，提高飞行事故分类的精确度，是提高分析准确性的一个重要的发展方向。

另外，文中飞行任务复杂度的确定等级仅仅是通过理论分析得到的，将来可以设计具体的飞行实验，对其进行验证和修正，以提高其精确度。

5　总结

飞行员执行任务的过程也是不断进行信息处理的过程，人的信息处理是信息获取、分析和执行的循环过程。本文主要研究了信息处理模型中信息获取、分析和执行三个阶段对飞行事故的影响。信息分析过程导致事故的可能性最大，其次为信息获取过程，而信息执行阶段发生事故的可能性比较小。因此，增强飞行员的信息获取和分析能力，是减少飞行事故的一个有效的途径。

参考文献

[1]　Wiegmann D，Shappell S. A human error analysis of commercial aviation accidents using the human factors analysis and classification system（HFACS）. DOT/FAA/AM - 01/3 ［R］，2001.

[2]　Earl L. Wiener，David C，Nagel. Human factors in aviation ［M］. California：Academic Press，1988.

[3]　NTSB/ARG - 09/01. Annual Review of Aircraft Accident Data U. S. General Aviation，Calendar Year 2005［R］.

[4]　Introduction to Aircraft Control. DSCB Home Page：http://dscb. larc. nasa. gov.

[5]　David R. Hunter. Measuring General Aviation Pilot Judgment Using a Situational Judgment Technique. The International Journal of Aviation Psychology ［J］，2003；13(4)，373 - 386.

大型客机飞行员操作程序综合评价

郭奋飞，王黎静，王昭鑫，何雪丽，向　维

北京航空航天大学航空科学与工程学院，北京 100191

摘要　在复杂的人-机系统中，成功的操作依赖于一套精心编制的操作程序，不合理或不合逻辑的操作程序可能导致飞行员的违规操作，进而引起飞行事故。参考波音 737 - NG 飞行机组操作手册，分析进近和着陆阶段的任务功能流程确定指标体系的层次结构。采用德尔菲法对现役飞行员进行两轮咨询，建立了飞行员操作程序综合评价指标体系。采用序关系分析法确定权重系数，对飞行员进行咨询得到各个指标的权重系数。应用模糊综合评价法对飞行员操作程序进行综合评价，得到了飞行员操作程序的评价等级和评价值。本文提出了一种将功能流程图法、德尔菲法、序关系分析法和模糊综合评价法相结合的飞行员操作程序综合评价方法，通过应用四种方法对波音 737 - NG 飞机的进近和着陆阶段的操作程序进近评价，说明了该综合评价方法的科学合理性。

关键词　飞行员操作程序，进近和着陆，评价指标，权重系数，模糊综合评价

Comprehensive Evaluation of Pilot Operation Procedures for Commercial Airliner

Guo Fenfei，Wang Lijing，Wang Zhaoxin，He Xueli，Xiang Wei

School of Aeronautic Science and Engineering，Beijing University of
Aeronautics and Astronautics，Beijing 100191

Abstract　In complex human-machine systems，successful operations depend on an elaborate set of procedures. Inconsistent or illogical procedures may lead to deviation operation procedures and flight accidents happen. Normal procedures of approach and landing phases which come from Flight Crew Operations Manual of the Boeing 737-NG airplane were divided into task function flow. The hierarchy of index system was confirmed by pilot interview and the comprehensive evaluation index system of POP was established by Delphi consulting in-service pilots. Index weight factors were calculated by Order Relation Analysis. POP was evaluated by Fuzzy Comprehensive Evaluation and evaluation scales and values of POP were calculated by weight factors and evaluation matrixes. A POP Comprehensive Evaluation method was proposed by the integration of Functional Flow Diagram, Delphi, Order Relation Analysis and Fuzzy Comprehensive Evaluation. The Boeing 737-NG airplane POP of approach and landing phases were evaluated by above four methods, which indicated the validity and rationality of the novel Comprehensive Evaluation method.

Keywords　Pilot Operation Procedures, Approach and Landing, Evaluation Index, Weight Factor, Fuzzy Comprehensive Evaluation

在飞机的设计中,设计生产单位制定出相关的机组使用手册(Flight Crew Operation Manual, FCOM)。手册分 4 部分[1, 2],其中最重要一部分为飞行操作(Flight Operation),即为通常所说的飞行员操作程序(Pilot Operation Procedures, POP)。飞行员操作程序指按照各种飞行工作状态模式,完整地规定驾驶员操作控制、显示画面和告警信息文件。当大型飞机公共航空运输承运人购买飞机后,结合FCOM,融合航空飞行的必需要素(包括飞行机组简令、客舱简令、陆空通信等多方面因素)制订出适用于本航空公司的机组标准操作程序(Standard Operating Procedures, SOPs)。SOP 中的 POP 是与飞行员操作直接相关的部分,如果 POP 出现了缺陷,将可能引起操作失误,导致航空事故。

在 1987 年,Lautman 和 Gallimore 完成了一项有关喷气式运输飞机事故报告的研究,他们分析了 1977～1984 年期间的 93 起坠机事件,其中事故最主要的因素是"飞行员背离基本的操作程序",占整个机组失误的 33%[3]。1994 年,美国国家运输安全委员会(National Transportation Safety Board, NTSB)通过对 37 起飞行事故研究表明:操作程序失误占整个机组失误的 24%,它是导致事故的最主要因素[4]。

在 20 世纪 90 年代,美国 NASA Ames 研究中心的 Degani 和 Wiener 通过分析正常操作程序及其功能、设计、长度、用法和与人交互的局限性,表明政府(尤其FAA 首席运行检查员)、厂商的理念,航空公司的文化以及飞行员都将影响操作程序的最终设计和使用[5, 6]。在 20 世纪八九十年代,俄罗斯国家航空系统研究院在操作程序的设计和评估研究方面,建立了人机工效数据库和驾驶员决策数据库;研制了一整套计算机仿真软件,着重于进行定量的计算分析,对飞行员操作程序的设计起到了验证和指导作用[7]。

本文以波音 737 - NG 飞机为例,在对其进近着陆阶段的飞行员操作程序分析的基础上建立了基于飞行安全的飞行员操作程序综合评价指标体系,然后确定各指标的权重系数,最后应用模糊综合评价对飞行员操作程序进行综合评价。所得结论符合实际情况,说明该方法可用于其他机型的飞行员操作程序的综合评价。

1 综合评价指标体系

指标体系是指由一系列指标构成的整体,它能全面真实地综合反映飞行员操作程序各个方面的情况,不仅是进行有效综合评价的前提,也是权重系数确定的依据。因此,建立一套完善、科学、合理的评价指标体系对飞行员操作程序综合评价至关重要。

1.1 进近着陆阶段综合评价指标体系模型

飞行的结束阶段是一个非常关键的飞行阶段,国际飞行安全基金会记录在案的飞行事故中有 56.2% 是发生在最后的进近和着陆阶段。因此,考虑到机组标准操作程序内容极其庞杂,本文选取进近和着陆阶段的正常操作程序为研究对象,对飞行员操作程序进行综合评价研究。

本文参考波音 737 - NG 飞行机组操作手册,将进近和着陆阶段作为一个整体目标来研究,根据实际飞行中的具体着陆情况,与飞行员讨论后,将其详细划分为以下 8 个阶段,说明各阶段流程关系的功能流程图如图 1 所示。其中,精密仪表进近和非精密仪表进近为两种可选的仪表进近方式,精密仪表进近指使用仪表着陆系统(Instrument Landing System,ILS)的进近,非精密仪表进近指使用垂直导航(Vertical Navigation,VNAV)的进近。此外,复飞也是进近和着陆过程中一个可选择的阶段,且十分重要。

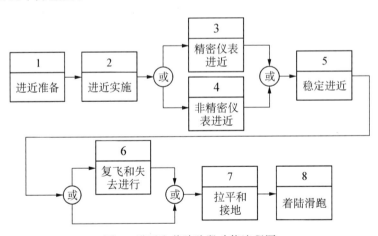

图 1   进近和着陆阶段功能流程图

依据综合评价指标体系的建立原则,通过对波音 737 - NG 现有的飞行员操作程序和相关文献进行研究和分析,并征求波音 737 - NG 机型现役飞行员的意见,确定出一个如表 1 所示的 4 层次评价指标体系模型。其中,O 表示评价总目标,$O_i(i=1,2,3)$ 表示评价子目标,O_{ii} 表示评价子子目标,U_k 表示评价指标集,指标集中的评价指标多达 68 个,由于篇幅有限就不一一列出。

表1 飞行员操作程序综合评价指标体系模型

总目标层	子目标层	子子目标层	指标集
O:飞行员操作程序综合评价指标体系	O_1:进近阶段	O_{11}:进近准备	U_1
		O_{12}:进近实施	U_2
		O_{13}:精密仪表进近	U_3
		O_{14}:非精密仪表进近	U_4
		O_{15}:稳定进近	U_5
	O_2:复飞阶段	O_{21}:复飞和失去进近	U_6
	O_3:着陆阶段	O_{31}:拉平和接地	U_7
	O_{32}:着陆滑跑	U_8	

1.2 德尔菲法筛选指标

本文采用德尔菲法进行专家咨询,接受咨询的专家都来自某航空公司波音737-NG机队的现役飞行员,且都在飞波音737-NG系列的机型。在第一轮咨询中,首先介绍了飞行员操作程序的研究目的以及相关的背景资料,供咨询专家参考。然后,要求每位专家根据已有的知识和经验判断出每个指标对飞行安全的影响程度。影响程度等级的划分采用5点Likert型标度,1~5分别表示的影响程度等级为"不大"、"一般"、"大"、"很大"和"极大"。同时,又要求专家给出在判断各指标对飞行安全的影响程度等级时的自信度。自信度等级的划分采用5点标度,1~5分别表示的自信度等级为"很低"、"低"、"一般"、"高"和"很高"。当专家组的自信度等级(The Group Confidence Ranking,GCR)不小于3时,即所有专家给出的自信度等级的均值,表示咨询结果比较接近于主观判断的真实情况。第二轮咨询增加了多数专家的反馈意见。对于改进德尔菲法,一般通过两轮咨询,专家的意见已基本趋于一致。因此,依据第二轮咨询形成的专家一致意见,筛选出评价指标体系。本文把专家一致意见定义为:不少于2/3(或67%)的专家判断等级为"很大"以上的判断结果,即Likert量表等级≥4[8]。当然,筛选指标的前提是该指标的自信度等级的均值大于3。

第一轮咨询后,从初步确定的68个指标中筛选出63个指标。筛除的5个指标中,有3个为指标专家一致意见的判断等级为"大",不满足≥4的条件;另外2个指标是根据咨询过程中多名飞行员反映指标不合理而筛除。第二轮咨询后,所有指标的专家一致意见判断等级都满足条件。

1.3 评价指标咨询结果统计分析

本论文采用目前最常用的克伦巴赫α信度系数法(Cronbach's Alpha)对指标体系进行信度分析。利用统计软件SPSS进行分析计算,对初选的68个指标的信度分析,可以看出,Cronbach's α信度系数值为0.964,量表的内在一致性程度非常高,说明量表的信度非常可靠。

对于飞行员操作程序评价指标的专家咨询结果,本研究采用以下几个统计参数进行频数统计处理和分析。

(1) 集中程度,用数值平均数和位置平均数来描述。

(2) 离散程度,用标准差(Std. Deviation)σ_i描述。它反映专家对第i个指标影响飞行安全程度评价的分散程度,该值越小,表明专家评价结果的分散程度越小。

(3) 协调程度,用变异系数(Coefficient of Variation,CV)V_i和协调系数(Kendall's Coefficient of Concordance)W表征。

(4) Kendall's W 协调系数检验,原假设为专家的判断标准不一致,用按泊松(Pearson)统计量的卡方检验进行非参数检验。

根据两轮咨询结果分别计算出协调系数及其显著性检验,如表2所示。其中,N表示样本数;$Kendall's\ W$表示肯德尔协调系数(或和谐系数);$Chi-Square$表示卡方值;df表示自由度;$Asymp.\ Sig.$表示概率p值。

表 2　协调系数非参数检验(Kendall's W Test Statistics)

检验参数	第一轮		第二轮
	筛除指标前	筛除指标后	
N	24	24	24
$Kendall's\ W$	0.059	0.073	0.341
$Chi-Square$	94.711	108.589	507.410
df	67	62	62
$Asymp.\ Sig.$	0.015	0.000	0.000

通过分析表2中的数据,可得出以下三点结论:

(1) 从表中可以看出,两轮咨询结果的概率p值都小于显著性水平0.05,则应拒绝原假设,说明专家评价的协调程度是可信的,咨询结果可取。

(2) 对第一轮咨询而言,筛除指标后,概率p值有较大提高,满足了严格的显著性水平0.01,说明筛除的5个指标影响专家对指标体系评价的协调程度,因此,按专家一致意见的判断等级$\geqslant 4$为筛选指标的标准是科学合理的。

(3) 对第一轮筛除指标后和第二轮而言,专家的意见集中,对整个评价指标体系的判断结果更加协调,概率p值趋于零。

(4) 对两轮咨询而言,由于第二轮咨询中给出了第一轮多数专家的反馈信息,因此,第二轮咨询结果的协调系数明显大于第一轮,说明专家对整个评价指标体系的判断结果更加协调,经过筛选后确立的指标体系更具有代表性。

2　指标权重系数确定

本文采用序关系分析法进行权重系数计算,其基本思想是:先对各评价指标按某种评价准则进行定性排序,然后再按一定标度对相邻指标间依次比较判断,进行定量赋值,并对判断结果进行数学处理,得出各个评价指标的权重系数[9]。

2.1 权重系数咨询

本文对建立的指标体系中的一级指标和二级指标进行了权重系数的专家咨询。接受指标权重系数咨询的专家都来自某航空公司波音 737 - NG 机队的现役飞行员,且都在飞波音 737 - NG 系列的机型。本次权重系数调查共送出 25 份调查表,收回 23 份有效问卷。应用序关系分析法迭代计算各指标的权重系数。飞行员操作程序一级指标权重系数的计算结果如表 3 所示。

表 3　指标集权重系数

指标集	权重系数
$\{w_{U1}, w_{U2}, w_{U3}, w_{U4}, w_{U5}, w_{U6}, w_{U7}, w_{U8},\}$	$\{0.14, 0.14, 0.11, 0.10, 0.16, 0.12, 0.13, 0.10\}$

2.2 权重系数统计分析

对不同专家人数下的 8 个一级指标 $\{w_{U1}, w_{U2}, w_{U3}, w_{U4}, w_{U5}, w_{U6}, w_{U7}, w_{U8}\}$ 权重系数统计如图 2 所示。通过分析图 2 可以得出以下两点结论:

(1) 随着专家人数的逐渐增加,权重系数趋于稳定,说明计算结果合理可信。

(2) 当专家人数到达 10 以上时,指标权重系数趋于平稳,说明所述"专家人数以 10～30 人为宜"是科学合理的。

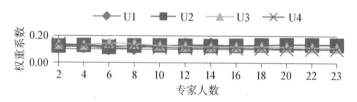

图 2　不同专家人数的指标权重系数比较

3 模糊综合评价

模糊综合评价(Fuzzy Comprehensive Evaluation,FCE)方法是一种用于涉及模糊因素的对象系统的综合评价法。在研究复杂系统中所遇到的问题时,不仅涉及多方面的客观因素,而且还涉及人的问题。因此,研究过程中就必然不能忽视客观外界事物在人脑中反映的模糊性以及事物本身的模糊性。所谓模糊评价,就是根据给出的评价标准和实测值,经过模糊变换后对事物做出评价,它是在模糊集合理论基础上形成的一种综合评价方法。

3.1 模糊综合评价应用

本文对飞行员操作程序进行模糊综合评价调查将飞行员操作程序中影响飞行安全的评价等级划分为 5 个等级,即优秀、良好、中等、较差、极差,其含义如表 4 所示。

表 4　飞行员操作程序评价等级及其含义

评价等级	分值	评语含义
优秀	5	不存在影响飞行安全的问题
良好	4	存在可忽略不计的影响飞行安全的问题
中等	3	存在一般的影响飞行安全的问题
较差	2	存在较高的影响飞行安全的问题
极差	1	存在极高的影响飞行安全的问题

本次对飞行员进行主观调查共收回 24 份有效问卷,对咨询结果进行统计分析,计算每个指标的评价等级频数,建立了模糊评价矩阵。

3.2　模糊综合评价结果分析

本文以指标体系中的一级指标"进近准备"(U_1)为例,对其进行模糊综合评价,具体步骤如下:

(1)评价对象:U_1 进近准备。评价指标集:$U = \{u_{11}, u_{12}, u_{13}, u_{14}, u_{15}, u_{16}, u_{17}, u_{18}, u_{19}\}$。

(2)拟定评语级:$V = \{优秀(v_1),良好(v_2),中等(v_3),较差(v_4),极差(v_5)\}$。

(3)指标权重系数:$W = \{w_{U1}, w_{U2}, w_{U3}, w_{U4}, w_{U5}, w_{U6}, w_{U7}, w_{U8}\}$。指标具体数值如表 3 所示。

(4)模糊评价矩阵:

$$
\boldsymbol{R} = \begin{pmatrix} r_{11} & r_{12} & r_{13} & r_{14} & r_{15} \\ r_{21} & r_{22} & r_{23} & r_{24} & r_{25} \\ r_{31} & r_{32} & r_{33} & r_{34} & r_{35} \\ r_{41} & r_{42} & r_{43} & r_{44} & r_{45} \\ r_{51} & r_{52} & r_{53} & r_{54} & r_{55} \\ r_{61} & r_{62} & r_{63} & r_{64} & r_{65} \\ r_{71} & r_{72} & r_{73} & r_{74} & r_{75} \\ r_{81} & r_{82} & r_{83} & r_{84} & r_{85} \\ r_{91} & r_{92} & r_{93} & r_{94} & r_{95} \end{pmatrix} = \begin{pmatrix} 0.42 & 0.54 & 0.04 & 0 & 0 \\ 0.50 & 0.46 & 0.04 & 0 & 0 \\ 0.54 & 0.42 & 0.04 & 0 & 0 \\ 0.54 & 0.42 & 0.04 & 0 & 0 \\ 0.54 & 0.42 & 0.04 & 0 & 0 \\ 0.33 & 0.54 & 0.13 & 0 & 0 \\ 0.54 & 0.42 & 0.04 & 0 & 0 \\ 0.63 & 0.33 & 0.04 & 0 & 0 \\ 0.54 & 0.46 & 0 & 0 & 0 \end{pmatrix} \tag{1}
$$

式中:r_{ij} 表示第 i 个评价指标对 j 个评价等级的隶属度,它反映了各评价指标与评价等级之间用隶属度表示的模糊关系。

其中,模糊评价矩阵的建立方法采用专家评议法,即由专家利用经验对因素集中各因素相对于其评语等级进行主观评价得到。确定评价矩阵中的元素 r_{ij} 的公式为

$$
r_{ij} = d_{ij}/d \tag{2}
$$

式中:d 表示参加评价的专家人数,d_{ij} 表示第 i 个评价因素做出第 j 评价等级的专家人数。

该矩阵中的元素值是根据飞行员主观评价咨询结果计算得到的。

(5) 模糊综合评价结果:利用权重系数和评价矩阵进行模糊复合运算可得到 $U \times V$ 上的模糊综合评价结果为

$$B = WR = \{0.51, 0.44, 0.05, 0, 0\} \tag{3}$$

(6) 模糊综合评价等级:按模糊综合判定的最大隶属度原则

$$B = \max_{1 \leqslant i \leqslant m} B_i \tag{4}$$

可得 $B_1 = 0.51$,该值位于评价矩阵的第一列,因此,判定一级指标"进近准备"的评价等级为"优秀"。

(7) 模糊综合评价值:

$$Y = B \cdot V^{\mathrm{T}} = \{0.51, 0.44, 0.05, 0, 0\} \cdot (5 \quad 4 \quad 3 \quad 2 \quad 1)^{\mathrm{T}} \tag{5}$$

可得 $Y = 4.46$,因此,一级指标"进近准备"的评价值为 4.46。

对每个一级指标重复上述步骤,得到的模糊综合评价结果如表 5 所示。

表 5　飞行员操作程序的模糊综合评价结果

指标名称	评价结果	评价等级	评价值
U_1:进近准备	{0.51, 0.44, 0.05, 0, 0}	优秀	4.46
U_2:进近实施	{0.80, 0.17, 0.03, 0, 0}	优秀	4.77
U_3:精密仪表进近	{0.83, 0.13, 0.04, 0, 0}	优秀	4.79
U_4:非精密仪表进近	{0.52, 0.48, 0, 0, 0}	优秀	4.64
U_5:稳定进近	{0.90, 0.10, 0, 0, 0}	优秀	4.90
U_6:复飞和失去进近	{0.87, 0.13, 0, 0, 0}	优秀	4.87
U_7:拉平和接地	{0.59, 0.41, 0, 0, 0}	优秀	4.59
U_8:着陆滑跑	{0.65, 0.35, 0, 0, 0}	优秀	4.65

由一级指标的评价结果可得到进近和着陆阶段飞行员操作程序这一总目标的评价矩阵为

$$R = \begin{pmatrix} 0.51 & 0.44 & 0.05 & 0 & 0 \\ 0.80 & 0.17 & 0.03 & 0 & 0 \\ 0.83 & 0.13 & 0.04 & 0 & 0 \\ 0.52 & 0.48 & 0 & 0 & 0 \\ 0.90 & 0.11 & 0 & 0 & 0 \\ 0.87 & 0.13 & 0 & 0 & 0 \\ 0.59 & 0.41 & 0 & 0 & 0 \\ 0.65 & 0.35 & 0 & 0 & 0 \end{pmatrix} \tag{6}$$

由式(6)和表 3 中的一级指标的权重系数$\{w_{U1}, w_{U2}, w_{U3}, w_{U4}, w_{U5}, w_{U6}, w_{U7}, w_{U8}\}$进行模糊运算,可得总目标的模糊综合评价结果为

$$B = W \cdot R = \{0.72, 0.27, 0.01, 0, 0\} \tag{7}$$

可得进近和着陆阶段飞行员操作程序模糊评价等级为"优秀",评价值$Y = 4.71$。

通过分析表 5 和式(7)中的模糊综合评价结果,可得到以下几点结论:

(1) 进近和着陆阶段飞行员操作程序的评价等级是优秀,这一结果说明波音 737 - NG 系列飞机的操作程序很好,不存在影响飞行安全的问题。

(2) 从总目标的评价值可以看出,进近和着陆阶段飞行员操作程序分值较高,说明波音 737 - NG 系列飞机的操作程序整体很好。

(3) 从每个次级阶段即一级指标的评价值可以看出,每个阶段的分值都较高且比较集中,说明每个阶段的操作程序都很完善。

4 结束语

本文以波音 737 - NG 系列飞机的进近和着陆阶段飞行员操作程序的综合评价研究为例,提出了一种功能流程图、德尔菲法、序关系分析法和模糊综合评价法相结合的飞行员操作程序综合评价方法。该方法是以飞行阶段为基础建立的飞行员操作程序评价指标体系,而对于不同类型大型客机而言,飞行阶段的划分是相似的,因此本方法可应用于其他大型客机的飞行员操作程序综合评价。

本文对飞行员操作程序的综合评价研究主要集中于操作内容,未考虑操作过程中飞行员认知和决策对飞行安全的影响。因此,在进一步的研究中可以对飞行员操作过程中脑力负荷和情景意识等进行分析研究。

致谢

感谢中国国际航空公司飞行总队第三大队的殷雪冬及其他飞行员的帮助和支持!

参考文献

[1] 《飞机设计手册》总编委会. 飞机设计手册:第 17 册. 航空电子系统及仪表[M]. 北京:航空工业出版社,2001.

[2] 中国民用航空总局飞行标准司. 机组标准操作程序[Z]. AC - 121 - 22,2007.

[3] Lautman L, Gallimore P L. Control of the Crew Caused Accident: Results of a 12 - operator Survey [J]. Boeing Airline. 1987(Apr-Jun): 1 - 6.

[4] National Transportation Safety Board. A Review of Flightcrew-involved Major Accidents of US Air Carriers, 1978 Through 1990[R]. Washington, DC: , 1994.

[5] Degani A, Wiener E L. Human Factors of Flight-Deck Checklists: The Normal Checklist [R]. NASA Contractor Report 177549,1990.

［6］ Degani A，Wiener E L． Procedures in complex systems The airline cockpit［J］． IEEE Transactions on Systems，Man，and Cybernetics Part A：Systems and Humans，1997，27(3)：302-312.

［7］ 杨治琰. 战斗机飞行员操作程序的设计和评估[J]. 电光与控制，2000(01)：9-19.

［8］ Jonathan O，Sue C，Mary R． What 'Ideas-about-Science' Should Be Taught in School Science? A Delphi Study of the Expert Community［J］． Journal of Research in Science Teaching，2003，40(7)：692-720.

［9］ 郭亚军. 综合评价理论、方法及应用[M]. 北京：科学出版社，2007.

机组工作负荷评价新方法及其应用

王黎静[1],向　维[1],高晓华[1],李润山[2]

1. 北京航空航天大学 航空科学与工程学院,北京 100191

2. 中国国际航空公司飞行总队,北京 100621

摘要　机组工作负荷的评估是适航审核的重要指标之一,FAA 适航标准 CFR - 25.1523 和中国适航标准 CCAR - 25.1523 中都对机组工作负荷的审核提出了要求。①研究适航标准、波音 737 和空客 320 机组操作手册及其他相关文献,通过与飞行员面谈建立评价体系基本框架。应用德尔菲法进行咨询确定评价体系指标,应用序关系分析法进行指标权重系数调查,最终建立机组工作负荷综合评价体系。②基于所建立的评价体系,应用模糊综合评价法对 B737 和 A320 机组工作负荷进行了综合评价,评价结果表明研究所建立的民机机组工作负荷评价体系合理,具有实用性。③所获得的机组工作负荷评估体系及评价结果可为 C919 等同类机型的机组工作负荷设计提供参考。

关键词　机组工作负荷,评价指标,权重系数,模糊综合评价,适航

A New Method to Evaluate Crew Workload and the Application

Wang Lijing[1] , Xiang Wei[1] , Gao Xiaohua[1] , Li Runshan[2]

1. School of Aeronautic Science and Engineering，Beijing University of Aeronautics and Astronautics，Beijing 100191

2. Flying Corps of Air China ，Beijing 100621

Abstract　Crew workload assessment is an important index of airworthiness certification. Wokload assess-ment is required by both CFR-25. 1523 of FAA airworthiness certification and CCAR-25. 1523 of Chinese airworthiness certification. The basic frame of evaluation index system was established by studying airworthiness certification, crew operation manual of Boeing 737 and Airbus 320, other correlated references and interviewing with pilots; the crew workload comprehensive evaluation index system was established by consultation with Delphi method, and index weight coefficients questionnaire with Order Relation Analysis method. The crew workload of Boeing 737 and Airbus 320 was evaluated with Fuzzy Comprehensive Evaluation method based on the built evaluation index and the result shows that this index is reasonable and practicable. The evaluation index system and evaluation results can provide design reference on crew workload for C919.

Keywords crew workload，evaluation index，weight coefficients，Fuzzy Comprehensive Evaluation，airworthiness

　　1981 年，美国出台了关于指导飞行机组定员任务的建议，即关于新一代商用运输机的两人制机组操作是否安全，其验证是否与秘书处的规定(1958 年联邦航空局法令中用来促进飞行安全的验证条款)一致的建议[1]。过去 30 年里，研究者们研究了大量的机组工作负荷的评价方法，但大多数方法和 AC25.1523 中的方法一样仅仅被谨慎地应用于可控的飞行模拟实验[2,3]。就目前检索的资料，能用于驾驶舱早期设计的负荷评估方法和工具少之又少。本文从早期介入设计的观点出发，研究一个基于 A320，B737(C919 竞争机型)机组工作负荷评估体系，之后基于所建立的评价体系，应用模糊综合评价法对 B737 和 A320 机组工作负荷进行了综合评价，在此基础上提出了 C919 驾驶舱机组工作负荷设计中应该关注的设计系统及设计参考原则。基于本文提出的机组工作负荷评价体系，设计人员可在驾驶舱设计中的各个阶段方便、清晰地评价驾驶舱机组工作负荷，使机组工作负荷的设计更合理。

1　机组工作负荷综合评价体系

1.1　建立民机机组工作负荷综合评价指标体系

　　通过对 FAA 适航标准 CFR - 25.1523、中国适航标准 CCAR - 25.1523[4]、B737 及 A320 现有飞行员机组使用手册[5,6]和相关文献[7]的研究，并与这两种机型现役飞行员多次面谈，从以下 11 个因素对机组工作负荷进行了细化：①操纵装置的可达性及操纵的简易程度；②仪表及警告装置的可视性及醒目程度；③燃油非正常操作程序；④消耗的负荷大小和持续的时间；⑤系统监控程度；⑥需机组成员离开原定职能岗位才能完成的动作；⑦飞机系统的自动化程度；⑧通信工作量；⑨导航工作量；⑩应急程序；⑪一名驾驶员丧失能力的操作程序。评估体系的部分内容如表 1 所示。

表 1　民机机组工作负荷综合评价指标体系模型

总目标层	子目标层	子子目标层	指标层
民机机组工作负荷	O_1：操纵装置的可达性及操纵的简易程度	$O_{11}(U_1)$：飞行操纵系统(人工飞行)	u_{11}：基本飞行操纵装置
		$O_{12}(U_2)$：发动机系统	……
		$O_{13}(U_3)$：自动飞行系统	……

（续表）

总目标层	子目标层	子子目标层	指标层
民机机组工作负荷	O_1:操纵装置的可达性及操纵的简易程度	$O_{14}(U_4)$:指示/记录系统	u_{41}:EFIS 面板和 ECAM 控制面板
			……
		$O_{15}(U_5)$:通信系统	……
		$O_{16}(U_6)$:导航系统	……
		$O_{17}(U_7)$:电气系统	……
		$O_{18}(U_8)$:燃油系统	……
		$O_{19}(U_9)$:液压系统	……
		$O_{110}(U_{10})$:气源系统	u_{101}:引气控制
			u_{102}:空调控制/通风控制
			u_{103}:增压控制
		$O_{111}(U_{11})$:防冰、防雨系统	……
		$O_{112}(U_{12})$:防火系统（发动机/APU 防火操纵）	……
		$O_{113}(U_{13})$:灯光系统	……
		$O_{114}(U_{14})$:起落架系统	u_{141}:起落架收放控制
			……
	O_2:仪表及警告装置的可视性及醒目程度	$O_{21}(U_{15})$:仪表	u_{151}:主飞行显示器（PFD）
			u_{152}:导航显示器（ND）
			……
		$O_{22}(U_{16})$:警告装置	……
			u_{164}:引导机组快速辨明故障类型的显示
	$O_3(U_{17})$:燃油非正常操作程序		……
	$O_4(U_{18})$:消耗的负荷大小和持续的时间		u_{181}:非正常情况下消耗体力负荷的大小和持续的时间
			u_{182}:非正常情况下消耗脑力负荷的大小和持续的时间
	$O_5(U_{19})$:系统监控程度		……
			u_{193}:监控增压系统信息
			u_{194}:监控飞行操纵系统信息
			……

（续表）

总目标层	子目标层	子子目标层	指标层
民机机组工作负荷		$O_6(U_{20})$：需机组成员离开原定职能岗位才能完成的动作	......
		$O_7(U_{21})$：飞机系统的自动化程度	......
		$O_8(U_{22})$：通信工作量	......
		$O_9(U_{23})$：导航工作量	......
		$O_{10}(U_{24})$：应急程序	......
		$O_{11}(U_{25})$：一名驾驶员丧失能力的操作程序	......

1.2 德尔菲法筛选指标

初步确定机组工作负荷评价指标体系后，采用改进德尔菲[8]法筛选指标确定了机组工作负荷的评价体系。两次接受咨询的专家都来自某航空公司飞行总队的现役飞行员，且均飞过 A320 系列及波音 737 系列机型。参考文献[8]把专家一致意见定义为：不少于 2/3（或 67%）的专家判断等级为"大"及"大"以上的判断结果，即 Likert 量表等级≥3。

1.3 确定机组工作负荷评价指标权重系数

研究中采用序关系分析法[9]确定各指标的权重系数。接受指标权重系数咨询的专家组同前。权重系数调查共送出 25 份调查表，收回 22 份有效问卷。对不同专家人数下 11 个一级指标中的部分权重系数的计算如图 1 所示。图中随着专家人数的逐渐增加，权重系数趋于稳定。研究中人数达 12 个以后，11 个一级指标的权重系数没再出现大的起伏，与文献[9]中所述"专家人数以 10～30 人为宜"是吻合的。计算所得的一级指标权重系数如表 2 所示。

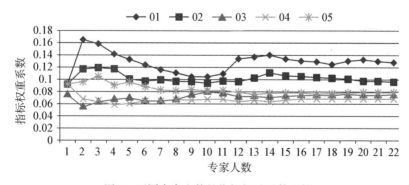

图 1 不同专家人数的指标权重系数比较

表 2 子目标层权重系数

子目标集	权重系数
$\{w_{O_1},w_{O_2},w_{O_3},w_{O_4},w_{O_5},w_{O_6},w_{O_7},$ $w_{O_8},w_{O_9},w_{O_{10}},w_{O_{11}}\}$	$\{0.13,0.1,0.08,0.07,0.08,0.07,0.09,$ $0.06,0.07,0.13,0.12\}$

2 机组工作负荷模糊综合评价

2.1 机组工作负荷的主观评价

利用主观评价法[10]对波音 737 和空客 320 机型的机组工作负荷评价体系中的指标进行了评价。评价专家组同前。调查中的评价等级及其含义如表 3 所示。发送空客 320 机型、波音 737 机型机组工作负荷综合评价调查问卷各 25 份,分别收回有效问卷 23 和 17 份。

表 3 民机机组工作负荷评价等级及其含义

评价等级	分值	评语含义
很大	0.9	给飞行机组带来极高的工作负荷
大	0.7	给飞行机组带来较高的工作负荷
一般	0.5	给飞行机组带来一般的工作负荷
小	0.3	给飞行机组带来较低的工作负荷
很小	0.1	给飞行机组带来忽略不计的工作负荷

主观综合评价值:按有界和、有界积算子计算原则 $X=R_i\cdot V^{\mathrm{T}}$,其中 $V^{\mathrm{T}}=(0.9\quad 0.7\quad 0.5\quad 0.3\quad 0.1)$。

矩阵 R 中的元素 r_{ij} 的公式为

$$r_{ij}=d_{ij}/d \tag{1}$$

式中:d 表示参加评价的专家人数,d_{ij} 表示第 i 个评价因素做出第 j 评价等级的专家人数。

2.2 机组工作负荷模糊综合评价

论文以指标体系中的子目标"操纵装置的可达性及操纵的简易程度"(O_1)为例,展示所进行的机组工作负荷模糊综合评价步骤:

(1) 评价对象:U_{10} 气源系统(A320)。评价指标集:$U_{10}=\{U_{101},U_{102},U_{103}\}$,指标具体名称如表 1 所示。

(2) 拟定评语级:$V=\{$很大(v_1),大(v_2),一般(v_3),小(v_4),很小$(v_5)\}$。

(3) 指标权重系数:

$$W=\{w_{u_{101}},w_{u_{102}},w_{u_{103}}\}=\{0.33,0.29,0.38\}$$

（4）模糊评价矩阵：

$$\boldsymbol{R} = \begin{bmatrix} r_{1011} & r_{1012} & r_{1013} & r_{1014} & r_{1015} \\ r_{1021} & r_{1022} & r_{1023} & r_{1024} & r_{1025} \\ r_{1031} & r_{1032} & r_{1033} & r_{1034} & r_{1035} \end{bmatrix} = \begin{bmatrix} 0 & 0.18 & 0.70 & 0.06 & 0.06 \\ 0 & 0.12 & 0.65 & 0.17 & 0.06 \\ 0 & 0.41 & 0.41 & 0.12 & 0.06 \end{bmatrix} \quad (2)$$

式中：r_{ij} 表示第 i 个评价指标对 j 个评价等级的隶属度，它反映了各评价指标与评价等级之间用隶属度表示的模糊关系。

（5）模糊综合评价结果：利用权重系数和评价矩阵进行模糊复合运算可得到 $U \times V$ 上的模糊综合评价结果为

$$\boldsymbol{B} = \boldsymbol{W} \cdot \boldsymbol{R} = \{0, 0.25, 0.58, 0.11, 0.06\} \quad (3)$$

（6）模糊综合评价等级：按模糊综合判定的最大隶属度原则

$$\boldsymbol{B} = \max_{1 \leqslant i \leqslant m} B_i \quad (4)$$

可得 $B = B_3 = 0.58$，该值位于评价矩阵的第三列，因此，指标集"气源系统"的评价等级为"一般"。

（7）模糊综合评价值：按有界和、有界积算子计算原则

$$\boldsymbol{Y} = \boldsymbol{B} \cdot \boldsymbol{V}^{\mathrm{T}} = (0 \quad 0.25 \quad 0.58 \quad 0.11 \quad 0.06) \cdot (0.9 \quad 0.7 \quad 0.5 \quad 0.3 \quad 0.1)$$
$$(5)$$

可得 $Y = 0.503$，因此，指标集"气源系统"的评价值为 0.503。对每个指标及子子目标、子目标集重复上述步骤，可获得空客 320 子目标模糊综合评价矩阵。

$$\boldsymbol{R} = \begin{bmatrix} 0.07 & 0.28 & 0.46 & 0.17 & 0.02 \\ 0.03 & 0.31 & 0.52 & 0.13 & 0.01 \\ 0.04 & 0.54 & 0.36 & 0.06 & 0 \\ 0.09 & 0.54 & 0.37 & 0 & 0 \\ 0.03 & 0.28 & 0.47 & 0.22 & 0 \\ 0.12 & 0.36 & 0.44 & 0.06 & 0.02 \\ 0.01 & 0.29 & 0.56 & 0.12 & 0.02 \\ 0.05 & 0.33 & 0.43 & 0.15 & 0.04 \\ 0 & 0.12 & 0.65 & 0.23 & 0 \\ 0.12 & 0.70 & 0.18 & 0 & 0 \\ 0.29 & 0.42 & 0.29 & 0 & 0 \end{bmatrix} \quad (6)$$

由式（6）和子目标的权重系数 $\{w_{o_1}, w_{o_2}, w_{o_3}, w_{o_4}, w_{o_5}, w_{o_6}, w_{o_7}, w_{o_8}, w_{o_9}, w_{o_{10}}, w_{o_{11}}\}$ 进行模糊运算，可得总目标的模糊综合评价结果为

$$\boldsymbol{B} = \boldsymbol{W} \cdot \boldsymbol{R} = \{0.09, 0.39, 0.41, 0.10, 0.01\} \quad (7)$$

$$Y = \boldsymbol{B} \cdot \boldsymbol{V}^{\mathrm{T}} = (0.09 \quad 0.39 \quad 0.41 \quad 0.10 \quad 0.01) \cdot (0.9 \quad 0.7 \quad 0.5 \quad 0.3 \quad 0.1)$$

$$\tag{8}$$

可得，空客 320 机组工作负荷模糊评价等级为"一般"，评价值 $Y = 0.590$。

同样的方法可计算出 B737 机组工作负荷评价值 $Y = 0.597$。

3 结果分析及讨论

用模糊综合评价方法对 B737 和 A320 机组工作负荷进行评价后，可获得如表 4、表 5、表 6 所示的系列评价结果。

表 4 320 和 B737 总目标层的模糊综合评价结果

指标名称	A320 模糊综合评价结果	评价等级	评价值	B737 模糊综合评价结果	评价等级	评价值
O	{0.09, 0.39, 0.41, 0.10, 0.01}	一般	0.590	{0.10, 0.41, 0.38, 0.09, 0.02}	大	0.597

表 5 部分 A320 和 B737 指标层的模糊综合评价结果

指标名称		A320 模糊综合评价结果	评价等级	B737 模糊综合评价结果	评价等级
O_{11}	U_1	{0.12, 0.59, 0.24, 0.06, 0.00}	大	{0.29, 0.41, 0.29, 0.00, 0.00}	大
O_{14}	U_4	{0.00, 0.06, 0.47, 0.29, 0.18}	一般	{0.00, 0.24, 0.24, 0.53, 0.00}	小
O_{15}	U_5	{0.04, 0.10, 0.61, 0.23, 0.02}	一般	{0.11, 0.45, 0.31, 0.10, 0.04}	大
O_{114}	U_{14}	{0.00, 0.22, 0.54, 0.24, 0.00}	一般	{0.12, 0.35, 0.34, 0.11, 0.05}	大
O_{21}	U_{15}	{0.04, 0.31, 0.46, 0.17, 0.02}	一般	{0.00, 0.47, 0.44, 0.09, 0.00}	大

表 6 部分 A320 和 B737 子子目标层的模糊综合评价结果

指标名称	A320 模糊综合评价结果	评价等级	B737 模糊综合评价结果	评价等级
u_{11}	{0.12, 0.59, 0.23, 0.06, 0.00}	大	{0.29, 0.41, 0.29, 0.00, 0.00}	大
u_{41}	{0.00, 0.06, 0.47, 0.29, 0.18}	一般	{0.00, 0.24, 0.24, 0.53, 0.00}	小
u_{141}	{0.00, 0.18, 0.47, 0.35, 0.00}	一般	{0.06, 0.41, 0.35, 0.18, 0.00}	大
u_{151}	{0.06, 0.59, 0.29, 0.06, 0.00}	大	{0.00, 0.71, 0.29, 0.00, 0.00}	大
u_{152}	{0.06, 0.35, 0.47, 0.12, 0.00}	一般	{0.00, 0.53, 0.35, 0.12, 0.00}	大
u_{164}	{0.06, 0.35, 0.53, 0.06, 0.00}	一般	{0.12, 0.35, 0.29, 0.24, 0.00}	大
u_{181}	{0.06, 0.41, 0.53, 0.00, 0.00}	一般	{0.21, 0.38, 0.35, 0.06, 0.00}	大
u_{182}	{0.12, 0.65, 0.24, 0.00, 0.00}	大	{0.18, 0.59, 0.24, 0.00, 0.00}	大
u_{193}	{0.00, 0.35, 0.47, 0.18, 0.00}	一般	{0.00, 0.47, 0.41, 0.06, 0.06}	大
u_{194}	{0.00, 0.47, 0.35, 0.18, 0.00}	大	{0.06, 0.35, 0.41, 0.18, 0.00}	一般

表 4 反映了总目标层的模糊综合评价结果，尽管表中 A320 机组工作负荷评价

等级为"一般",B737 机组工作负荷评价等级为"大",但是两者模糊综合的评价值相差不大,A320 是 0.590,B737 是 0.597,由此可见 A320 的机组负荷与 B737 的机组负荷差异不大。

表 5 反映了部分指标层的评价结果。这些评价结果反映了系统设计对机组工作负荷的影响。

O_{11}-飞行操纵系统和 u_{11}-基本飞行操纵装置,在 A320 和 B737 中负荷评价等级均为"大"。原因是在飞行中经常使用,而且飞行操纵与气压、温度、飞机重量、重心等诸多因素相关,在起飞和着陆时操纵尤为重要。

O_{14}-指示/记录系统,u_{41}-EFIS 面板和 ECAM 控制面板,在 A320 和 B737 的负荷评价等级分别为"一般"、"小"。原因是 A320 比 B737 多了 ECAM 控制面板,增加了飞行员操作的按钮。

O_{15}-通信系统,在 A320 和 B737 的负荷评价等级分别为"一般"、"大"。在使用"内话/无线电开关"时,当 A320 不发射时,无需一直按着按钮,而 B737 无论是发射还是不发射都需要一直按着按钮,给飞行员带来了负荷。

O_{114}-起落架系统,u_{141}-起落架收放控制,在 A320 和 B737 中的负荷评价等级为"一般"、"大"。因为在 A320 中,起落架的收放控制只有"收上"和"放下",但是在 B737 中除此之外,还有"关断位(切断液压)"的操作。

O_{21}-仪表的可视性及醒目程度,u_{152}-导航显示器(ND),u_{164}-引导机组快速辨明故障类型的显示,在 A320 中负荷评价等级为"一般",在 B737 中负荷评价等级为"大"。u_{151}-主飞行显示器(PFD),在 A320 和 B737 中负荷评价等级均为"大"。因为系统集成导致 PFD 显示的信息量很大,这对于飞行员来说在众多信息中识别有效信息、关键信息等来说负荷是很大的,因此无论是 A320 还是 B737,PFD 带给飞行员的负荷都很大。而对于 ND,在 A320 中是自动调节的,而在 B737 中手动调节的,这导致在 B737 中飞行员的负荷要大些。而对于引导机组快速辨明故障类型的显示,由于 A320 中多了 ECAM(电子中央飞机监控),飞行员只需按照 ECAM 的指示来执行各种操作,但是 B737 中只告诉信息的位置,还需要飞行员自己判断操作。

表 6 给出了部分子子目标层的评价结果。这些评价结果都反映了具体设计给机组工作负荷带来的影响。

u_{182}-非正常情况下消耗脑力负荷的大小和持续的时间,在 A320 和 B737 中负荷评价等级均为"大"。因为在非正常情况下,飞行员需要大量的思考,加上心理的紧张,导致脑力负荷极大。

u_{181}-非正常情况下消耗体力负荷的大小和持续的时间,在 A320 和 B737 中负荷评价等级分别为"一般"、"大"。因为 A320 是电传系统,没有很多操作需要使很大的力,而 B737 是机械传动,很多操作需要很大的体力(比如双液压失效),而且 A320 的计算机系统会减少飞行员很多的体力负荷,比如单发失效时,A320 会自动补偿,而 B737 不能。

u_{193}-监控增压系统信息,在 A320 和 B737 中负荷评价等级分别为"一般"、"大"。因为 A320 中,增压信息的调节显示在人工位时,只需判断"上升"和"下降"来调节上升和下降,信息比较直接。而在 B737 中,需通过调节排气活门来调节上升和下降,信息的获取不是很直接。u_{194}-监控飞行操纵系统信息,在 A320 中负荷评价等级为"大",在 B737 中负荷评价等级为"一般"。原因是在 A320 中,飞行操纵由计算机完成,如 ELAC(升降舵副翼计算机)、SEC(扰流板升降舵计算机)、FAC(飞行增稳计算机)等来完成,因此飞行员需监控的飞行操纵信息比较多,而在 B737 中飞行操纵信息的监控一部分是靠仪表,一部分是靠感觉,来判断飞机是否符合当时的状态。

4　结论

基于空客 320 和波音 737,确定了民机机组工作负荷综合评价体系,包括评价指标体系及其权重系数。应用模糊综合评级方法对空客 320 和波音 737 机组工作负荷进行了综合评价后发现:

(1) 基本飞行操纵装置、主飞行显示器(PFD)、非正常情况下消耗脑力负荷的大小和持续的时间三个指标的工作负荷等级均为大,建议 C919 在进行与这些指标相关的系统设计时应着重考虑其引起的机组工作负荷问题。

(2) EFIS 面板、ECAM 控制面板、通信系统、监控飞行操纵系统信息四个指标,A320 的工作负荷等级比 B737 的高,建议 C919 在进行与这些指标相关的系统设计时参考 B737 的设计。

(3) 起落架收放控制、导航显示器(ND)、引导机组快速辨明故障类型的显示、监控增压系统信息、非正常情况下消耗体力负荷的大小和持续的时间共五个指标,B737 的工作负荷等级比 A320 的工作负荷等级高,建议 C919 在进行与这些指标相关的系统设计时参考 A320 的设计。

参考文献

[1] Federal Aviation Administration. Minimum Flight crew[R]. AC 25. 1523 - 1,1993,2005.

[2] Roscoe A H. Assessing pilot workload [R]. AGARDograph No. 233, Neuilly Sur Seine: 1978.

[3] Williges R C, Wierwille W W. Behavioral measures of aircrew mental workload [J]. Human Factors,1979,21:549 - 574.

[4] 中国民用航空总局. CCAR - 25 - R3,中国民用航空规章第 25 部运输类飞机适航标准[S].

[5] The Boeing Company. Boeing 737 flight crew training manual [R]. FCT 737 (TM),2006.

[6] The Airbus Company. A318/A319/A320/A321 flight crew operating manual [R]. FCOM VOL. 1 REV041E, 2008.

[7] 《飞机设计手册》总编委会. 飞机设计手册[M].北京:航空工业出版社,2001.

[8] Jonathan O, Sue C, Mary R et al. What "Ideas-about-Science" School Be Taught in School

Science? A Delphi Study of the Expert Community [J]. Journal of Research in Science Teaching，2003：692－720.

[9] 郭亚军. 综合评价理论、方法及应用[M]. 北京：科学出版社，2007.

[10] 苊垆. 实用模糊数学[M]. 北京：科学技术文献出版社，1989.

[11] 王黎静，郭奋飞，何雪丽，等. 大型客机飞行员操作程序综合评价，北京航空航天大学学报，2010，36(11)：1436－1439.

飞行员视觉信息流强度模拟及适人性分析

张慧姝，庄达民

北京航空航天大学航空与工程学院，北京 100191

摘要 针对飞行员在执行飞行任务时，需要快速辨认和处理很多视觉信息问题，开发了用于工效学实验的动态模拟飞行座舱显示界面的仿真模型，包括执行飞行任务信息，设置天气、机场、跑道等。实验基于该模型，运用工业设计学、语义学和人机工程学的理论和方法设计目标图符编码和战术显示界面，以正确反应率和反应时间来评价不同的视觉信息量对目标辨识和绩效的影响，以及不同的视觉信息量在白天和夜晚对目标辨识和绩效的影响，为飞机人机界面的工效学设计提供科学依据和评价方法。在上万人次的实验和实验分析的基础上，得到以下结果：所建立的视觉信息流模型中视觉信息流强度曲线符合正态分布，正态曲线的峰值不超过1，左右端点在 600 ms 和 900 ms 之间。不同的视觉信息量完成任务差异显著，随着信息量增加，最初辨识较快较准确，达到一定程度时辨识变慢，出现错误的几率升高；在相同视觉信息量情况下，夜晚与白天、动态与静态执行任务辨识和绩效差异显著，白天要比夜晚、静态要比动态辨识效率和绩效更好；不同信息量 900 ms 内基本可清晰辨认目标。

关键词 动态模拟，显示界面，编码设计，视觉信息流，辨识性和绩效

Simulation and Ergonomics Analysis of Pilot Visual Information Flow Intensity

Zhang Huishu, Zhuang Damin

School of Aeronautics Science and Engineering, Beijing University of Aeronautics and Astronautics, Beijing 100191

Abstract Aiming at the problem that the pilot needs to recognize and process many pieces of visual information rapidly when operating an aircraft, a simulation model was developed, which can dynamically simulate the aircraft cockpit display interface, including task-relevant information, weather settings, airdrome, flight path, etc.. Based on this model and in the experiment, the target icon coding and display interface of tactics were designed by applying the theories and approaches of industrial design, semantics and ergonomics. The influence of different visual information flows on the target identification and performance, as well as the influence of different visual information flows in the daytime and at night were evaluated by correct reaction rate and reaction time. The experiment results provide scientific reference and evaluation methods for ergonomic design of aircraft cockpit display interface. On the basis of ten thousand person-times of experiments and the analysis, the results are as follows. The intensity curve of visual information flow is consistent with the normal distribution in the established visual information flow model. The peak amplitude of the normal distribution curve goes beyond 1, and the left and right points are between 600 ms and 900 ms. The task performance was greatly influenced by different information flow. At the very beginning, the identification is relatively rapid and accurate. When the visual information flow increases to a certain amount, the identification becomes slow and the possibility of making mistakes increases. Under the same information flow, there are significant differences of identification performance between daytime and night, as well as between statically and dynamically. The identification performance is better in the daytime than at night, and is better at static state than dynamic state. Target identification can be completed within 900 milliseconds under different information flow.

Keywords dynamic simulation, display interface, coding design, visual information flow, identification and performance

视觉显示是飞机向飞行员传递信息的主要形式。随着科技的发展,人机系统的复杂程度越来越高,对视觉显示的设计要求越来越高。这是因为飞行员通过座舱显示系统接收到的飞行信息大量的增加,如 F-18 座舱内,一个 HUD,三个 MFD 上共计有 62 幅显示画面,675 个缩写符号,其中 177 个符号有四种不同大小,总信息量达 1,000 多个,导致飞行员脑力负荷加重,心理疲劳增多[1, 2]。如何通过视觉快速地获取和正确的选取这些信息,做出合理的决策,使飞行员准确有效地操作,就需要对飞机座舱显示信息进行规划和整合,对飞机座舱显示界面进行良好的设计[3, 4]。本文的目的是通过理论分析和实验来建立视觉信息流强度模型和评价不同信息量、不同时间执行飞行任务的辨识和绩效,为飞机座舱显示界面的工效学设计提供科学依据和评价方法。

1 实验设计

1.1 被试

实验被试为 22 名北京联合大学工业设计专业的本科学生(男 12 名,女 10 名),被试了解飞机座舱显示系统和控制系统的相关知识(在实验前学习 60 学时人机工程学课程,对飞机座舱显示系统和控制系统进行了充分调研,并掌握了相应的理论知识),年龄在 19~20 岁之间,无色盲和色弱,矫正视力 1.0 以上。

1.2 实验设备及环境

实验设备包括惠普 xw8600 图形工作站 8 台;遥杆 5 个(油门杆、操纵杆、方向舵、升降舵和飞行动力系统)能完成飞机的加速、减速、俯仰、偏航和滚转等基本控制;音响 1 个和飞行仿真系统 1 套。在动态模拟飞行座舱显示界面的仿真模型上开发相应的实验程序来开展被试实验并进行工效评价。飞行仿真系统显示在 hp 工作站 19′显示屏上,分辨率为 1,280×1,024 像素。人机信息交互采用遥杆、键盘和鼠标。坐姿忽略,被试和显示屏的距离为 600 mm。由于照明分布和光源对作业绩效会产生一定的影响,要求室内照明分布稳定,光源采用白光照射。实验从四月份持续到七月份,该期间为防止室内温度发生急剧变化而采用空调来控制环境温度。

1.3 实验任务

实验分动态和静态两种情况,动态意指被试在操纵飞机驾驶同时完成确认任务。静态意指被试无须操纵飞机飞行去完成确认任务。根据动态和静态及白天和晚上等共设以下三种实验任务。

(1) 实验要求在模拟系统中设置任务场景,起飞机场为 San Francisco 国际机场,跑道为 10 L 跑道,季节为夏天,白天中午时间,天气晴朗,在该机场沿跑道起飞,飞行高度达到 5,000 ft(1,524 m)时进行平飞。舱内照度 872 lx,无眩光。任务设定为:①在战术信息显示界面中分别完成确认敌机 1 个目标。②确认敌机和友机 2 个目标。③确认敌机、友机和僚机 3 个目标。④确认敌机、友机、僚机和不明飞机 4 个目标。如图 1 所示,为战术信息界面及目标呈现状况。这里完成各项任务同时需要监视飞机姿态(高度,空速,俯仰,滚转)。分别把确认一个目标定义视觉信息量为 1 个,确认 2 个目标定义视觉信息量为 2 个,确认 3 个目标定义视觉信息量为 3 个,确认 4 个目标定义视觉信息量为 4 个。在雷达范围外出现的目标无需进行反应,在雷达扫视范围内出现的目标通过辨认其是友机或是敌机,是敌机的情况通过点击该目标来加以确认。

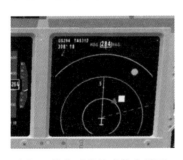

图 1　简化后的战术信息界面

(2) 实验要求在模拟系统中设置任务场景,起飞机场为 San Francisco 国际机场,跑道为 10 L 跑道,季节为夏天,午夜时间,天气晴朗,在该机场沿跑道起飞,飞行高度达到 5,000 ft 时进行平飞,舱内照度 3 lx,无眩光。任务设定同(1)。

(3)飞机在地面上静态时完成任务同(1)。

(1),(2)和(3)要求被试爬升时仰角尽量保持在 13°。

1.4 战术显示界面设计

1.4.1 完成任务时的视觉信息加工过程

如图 2 所示,飞机显示界面信息源通过视觉通道把来自于外部的战术情况和飞机的航姿情况等信息,经过大脑加工,做出判断和决策并向效应器——手和脚对飞机进行操作,改变飞机的运动状态和攻击状况。这种过程不断的循环,直至达到人机系统的预定目的——完成战术任务。在显示界面设计中要充分考虑视觉信息加工过程,并进行评估。

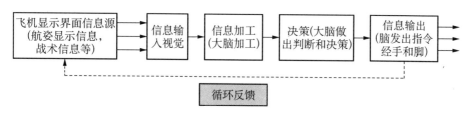

图 2 飞行员完成任务时的视觉信息加工过程

1.4.2 飞行员的视觉行为过程

总的来说这个过程分为做某事和检查所做的事的结果这两个阶段。所以在设计的时候要考虑如图 3 所示行为目标、行为、行为对象和行为结果 4 个方面的问题。如飞行员视觉行为过程,首先确立了作战任务的目标,按照作战任务在飞机座舱显示界面进行视觉行为,经过大脑处理再进行操作,座舱界面又通过视觉将行为结果反馈给人[5]。

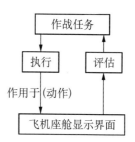

图 3 飞行员视觉行为过程

1.4.3 目标图符设计

图符设计包括图的形状、颜色和大小。根据视觉信息加工过程和飞行员的视觉行为过程特点,设计信息的信道尽量越少越好,以减轻脑力负荷,采用色彩、形状和界面信息几个信道。由于容易吸引视觉注意的视觉特征有大的、亮的、彩色的和变化的(或者闪动的),但是变动的或尽量减少变动和闪动的应用。图形标志的设计原则参照国标《图形符号表示规则·标志用图形符号》[6],优先采用对称图形和实心图形,长宽尽量接近,醒目清晰等。目标的形状和颜色也是影响辨认效率的因素。其形状的优劣次序为:三角形、圆形、梯形、方形、长方形、椭圆形、十字形。当干扰点强度较大时,方形目标优于圆形目标[7]。如图 1 中被确认目标中的图符设计采用了语义学[8]、心理学[9]、工业设计学[10]和人机工程学的原理和方法设计了 11 套方案,根据色彩和形状的变化共设计了表 1 所示的 28 个图符。为了避免构成图符的视觉要素发生混淆,四个目标图符的形状采用不同的形式,设计时采用具象型特征与使用者所熟悉的庶务关联起来,同时设计采用具有语义指向性的图形和色彩[11, 12]。

表 1　图符设计方案

飞机类型 ＼ 方案序号	1	2	3	4	5	6	7	8	9	10	11
敌机	△	▲	△	▲	△	▲	△	△	▲	△	⚡
友机	○	●	○	●	✚	✚	✚	✚	♥	♥	♥
僚机	◎	◉	◎	◉	●	●	◎	◎	◉	○	○
不明飞机	□	■	□	■	◆	◆	◇	◇	◆	◆	◆

色彩设计也参考了 F-22 猛禽战斗机的图符设计。如敌机被显示成红色三角形,友机被显示成绿色三角形,未知敌友的飞机则显示成黄色方块,用蓝色 F-22 图标代表自己的编队飞机。

参照国际《标志用图形符号的制定和测试程序》,采用适当性排序测试和匹配测试评价测试方法,用统计学进行分析,最后选定的方案为表 1 中的 2 号方案[13]。敌机为红色实心三角形,友机为绿色实心圆形,僚机为蓝色圆环实心,不明飞机为黄色实心方块。

1.4.4　战术显示界面设计

战术显示界面配色采用原始的配色,界面其他元素不变。真正的战术显示界面是比较复杂的,为了研究需要,将界面进行了适当的简化。只设计了雷达范围和必要的标识,可在动态情况下进行操作。该显示界面设计了三套方案,黑色背景、黄色字符和雷达范围线;黑色背景、绿色字符和雷达范围线;蓝色背景、白色字符和雷达范围线。

1.5　实验方法

实验前被试进行 3 周的飞行训练(包括起飞、爬升、巡航、降落)和 1 周实验过程的熟悉(包括实验操作方法、实验要求、数据记录等),被试与显示器屏幕之间的距离为 600 mm。为消除疲劳效应采用交替方式进行,中间休息时间 5 min。实验时间为 120 s,允许稍微多点或少点时间,计算机每隔 1 s 记录 1 次数据,记录内容包括通过点击确认的敌机次数,误点击到友机、僚机和不明飞机的次数,误点击到雷达范围外飞机的次数,对目标没有作出反应的次数,没有击中到任何目标的次数,共 6 个指标,以及它们相应的反应时间。当失速时发出警报,并进行记录。每次测试时间为 120 s,数据取 110 s 的数据。

在实验之前进行了小范围的实验,当目标出现间隔时间为 500, 800 和 1,000 ms 时,间隔时间为 500 和 800 ms 的确认正确率较低,未反应率较高。同时根据参考文献[14],当一个具有颜色、形状等多维目标时,以持续时间 1 s,间隔 1 s 的形式呈现,人对其辨认绩效较好,故选择目标出现间隔时间为 1,000 ms。

2　实验分析及讨论

将实验结果进一步分为[情况 1]为完成夜间动态飞行任务、[情况 2]为完成白天动态飞行任务、[情况 3]为完成白天静态飞行任务的平均反应时和平均正确率。

2.1　不同视觉信息量下的辨认性分析

采用 Spss 软件进行数据处理。对 4 种视觉信息量完成任务的反应时进行重复测量方差分析的检验,方差结果显著,$(F(3, 171) = 9.968, p = 0.000)$,也就是被试内因素"视觉信息量"的因素主效应显著,说明不同的视觉信息量对辨识目标的快慢有显著的差异。对其进行 LSD 方法的验后多重比较检验,结果如表 2 所示,发现信息量为 1 个时与 2 个、3 个和 4 个反应时有显著性差异(有 * 为差异显著)。说明视觉信息量为 1 时与视觉信息量为 2,3 和 4 时完成任务辨识快慢有很大的差别(见表 2)。

表 2　视觉信息量主效应验后多重比较结果

评价指标	视觉信息量 I	视觉信息量 J	平均误差	标准误差	p
反应时间	1	2	-23.067^*	4.743	.000
		3	-22.067^*	4.323	.000
		4	-16.867^*	5.654	.004
	2	1	23.067^*	4.743	.000
		3	1.000	4.012	.804
		4	6.200	4.878	.209
	3	1	22.067^*	4.323	.000
		2	-1.000	4.012	.804
		4	5.200	4.940	.297
	4	1	16.867^*	5.654	.004
		2	-6.200	4.878	.209
		3	-5.20	4.940	.297

对实验设计中所定义的三种情况下完成任务的反应时进行方差分析,发现 $(F(4, 302) = 29.113, p = 0.000, p < 0.005)$,说明不同情况下完成任务反应时差异较大。事后检验结果如表 3 所示,此时,[情况 1]与[情况 2]和[情况 3]的反应时有显著差异,说明夜间完成任务和白天完成任务的反应时差异显著,[情况 2]和[情况 3]的反应时有差异显著,说明动态完成任务和静态完成任务差异显著。图 4 为三种情况下确认目标的辨识性比较,由图可知,不同情况下随着视觉信息量的增加,反应时也在增加,增加到一定值时下降;夜间动态完成任务的反应时明显高于白天动态完成任务的反应时,同时也高于静态完成任务时的反应时;动态完成任务的反应时高于静态完成任务的反应时(见表 3)。

表 3 各[情况]主效应验后多重比较校验结果

评价指标	[情况]I	[情况]J	平均误差	标准误差	p
反应时间	1	2	49.750*	6.418	.000
		3	66.625*	7.123	.000
	2	1	−49.750*	6.418	.000
		3	16.875*	5.830	.005
	3	1	−66.625*	7.123	.001
		2	−16.875*	5.830	.005

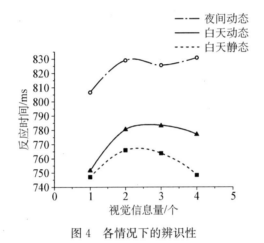

图 4 各情况下的辨识性

2.2 不同视觉信息量下的绩效分析

对 4 种视觉信息量完成任务的正确率进行重复测量方差分析,发现($F(3, 171) = 2.202$,$p = 0.009$),$p > 0.05$,说明不同视觉信息量完成任务的绩效差异不显著,稍有差别。对各情况下完成任务的正确率进行方差分析,发现($F(4, 302) = 29.113$,$p = 0.000$),$p < 0.005$,有统计学意义,说明不同情况下确认目标的绩效差异较大。对其进行 LSD 方法的验后多重比较检验结果如表 4 所示。

表 4 各[情况]主效应验后多重比较校验结果

评价指标	[情况]I	[情况]J	平均误差	标准误差	p
正确率	1	2	−6.637*	1.634	.000
		3	−5.575	2.014	.007
	2	1	6.637*	1.634	.000
		3	1.063	1.787	.554
	3	1	5.575	2.014	.007
		2	−1.063	1.787	.554

情况 1 和情况 2 差异较大,说明夜间确认目标和白天确认目标的绩效差异显

著。如图 5 所示,夜间确认目标的正确率比白天低,白天静态情况下确认目标的正确率比动态情况下要高;随着视觉信息量增加,最初正确率在上升,达到一定值后开始下降。

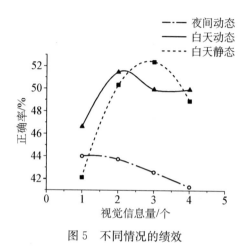

图 5   不同情况的绩效

2.3   视觉信息量与辨识性和绩效之间的关系

如图 6 所示,随着视觉信息量增加,正确率上升,反应时也增加,当反应时和正确率达到一定值时,随着信息量的增加,正确率在下降,反应时稍有增加,但变化不大。

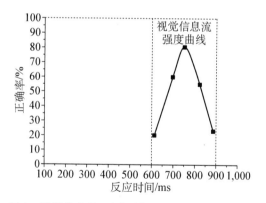

图 6   视觉信息量、反应时间和正确率之间的关系

2.4   视觉信息流强度模型讨论

对多次实验进行统计学正态检验,视觉信息量曲线符合正态分布。由上面的实验分析可以得出信息流强度模型如图 7 所示,视觉信息流强度曲线为正态曲线,正态曲线的峰值不超过 1,左右端点在 600 ms 和 900 ms 之间。根据视觉信息流强度模型可评价界面设计的好坏,包括图符(图形、颜色和大小)、字符(字体、颜色和大小)和信息等设计对信息流的影响。

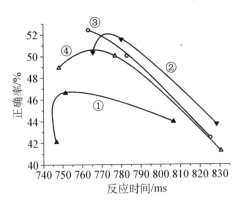

图 7 视觉信息流强度模型

2.5 实验中所测各项分析

实验中共测 6 个指标,每 1 s 记录一次,各项指标统计百分率的平均值如表 5 所示。由表 5 可知,随着视觉信息量的增加,确认目标的正确率在下降,正确判断率略有升高,错误判断率先升高,后下降,误确认情况也在增加。说明由于视觉信息量的增加,视觉信息加工过程的时间变长,出错和没反应的几率在上升。

表 5 实验中各项数据统计(%)

视觉信息量	击中目标(正确率)	在雷达外没有点击(正确判断)	在雷达外点击的(错误判断)	点击了什么都没点中的	没有任何反应的	点中友机或僚机或不明飞机的
1	44	7	23.5	21.5	4	0
2	43.75	7.85	24	16.9	4.8	2.7
3	42.5	7.8	23.6	18	5.4	2.7
4	41.3	7.9	21.9	18	5.5	5.4

2.6 不同视觉信息量的反应时

不同视觉信息量在完成任务时的平均反应时相差不多,测得的视觉平均反应时基本都超过 800 ms,但未超过 900 ms,说明这个时间可清晰辨认目标。

3 结论

(1) 所建立的视觉信息流强度模型表明视觉信息流强度曲线为正态分布,可用于评价界面设计的好坏。

(2) 不同的视觉信息量完成任务差异显著,随着信息量增加开始辨识较快较准确,达到一定时辨识变慢,出现错误的几率升高。

(3) 在相同视觉信息量情况下,夜晚与白天执行任务辨识差异显著,白天要比夜晚辨识更快更好。

（4）在相同的视觉信息量下，动态和静态完成任务辨识差异显著，静态要比动态情况辨识更快。

（5）白天动态完成任务的绩效好于夜晚，差别较明显，白天动态和静态情况下完成任务绩效相差不明显。

（6）不同视觉信息量情况下 900 ms 内基本可清晰辨认目标。

本实验的方法和结论对于人机显示界面设计的适人性研究有一定参考价值。

参考文献

［1］ 郭小朝，刘宝善，马雪松，等. 战术导航过程中新歼飞行员的信息显示需求[J]. 人类工效学，2003，9(1)：5-22.

［2］ 熊端琴，郭小朝，马雪松，等. 新型歼击机平显地空数传指令显示方案的工效学实验研究[J]. 人类工效学，2006，12(3)：7-9.

［3］ 张磊，庄达民，颜吟雪. 飞机座舱显示界面编码方式[J]. 南京航空航天大学学报，2009，41(4)：466-469.

［4］ Zhang H S, Zhuang D M, Wu F. The Study on Pleasure and Ergonomics of Cockpit Interface Design[C] //. Proceeding 2009 IEEE 10th International Conference on Computer-Aided Industrial Design & Conceptual Design. Wenzhou：IEEE Press，2009：1400-1402(in Chinese).

［5］ 赵江洪. 人机工程学[M]. 北京：高等教育出版社，2008：174-211.

［6］ GB/T 16901.1—1997,图形符号表示规则·标志用图形符号[S].

［7］ 颜声远，许彧青. 人机工程与产品设计[M]. 哈尔滨：哈尔滨工程大学出版社，2005：27-43.

［8］ 陈慎任，等. 设计形态语义学[M]. 北京：化学工业出版社，2005：392-482.

［9］ Fredrickson B L, Joiner T. Positive emotions trigger upward spirals toward emotional well-being [J]. New York：Psychological Science，2002，13(2)：172-175.

［10］ Donald A N. Emotion design：Why we love (or hate) everyday things[M]. New York，A member of the Perseus Books Group，2005：161-194.

［11］ 张宪荣，张萱. 设计色彩学[M]. 北京：化学工业出版社，2003：112-115.

［12］ 薛澄崎. 产品色彩设计[M]. 南京：东南大学出版社，2007：4-51.

［13］ GB/T12103—1990,标志用图形符号的制定和测试程序[S].

［14］ 曾庆新，庄达民，马银香. 脑力负荷与目标辨认[J]. 航空学报，2007，8(28)：76-80.

复杂人机智能系统功能分配方法综述

汤志荔,张　安,曹　璐,刘跃峰

西北工业大学电子信息学院,西安 710072

摘要　功能分配是复杂人机智能系统设计进程中的重要内容,它需要应用系统的分析方法,合理地进行人、机两者的任务分配和科学地设计两者的功能结合。分析了国内外功能分配的研究现状和存在的问题。针对复杂人机智能系统的设计需求,指出了系统功能分配方法的研究方向。

关键词　复杂人机智能系统,系统工程,功能分配

Survey of Functions Allocation Methods of Complex Human-Machine Intelligent System

Tang Zhili，Zhang An，Cao Lu，Liu Yuefeng

School of Electrical Information，Northwestern Polytechnical University，Xi'an 710072

Abstract　Functions Allocation(FA)，which applies system analytic methods to properly allocate the tasks of system to human or machine，is an important issue in the research of complex human-machine intelligent system. Based on the principles of FA，the status and problems of FA are discussed. Furthermore，aiming at the demands of system design，the research trends of FA methods are presented.

Keywords　Complex Human-Machine Intelligent System，System Engineering，Functions Allocation

1　引言

在复杂智能人机系统中,存在两个智能个体——人与智能机器。这样的人机系统与传统的人机系统无论在组成结构,还是内在机理上都十分不同,它是一个人机

结合的智能系统,具有智能性、开放性、复杂性、突现性等特点。为了研究这种系统,美国著名学者 Lenat 和 Feigenbaum 提出了人机智能系统(Man-machine Intelligent System)的概念[1],与此类似的还有 Sheridan 提出的人与自动化(Human-automation)概念[2]等。我国著名科学家钱学森以及中科院的戴汝为等人提出了人机综合集成思想[3,4]。浙江大学的路甬祥、陈鹰等提出了人机一体化理论[5,6]。龙升照等人提出了人-机-环境系统工程理论[7],这些都为研究人机智能系统提供了大量的理论依据。虽然目前还没有一套统一的严格的研究复杂人机智能系统的理论体系,但是从上述各种方法可以发现共同点,即必须在系统设计之初到系统实现的整个生命周期过程中考虑人与机器的关系,人机功能的合理分配是体现复杂人机系统智能化的关键。

2 功能分配的概念

早在 1951 年,Fitts 就第一次明确提出功能分配的概念[8]。它是指将系统中的功能或任务分派给人或者机器的过程。这种功能分配活动发生于系统综合和评价过程中的早期阶段,因此属于系统工程的研究范畴,这里的功能分配主要是强调系统组成成分之间的功能分解。后来由于机械、自动化以及计算机技术的发展,绝大部分系统都属于人机系统,这类系统设计任务中的功能分配需要越来越多地考虑到系统中人的因素,所以将它专门作为人因工程的一项重要研究内容,而称之为人机功能分配。自从功能分配的概念提出以来,它在工业与自动化领域尤其是在核电厂监控自动化系统、空中交通管制系统、飞机座舱、载人航天器等复杂系统的设计中得到了广泛的应用。

3 功能分配方法的研究现状

3.1 国外的研究情况

功能的分配原则与一定的应用背景相结合,逐渐形成了各种功能分配方法。目前在国际比较有影响力的几种功能分配方法有:人机能力比较分配法、Price 决策图法、Sheffield 法、自动化分类与等级设计法、York 法等。

3.1.1 人机能力比较分配法

它是最初的功能分配方法,例如著名的 Fitts Lists 分配方法,也是迄今为止应用最为普遍的方法,在早期的简单工业自动化监控系统中得到大量的应用。表 1 即为 Fitts 所列出人机各自的优势特性,也称为 MABA - MABA 方法。

表 1 Fitts 人机能力对比表

人擅长于:(Men Are Better At)
能够探测到微小范围变化的各种信号
对声音或光的模式感知

（续表）

创造或运用灵活的方法
长期存贮大量的信息并在适当的时候运用
运用判断能力
归纳推理能力

机器擅长于：(Machines Are Better At)
对控制信号的快速反应
能够精确和平稳的运用能量
执行重复、程序性的任务
能够存储简短的信息，并能完全删除它们
计算和演绎推理能力
能够应付复杂的操作任务

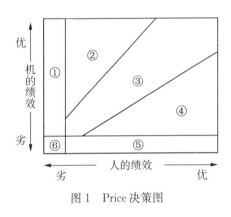

图 1　Price 决策图

3.1.2　Price 决策图法

Price 决策图法[9]对任意一个功能，从人机两方面的特性做出比较，然后根据效能、速度、可靠性、技术可行性等做出评估，评估结果为一个复数值（人的绩效值为实部，机器的绩效值为虚部）。这个复数值落在决策图的某一区域，如图 1 所示。

决策图由 6 个区域组成，每个区域对应于不同的人机绩效和分配方案：①表示将功能分配给机器；②表示将功能分配给机器；③表示既可分配给人也可分配给机器，存在一个最佳分配点；④将功能分配给人；⑤将功能分配给人；⑥采用其他的方法重新设计。

Price 决策法虽然在 Fitts Lists 分配法的基础上更进一步的明确了人机功能分配的过程，但是它对于如何计算绩效却没有明确的描述，并且客观上计算人和机器的绩效相当困难。

3.1.3　Sheffield 法

Sheffield 法[10]是由英国 Sheffield 大学在对海军的舰艇控制系统进行设计时所开发的一种功能分配方法。它在分配过程中共需要考虑 100 多项决策准则，将其分为 8 组，其中不仅考虑了人机的能力特性，还从人因工程的角度考虑了人员的作业设计、社会性、训练、安全等因素，另外还包括自动化的精度、费用等。它的主要流程如图 2 所示。

Sheffield 法的优点是考虑的因素比较全面，而且包含了系统的静态和动态功能分配过程。它主要是针对一个海军舰艇控制系统的设计，所以同时还考虑了舰艇操作人员之间的功能分配。但它也有明显的不足，首先由于考虑的因素太多，反而使

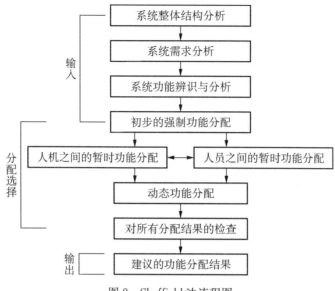

图 2　Sheffield 法流程图

得设计任务由于缺乏相关信息而无法操作；其次由于 Sheffield 法必须将功能分解到能够完全分配给人或机器，也就是足够细的粒度才能实施操作，但这在一个复杂系统中往往是不可能的。

3.1.4　自动化分类与等级设计法

该方法由英国科学家 Parasuraman 和 Sheridan 提出[11]，主要应用于工业自动化系统如核电站监控中。该方法认为任何人机自动化系统的工作过程类似于人类的信息处理，可分为四个步骤即：获取、分析、决策、行动。而机器的自动化程度分为连续的 10 个级别如表 2 所示。

表 2　自动化等级

1. 计算机不提供任何帮助
2. 计算机提供整套的决策或行动方案
3. 缩小选择范围
4. 建议一个方案
5. 如果人同意则执行这个方案
6. 在执行前允许人在短时间内否决
7. 自动执行，仅在必要时通知人
8. 如果人需要则告知他
9. 是否通知人全由计算机决定
10. 计算机决定所有的工作，拒绝人的干预

在此基础上对系统功能分别按上述的四个步骤进行分类，并对属于每一个分类的功能确定其自动化程度，然后建立多级评价准则逐步对分配结果进行修改，直到

最终确定系统应该采用的自动化类型和等级。其流程图如图 3 所示。

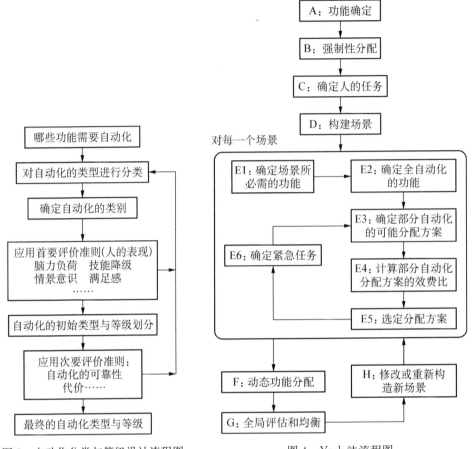

图 3　自动化分类与等级设计流程图　　图 4　York 法流程图

3.1.5　York 法

它是由英国 York 大学的 Dearden 等人提出一种基于场景(Scenario)的功能分配方法[12]，最初是为海军舰艇的设计而开发的，由于取得了比较好的效果，之后又被成功的用于单座飞机的功能分配设计中。York 法的基本步骤如图 4 所示。

这种设计方法将功能运行时的环境因素也考虑进去，因而是一种较为完善的功能分配方法。但是它没有考虑系统中人员之间的功能分配。

3.2　国内的研究情况

国内在功能分配的研究还处于起步阶段。西北工业大学的张炜等人对作战无人机系统的人机功能动态分配进行了研究[13]，提出无人机地面控制系统人机功能分配的原则和方法，主要针对远程无人机监控系统，在功能分配时所考虑的因素还比较简单，并不适合复杂系统的设计和研究。周家祥等人对载人航天器的功能分配进行了研究[14]，为载人航天器的人机最优功能分配优化设计提供了一定的依据。

但是模型的一些初始输入数据很难获得,这对最终分配结果有着直接的影响。

4 存在的问题

目前在系统功能分配方法及其应用上,主要存在的问题如下[15]:

(1)分配方法通用性较差。功能分配应用范围极其广泛,在各自的领域由于应用的环境、任务性质以及涉及的技术都不相同,分配标准也不统一,造成各领域的功能分配方法相互之间不能很好地兼容和共用,大大限制了功能分配在工程上的应用。

(2)分配标准单一化,分配过程较简单。在实际工程应用中功能分配往往是最容易忽视的一个环节,即使在设计之初考虑功能分配,也只是选取了某个单一的标准,如系统的负荷,或者费用等。分配标准的单一化必然也会造成分配过程的简单化。在这种情况下设计出来的系统可能会造成整个系统的某一单项指标较高,而其他指标却较低,因而综合性能往往达不到设计标准。

(3)功能分配过程和设计过程结合不够紧密,没有形成工程化的方法,并且对环境因素缺乏足够的考虑。传统的功能分配方法将功能分配作为单独的一个过程来考虑,与系统工程设计结合不够紧密,且需要功能分配专家的参与,而普通的设计人员要想参与进来是比较困难的,这就造成功能分配和工程设计的脱节,提高了设计成本。另外,由于对环境因素缺乏足够的考虑,当系统投入使用后,系统没有足够的动态调整能力,并有可能导致系统崩溃或失败。

5 复杂人机智能系统功能分配方法研究方向

复杂人机智能系统最主要的特征是人机之间存在着感知、决策和执行这三个层次的信息耦合。它要求在系统运行过程中人、机功能能够动态调整,在任务的不同阶段体现不同程度的智能。因此,应用单一的功能分配标准和方法很难满足复杂人机智能系统的设计需求。根据已有的方法,对复杂人机智能系统功能分配方法提出以下几点建议。

(1)必须分析系统中人、机的特性,以确定功能分配的影响因素。复杂人机智能系统中的"机"是广义的概念,包括一切与系统任务有关的机械、自动化、计算机等除人以外的各种软硬件设施。系统中的人应该是指一类群体,首先他具有一般人的生理限制特性;其次由于训练及学习,他具有专业人员的特性;最后不同的人员之间存在个人差异的特性。所以需要对人的特性进行重点分析。在人、机特性分析的基础上,从其中选择影响功能分配的因素,即特定的限制条件,例如对机器而言有可靠性、费用、性能等,对人而言有工作负荷、心理压力、生理极限、可靠性等。

(2)在进行功能分配之前,首先必须结合系统任务场景进行功能分析,它是一切后继分配活动的基础。功能分析的目的是确定具体设计准则,它向下拆分系统层次的要求为子系统,以及辨识输入设计准则和/或系统各种单元的约束所需的递阶结构。最终我们能够确定在每一个任务场景下功能分配的对象集。显然不同的任务场景,功能分配对象集也不相同。对于一个智能系统,应该对每一项功能进行分类,并确定每一类功能可能的智能程度,它们的组合构成了功能分配的基本方案。

(3) 功能分配既包含静态功能分配,更重要的是要包含动态的功能分配过程。在静态功能分配阶段,缺乏与分配准则相关的必要的初始数据(这些数据用来评价人或机器完成该项功能的表现),而且也无法构造任务的真实环境。因此只能采取某些预先估计的方法,对特定任务环境中每一种分配方案的假设,做出人和机器绩效以及其他影响因素的预测,然后评价哪一个方案最合适,进而描述当时的人机关系。而在动态分配阶段,可以借助仿真手段或原型系统,通过人机绩效实验的方式来评价功能分配方案的优劣,并做出修改和调整。

6　结语

功能分配的理论和方法经过几十年的发展,在工业自动化领域取得了卓有成效有效的成果。随着科学技术的进步,自动化系统逐渐向复杂人机智能系统发展,这对系统的设计和综合提出了更高的要求,尤其对系统的功能分配提出新的挑战。为此,系统研究人员必须从系统论的角度运用系统工程的思维、数学方法、建模理论对包含人在内的整个大系统进行设计、分析、建模和优化,对人机功能实施合理的分配,充分利用人机各自的优势,从而建立真正的复杂人机智能系统。

参考文献

[1]　Lenat D B, Feigenbaum E A. On the Threshold of Knowledge[J], Artificial Intelligence, 1991, 13, 1, 3 - 10.

[2]　Sheridan T B. 人与自动化——系统设计和研究问题[M]. 胡保生译. 西安:西安交通大学出版社, 2007.

[3]　王寿云,于景元,戴汝为,等. 开放的复杂巨系统[M]. 杭州:浙江科技出版社, 1996.

[4]　钱学森,于景元,戴汝为. 一个科学新领域——开放的复杂巨系统及其方法论[J]. 自然杂志, 1999, 13(1), 3 - 10.

[5]　路甬祥. 人机一体化系统与技术立论[J]. 机械工程学报, 1994, 30(6), 1 - 9.

[6]　陈鹰,杨灿军. 人机智能系统理论与方法[M]. 杭州:浙江大学出版社, 2006.

[7]　龙升照,黄端生,陈道木,等. 人-机-环境系统工程理论及应用基础[M]. 北京:科学出版社, 2004.

[8]　Fitts P M. Human Engineering for an Effective Air Navigation and Traffic Control System [M]. Washington, DC: National Research Council, 1951.

[9]　Price H E. The Allocation of Functions in Systems [J]. Human Factors, 1985, 27, 1, 33 - 45.

[10]　Older M, Clegg C, Waterson P. Report on the Revised Method of Function Allocation and Its Preliminary Evaluation [R], Institute of Work Psychology, University of Sheffield, 1996.

[11] Parasuraman R，Sheridan T B. A Model for Types and Levels of Human Interaction With Automation [C]，IEEE Transactions on Systems，Man，and Cybernetics — Part A：Systems and Humans，2000，30，3，286 - 297.

[12] Dearden A，Harrison M，Wright P. Allocation of Function：Scenarios，Context and the Economics of Effort [J]，International Journal of Human-Computer Studies，2000，52，289 - 318.

[13] 张炜，李道春，宋笔锋. 作战无人机系统的人、机功能动态分配模拟仿真[J]. 人类功效学，2005，11(1)，5 - 7.

[14] 周前祥，周诗华. 一种用于载人航天器人机功能分配的模型[J]. 人类功效学，2003，9(2)，3 - 6.

[15] 周诗华，周前祥，曲战胜. 复杂系统人机功能分配方法的研究进展[J]. 2004 中国控制与决策会议论文集，955 - 958.

基于非任务相关 ERP 技术的飞行员脑力负荷评价方法

完颜笑如[1]，庄达民[1]，刘　伟[2]

1. 北京航空航天大学 航空科学与工程学院，北京 100191
2. 北京邮电大学 自动化学院，北京 100876

摘要　为研究不同飞行脑力负荷水平对被试脑功能的影响，在飞行模拟任务中，要求被试对平视显示器（HUD）仿真模型上所呈现的目标信息状态进行监视，发现异常信息后按指定键进行消除。实验通过设定所需监视的目标信息数量及刷新频率来控制被试的脑力负荷水平。在进行飞行模拟任务的同时，对被试双耳输入oddball 模式下的听觉刺激，要求被试忽略该声音刺激。记录和分析不同脑力负荷下的听觉失匹配负波（MMN）。结果表明，在额中央区，与低脑力负荷相比，高脑力负荷下的 MMN 平均波幅增强，且与被试对异常信息的正确探测率呈正相关。本研究表明，听觉 MMN 对飞行脑力负荷具有较好敏感性，可为复杂飞行任务的脑力负荷评价提供一定的客观依据。

关键词　脑力负荷，事件相关电位，失匹配负波，警觉度，人机工效

Evaluation Method of Pilot Mental Workload Based on Task-irrelevant ERP Technology

Wanyan Xiaoru[1], Zhuang Damin[1]*, Liu Wei[2]

1. School of Aeronautic Science and Engineering, Beihang University, Beijing 100191
2. School of Automation, Beijing University of Posts and Telecommunications, Beijing 100876

Abstract　In order to investigate the effects of different mental workloads on the brain functions, the flight simulation task was performed in the experiment, and the subjects were asked to monitor the information targets presented on the simulation model of head-up display (HUD). When abnormal information was detected, the subjects were required to eliminate abnormity by pressing the corresponding key. According to setting the quantities and refresh frequencies of the information targets, the high and low mental workload levels were manipulated. During the process of flight simulation, auditory stimuli were presented binaurally through headphones with an oddball paradigm, and the subjects were instructed to ignore the auditory probes. Mismatch negativity (MMN) was recorded and analyzed under different mental workload levels. The results revealed that the MMN amplitudes were increased under the high mental workload level compared with the low mental workload level, and that the average amplitude of the fronto-central MMN was positively correlated with the accuracy rate of detecting abnormal information. The present study suggests that the auditory MMN is sensitive to flight mental workload and can provide effective electrophysiological evidence for flight mental workload assessment in complex tasks.

Keywords   mental workload, event-related potentials, mismatch negativity, alertness, ergonomics

脑力负荷指作业人员为达到业绩标准而付出的注意力的大小,其涉及完成某项任务时的工作要求、时间压力、作业人员的能力和努力程度以及任务不顺利时的挫折感等[1]。飞机驾驶舱属于信息高度密集的特殊作业环境,近年来,随着飞行智能化、信息化程度的提高,飞行员在执行任务时往往需要同时关注多个信息,当遇到紧急情况时,由作业负荷过高而导致脑力负荷超载的情况时有发生,严重影响到飞行安全,因此,对由飞行员作业负荷变化而引发的脑力负荷进行测量评价具有重要意义[2]。目前国内外普遍应用的飞行员脑力负荷测量方法包括主观评价法、主任务测量法、辅助任务测量法以及生理测量法。这些测量方法各有优点及应用局限性,虽然通过使用这些方法已取得一系列研究成果,但仍远不能满足系统设计对于脑力负荷测量的要求,因此有必要不断改进现有的脑力负荷测量方法并开创新的方法[3]。

事件相关电位(Event-Related Potentials,ERPs)测量是反映大脑信息加工活动的敏感而有意义的技术手段,由于脑力负荷与大脑的信息加工能力密切相关,因此将 ERPs 作为脑力负荷状况的评价指标是合适的[4]。随着脑电技术设备的发展,更多的脑电位可被采集记录到,从而促使 ERP 成为目前在脑力负荷、疲劳评价领域很有发展前景的指标之一。近年来,在简单抽象任务、复杂模拟任务以及各种作业环境中探讨 ERPs 在脑力负荷评价方面的可用性成为国际认知脑科学界及人机工效学界的研究热点[5—6]。

听觉失匹配负波(Mismatch Negativity,MMN)成分由 Naatanen 等人于 1978 年所发现,其产生机制及认知学意义主要通过记忆痕迹假说获得解释,不断重复的规律性标准刺激的物理特征被编码储存于大脑后,成为记忆痕迹,每一个输入的听觉刺激都自动与之相比较,如果有偏差刺激在标准刺激记忆痕迹存续期内出现,便会产生神经失匹配过程,诱发出 MMN[7]。由于 MMN 的产生是大脑基于非注意条件下对偏差刺激的自动反应,因此反映了大脑对外界信息的自动加工过程[8]。目前,在作业负荷评价领域,基于 MMN 成分所开展的相关研究主要为在单一作业任务或双作业任务下,探讨任务难度变化对于被试听觉 MMN 的影响[9—11]。由于这些研究所取得的结论并不一致,且多数研究是基于抽象的心理学实验所展开的,与飞行员在实际飞行时从显示器获取信息并做出反应判断的作业过程还存在较大差异,考虑到飞行任务的复杂性与特殊性,故这类研究结论能否直接应用于飞行任务的评价有待进一步的验证。

本研究基于前期对信息获取与脑力负荷之间关系的研究基础上[12—13],设定高、低两种作业负荷来控制被试的脑力负荷水平,通过研究高、低脑力负荷水平对被试听觉 MMN 的影响,检验 MMN 成分的敏感性,以期提出一种用于飞行任务脑力负荷评价的新的客观评价指标。

1 方法

1.1 被试

被试为北京航空航天大学 13 名在飞行模拟器上受过培训的模拟飞行员(男 9 例,女 4 例,22～28 岁,平均年龄 25.4 岁),右利手,能够熟练完成多种飞行模拟任务,视力或矫正视力正常,无精神疾患,听力正常,实验前对实验内容均知情同意。

1.2 飞行模拟任务

被试需要在"飞机座舱人机工效评定实验台"[14]上完成以巡航任务为主,包括起飞及降落在内的完整的飞行过程。在飞行中,被试需要对平视显示器仿真模型中的目标信息的状态进行监视,发现状态异常时,通过按指定键消除异常,不同的目标信息对应不同的按键。平视显示器仿真模型可显示空速、气压高度、雷达高度、航向角、俯仰角、滚转角、方向舵状态、起落架状态、发动机状态等多种飞行信息。实验通过设定需要被试保持监视的目标信息的数量及刷新频率来改变被试的作业负荷,从而控制被试的脑力负荷水平[12]。实验设定在高脑力负荷下,目标信息数量为 9 个,信息异常状态的平均呈现时间与间隔时间(Inter-Stimulus Interval, ISI)分别为 1 s 与 0.5 s;在低脑力负荷平下,目标信息数量为 3 个,信息异常状态的平均呈现时间与间隔时间均为 2 s。

1.3 Oddball 任务

一个完整的听觉刺激序列包含 1,200 个标准刺激(1,000 Hz, 65 dB SPL,出现概率 80%)以及 300 个偏差刺激(1,100 Hz, 65 dB SPL,出现概率 20%),刺激呈现时间为 50 ms,刺激间隔(Stimulus Onset Asynchrony, SOA)为 500 ms。每个偏差刺激前至少出现两个标准刺激。要求被试关注飞行模拟任务,忽略听觉刺激。

1.4 实验程序

13 名被试均参与高、低两种脑力负荷下的 2 次飞行模拟任务,一次飞行模拟任务约需要 13 min,每个被试的两次飞行模拟任务之间间隔 3 d,同一被试的两次实验均在上午或下午进行。高、低脑力负荷的实验顺序在被试中交叉平衡。整个飞行模拟过程中通过耳机对被试双耳输入一个完整的听觉刺激序列。图 1 为"飞机座舱人机工效评定实验台"中刚完成实验的某被试者。

图 1　飞机座舱人机工效评定
实验台及某被试者

1.5 数据记录

计算机自动记录被试对异常信息的正确探测率和反应时间作为后期作业绩效的评价指标。脑电信号采用 Neuroscan 系统(40 导,美国 Compumedics NeuroScan 公司)记录 FZ,FCZ,CZ 位置的 EEG。以鼻尖为参考电极,同时记录水平眼电和垂直眼电。电极与皮肤接触阻抗小于 5 kΩ,记录带宽为 0.1～100 Hz,采样率为 500 Hz/导。实验结束后存储脑电数据备离线分析。

1.6 数据处理

采用 Scan 4.3 软件分析 EEG 数据,利用 EOG 信号相关法去除垂直眼电和水平眼电对 EEG 信号的影响,排除有明显伪迹的数据。以听觉信号为触发,分别得出每个被试在每种条件下的平均 ERPs。分析时程(epoch)为 450 ms,包含刺激前的 50 ms 为基线矫正,波幅大于 ±70 μV 视为伪迹予以剔除,所得 ERPs 经 1～20 Hz 的无相移带通数字滤波器滤波。用偏差刺激的 ERPs 减去标准刺激的 ERPs,得到 MMN。对 MMN 平均波幅进行重复测量的两因素方差分析(Repeated Measures ANOVA):脑力负荷(2 水平:高、低)×电极(3 水平:FZ,FCZ,CZ)。通常,MMN 的测量时间窗为刺激后 150～250 ms,但对于复杂的听觉或认知加工,需要考虑更宽的时间窗[15]。根据总平均图,本实验设定 MMN 的测量时间窗为刺激后 150～300 ms。对高、低脑力负荷下被试对异常信息的正确探测率及反应时间进行统计比较,并对 MMN 与作业绩效之间的相关性进行检验。

2 结果

2.1 作业绩效结果

被试在高、低脑力负荷下对异常信息的平均正确探测率和反应时间如表 1 所示。单因素重复测量的方差分析表明,在高脑力负荷下,被试对异常信息的正确探测率显著降低($F(1, 12) = 56.43$;$P < 0.001$),且反应时间出现延长的趋势($F(1, 12) = 4.05$;$P = 0.067$),但未达到显著性水平。

表 1　高、低脑力负荷下的正确探测率及反应时间

	正确探测率/%	反应时间/ms
高脑力负荷	64.99±15.02	738.93±66.81
低脑力负荷	95.98±3.62	699.93±87.09

2.2　ERP 结果

在高、低脑力负荷下,FZ,FCZ,CZ 电极记录到的 MMN 总平均图如图 2 所示。在 150～300 ms 测量时间窗,两因素重复测量的方差分析表明脑力负荷对 MMN 平

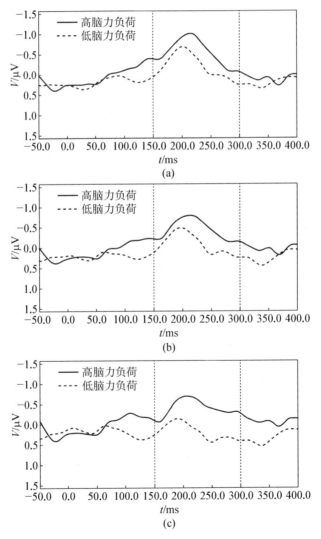

图 2　高、低脑力负荷下的 MMN 总平均图
(a) FZ 电极;(b) FCZ 电极;(c) CZ 电极

均波幅的主效应显著（$F(1,12) = 5.35$；$P = 0.039$），表现为高脑力负荷下的 MMN 平均波幅（$-0.70\,\mu\text{V}$）显著高于低脑力负荷下的 MMN 平均波幅（$-0.23\,\mu\text{V}$）。电极主效应显著（$F(2,24) = 10.37$；$P = 0.001$），表现为额区电极位置（FZ）的 MMN 平均波幅（$-0.62\,\mu\text{V}$）显著高于额中央区电极位置（FCZ）的 MMN 平均波幅（$-0.50\,\mu\text{V}$），额中央区电极位置的 MMN 平均波幅显著高于中央区电极位置（CZ）的 MMN 平均波幅（$-0.34\,\mu\text{V}$）。

在 $150\sim300\,\text{ms}$ 测量时间窗，脑力负荷与电极的交互作用显著（$F(2,24) = 4.02$；$P = 0.031$），多重比较检验结果表明，在低脑力负荷下，FZ 电极的 MMN 平均波幅显著高于 CZ 电极（$P < 0.001$），FCZ 电极的 MMN 平均波幅显著高于 CZ 电极（$P < 0.001$），FZ 电极的 MMN 平均波幅的均值高于 FCZ 电极的 MMN 平均波幅的均值，但两者间的差异未达到显著性水平（$P = 0.101$）；在 CZ 电极，高脑力负荷下的 MMN 平均波幅显著高于低脑力负荷下的 MMN 平均波幅。被试在高、低脑力负荷下的 MMN 平均波幅如表 2 所示。

表 2　高、低脑力负荷下的 MMN 平均波幅

电极	平均波幅/μV	
	高脑力负荷	低脑力负荷
FZ	-0.79 ± 0.58	-0.45 ± 0.42
FCZ	-0.68 ± 0.63	-0.32 ± 0.50
CZ	-0.63 ± 0.74	-0.05 ± 0.51

2.3　相关分析

为验证作业绩效与 MMN 成分之间的相关程度，分别对 FZ，FCZ，CZ 电极的 MMN 平均波幅与正确探测率和反应时间的相关性进行检验。结果表明：在 FZ 电极，MMN 平均波幅与正确探测率呈正相关（$r = 0.46$，$P < 0.05$），在 FCZ 电极，MMN 平均波幅与正确探测率呈正相关（$r = 0.45$，$P < 0.05$），在 CZ 电极，MMN 平均波幅与正确探测率呈显著正相关（$r = 0.52$，$P < 0.01$）。但 3 个电极的 MMN 平均波幅与反应时间的相关性均不显著，具体表现为在 FZ 电极，MMN 平均波幅与反应时间的相关性为（$r = -0.18$，$P > 0.05$）；在 FZ 电极，MMN 平均波幅与反应时间的相关性为（$r = -0.25$，$P > 0.05$）；在 FCZ 电极，MMN 平均波幅与反应时间的相关性为（$r = -0.28$，$P > 0.05$）。

3　讨论和结论

有关研究表明，作业人员在高脑力负荷下因面临更高的任务要求及时间压力，因此很容易产生紧张、焦虑等情绪，而这些情绪因素均会增加生理唤起，提高作业人员的警觉度，并且可能调节注意的方向，使注意焦点以外的信息获得关注[16]。由此可见，在本研究中，高脑力负荷水平下的 MMN 成分诱发增强，可能与被试在面临困

难任务时,因担心任务失败而使整体警戒水平提高,从而促使更多的脑力资源被唤醒,对任务非相关的听觉信息加工增强有关。相反,在持续性警戒作业中,当作业任务较为容易单调而又需要长时间保持注意时,作业人员的警觉水平容易发生下降,反映在本研究的实验结果中,则为当脑力负荷降低时,MMN 的诱发削弱。

与本研究较为接近的工作为 Kramer 等通过模拟实验验证了 MMN 平均波幅在评价海军雷达作业人员脑力负荷时的敏感性[5]。但 Kramer 的实验结果显示,随着脑力负荷的提高,作业人员对非任务相关信息的自动加工能力减弱,与本研究结果不一致。可能的原因在于本研究的单次实验时间较短,约为 13 min,而在 Kramer 的研究中,单次实验时间为 45 min,有研究表明,MMN 诱发实验的时间不宜过长,当被试接受实验 30 min 左右时,会因疲劳而导致 MMN 诱发电位的幅度出现明显下降(超过 40%)[17],而该幅度的降低与脑力负荷水平无直接关系。张朋等人研究了在视觉追踪任务中,不同注意负荷对被试听觉 MMN 的影响,结果表明额中央区 MMN 的平均波幅随任务负荷的增加而增高[11],与本研究结果一致。

本研究在较真实的飞行模拟环境下,讨论了不同脑力负荷对被试大脑的信息自动加工能力的影响。结果发现,与低脑力负荷相比,高脑力负荷下的 MMN 平均波幅增强,且 MMN 平均波幅与被试在飞行期间对异常信息的正确探测率呈显著正相关。本研究清楚地表明 MMN 成分对飞行任务的脑力负荷具有较好的敏感性,可为飞行员的脑力负荷评价提供了一定的电生理依据,并将可能应用于航空航天领域的警觉度自动估计与监测[18]。

值得注意的是,本研究所采用的非注意条件下的 MMN 诱发实验方法,无需被试在实验过程中主动识别刺激,可避免考虑被试能否配合实验的因素,从而将对飞行模拟主任务的侵入性降到最低,为其应用的客观性提供了保障。

致谢

感谢视觉艺术与脑认知研究中心对本研究的技术支持及建议。

参考文献

[1] Young M S, Stanton N A. Mental workload: theory, measurement, and application. In: Karwowski W, eds. International Encyclopedia of Ergonomics and Human Factors [M]. London: Taylor & Francis. 2001.

[2] Noel J B, Bauer K W, Lanning J W. Improving pilot mental workload classification through feature exploitation and combination: a feasibility study [J]. Computers & Operations Research, 2005, 32(10): 2713 - 2730.

[3] 柳忠起,袁修干,刘涛,等. 航空工效中的脑力负荷测量技术[J]. 人类工效学, 2003, 9(2): 19 - 22.

[4] 宋健,苗丹民. 脑力疲劳客观评定方法研究进展[J]. 中华航空航天医学杂志, 2006, 17(1): 74 - 76.

[5] Kramer A F, Trejo L J, Humphrey D. Assessment of mental workload with task-irrelevant auditory probes [J]. Biological Psychology, 1995,40(1−2):83−100.

[6] 曹雪亮,苗丹民,刘练红. 脑力疲劳评定方法现状[J].第四军医大学学报,2006,27(4):382−384.

[7] 马俊,胡斌,陈宪生,等.心境障碍患者的非匹配负波[J].神经病学与神经康复学杂志,2008,5(4):220−222.

[8] Naatanen R, Paavilainen P, Rinne T, et al. The mismatch negativity (MMN) in basic research of central auditory processing:A review [J]. Clinical Neurophysiology, 2007,118(12):2544−2590.

[9] Otten L J, Alain C, Picton T W. Effects of visual attentional load on auditory processing [J]. Neuro Report, 2000,11(4):875−880.

[10] Yucel G, Petty C, McCarthy G, et al. Visual task complexity modulates the brain's response to unattended auditory novelty [J]. NeuroReport, 2005,16(10):1031−1036.

[11] Zhang P, Chen X C, Yuan P, et al. The effect of visuospatial attentional load on the processing of irrelevant acoustic distractor [J]. Neuroimage, 2006,33(2):715−724.

[12] 张磊,庄达民,完颜笑如. 基于不同脑力负荷和任务类型的信息颜色编码[J]. 兵工学报,2009,30(11):1522−1526.

[13] 曾庆新,庄达民,马银香.脑力负荷与目标辨认[J].航空学报,2007,V28(4):S76−S80.

[14] 张磊,庄达民,邓凡,等.飞机座舱人机工效评定实验台研制[J].飞行力学,2009,27(1):81−84.

[15] 赖永秀,田银,尧德中.音乐速度变化感知的失匹配负波[J].中国生物医学工程学报,2010,29(2):277−282.

[16] Wickens C D, Lee J D, Liu Y L, et al. Introduction to Human Factors Engineering [M]. (2nd Edition). New Jersey:Prentice Hall, 2003.

[17] 丁海艳,叶大田. MMN 提取过程中关键问题的讨论[J].北京生物医学工程,2006,25(1):79−84.

[18] 傅佳伟,石立臣,吕宝粮. 基于 EEG 的警觉度分析与估计研究综述[J].中国生物医学工程学报,2009,28(4):589−595.

CREAM 失误概率预测法在驾驶舱机组判断与决策过程中的应用

孙瑞山,王　鑫

中国民航大学民航安全科学研究所,天津 300300

摘要　CREAM 强调人在生产活动中的绩效输出不是孤立的随机性行为,而是依赖于人完成任务时所处的环境或工作条件,它通过影响人的认知控制模式和其在不同认知活动中的效应,最终决定人的响应行为。在驾驶舱内,机组的绩效输出不仅仅是人的自身行为,还依赖于其完成任务时所处的情景环境,所以 CREAM 方法能够结合驾驶舱环境对机组的认知差错进行分析。在飞行中,驾驶舱内机组非常重要的一个环节是判断与决策过程,这一过程中包括询问、讨论、确定方案、执行、反馈五个环节。本文将通过分析这五个环节的相互关系及影响,以明确这种讨论过程是减少机组人为差错发生的一种有益方式,然后应用 CREAM 的预测法对这五个环节进行定量化分析,得出机组判断与决策过程的失误概率,完成对机组认知行为的客观评价。并为以后能够定量化研究驾驶舱内飞行员认知差错提供方法的借鉴。

关键词　认知可靠性与失误分析方法(CREAM),判断与决策,人为差错,机组

The Application of Judgement and Decision-making in the Cockpit Based on Failure Probability of CREAM

Sun Ruishan，Wang Xin

Research Institute of Civil Aviation Safety，Civil Aviation
University of China，Tianjin 300300

Abstracts　CREAM emphasizes that the performance output is not the isolated random behavior but dependents on working conditions and the environment when human complete tasks in the production activities. And it has the different response behavior by affecting human's cognition control mode and the effect in different cognitive activities. In the cockpit, the performance output of the crew is not only the independent behavior, but also depended on its environment.

Therefore, CREAM can combine with the cockpit environment to analysis the crew's cognitive errors. The process of judgement and decision-making in the cockpit by fight crew is a very important part. It includes questioning, debate, identify solutions, implement and feedback. This paper intends to analyze the relationship and impact between these five aspects that to clarify the discussion process is a useful way to reduce the human error. The CREAM method is applied to the quantitative analysis in these five aspects. It will be obtained the error probability in the judgement and decision-making and completed cognitive behavior of judgement for flight crew. It will provide a method for pilot in the cockpit that can quantify whose cognitive errors.

Keywords   cognitive reliability and error analysis method(CREAM), judgement and decision-making, human error, flight crew

从我国近 10 年民航飞行事故中分析,由机组原因引发的事故达到 67.86%(见表 1)。而据 NTSB(美国国家运输安全委员会)对 1950 年至 1999 年的 1,286 起[2]事故的统计中也可以发现,飞行机组的差错占到了 60% 以上,因此,对机组人员人为因素的研究已经不可代替地成为航空安全的重要环节。而在我国,随着航空市场新一轮的扩张、兼并和超乎寻常的快速增长,飞行安全已经成为公众关注的热点问题。

表 1   2000~2009 年国内飞行事故按主要原因统计表[1]

主要原因	事故次数		事故主要原因百分比/%	
	近 10 年	近 5 年	近 10 年	近 5 年
	2000~2009 年	2005~2009 年	2000~2009 年	2005~2009 年
机组	19	8	67.86	57.14
机务	2	2	7.14	14.29
机械	1	0	3.57	0
空管	0	0	0	0
地面保障	0	0	0	0
天气/意外	1	1	3.57	7.14
其他	0	0	0	0
待定	5	3	17.86	21.43
总计	28	14	100	100

本文将对机组的判断与决策过程进行分析,然后使用认知可靠性与失误分析方法(CREAM)中的失误概率预测法对这一过程进行定量化分析,从而完成对机组客观的评价,为管理者及理论研究人员提供一种方法的借鉴。

1　驾驶舱内机组判断与决策过程[3]

在飞行中,机组的判断与决策过程是保证飞行安全的重要因素。两人制机组或多人制机组主要依靠充分发挥机组团队和群体优势预防人为差错。而这一过程是建立在良好的驾驶舱沟通基础之上,良好的沟通可以帮助机组人员交流看法,做到相互理解,从而配合默契,行动一致。机组判断与决策过程包括询问、讨论、确定方案、执行和反馈。

1.1　询问

在飞行过程中,机组成员发现可疑迹象,及时发问,可启动机组的情况判断过程,帮助及早发现问题。通常机长应明确说明自己的发现和判断,其他成员有疑问的应鼓励他们大胆询问。若机长的判断正确,询问可使其他成员理解机长的意图,从而使以后的配合更加主动默契;若机长的判断不正确,询问又可以帮助改正错误,得出正确的结论,因此,驾驶舱内应当鼓励直截了当的询问。

1.2　讨论

这一阶段机组成员需要运用已有的知识和经验,对发现的问题发表意见,然后供机组其他成员讨论,在讨论过程中,提倡为自己的立场辩护,这是保证机组做出最佳决策的重要手段。

1.3　确定方案

飞行中往往情况非常复杂,时间非常紧迫,要把问题讨论明白,选出最优方案,常常是不现实的。因此,对于机组确定的方案不要求是最优的,但要求在安全上是可行的。机长在综合机组其他成员的意见后,需要果断地确定实施方案。

1.4　执行

确定实施方案后,全体机组成员就要严格按照机长的要求,分工协作,采取行动,以求解决问题。

1.5　反馈

反馈就是要对执行后的效果发表意见。反馈的目的,是要借助机组的集体智慧,来对所采取的行动是否有效,问题是否解决,先前的判断与决策是否正确做出更好的回答。

2　驾驶舱内机组判断与决策过程的 CREAM 定量化分析

2.1　适用性分析

1998 年,Erik Hollnagel 在其著作 *Cognitive Reliability and Error Analysis Method* 中正式提出 CREAM 方法。该方法是第二代可靠性分析中的一种代表性方法。CREAM 强调人在生产活动中的绩效输出不是孤立的随机性行为,而是依赖于人完成任务时所处的环境或工作条件,它们通过影响人的认知控制模式和其在不同认知活动中的效应,最终决定人的响应行为。而在驾驶舱内,飞行员的绩效输出就是依赖于其完成任务时所处的情景环境,所以该方法能够结合驾驶舱环境对机组的差错进行分析。

2.2 CREAM 预测分析方法[4—10]

CREAM 预测分析的主要功能就是对某一项人的认知活动的任务可能发生失效的概率进行预测,它包括基本法和扩展法两种方法。本文采用的方法是基于基本法之上更先进的扩展法。

扩展法预测分析的基本思想是分析人在完成任务过程中的认知活动和可能的认知功能失效,在得到认知功能失效概率基本值基础上,研究所处的情景环境的CPC 因子水平对基本值进行修正,从而对人在完成任务时可能发生失效的概率进行预测。扩展法预测分析过程有四个步骤:分析任务,建立认知需求剖面;评价共同绩效条件;识别最可能的认知功能失效;预测失效概率。

2.2.1 分析任务,建立认知需求剖面

建立事件序列,基于驾驶舱机组判断与决策的认知行为过程分析,并对该过程的发展进程进行系统化的任务描述,识别并确认驾驶舱机组判断与决策过程中任务的细节,包括操作的过程及相应的情境环境条件。CREAM 将认知功能归纳为观察、解释、计划和执行四类,每类功能有若干个失效模式。认知活动包括协调、通信、比较、诊断、评估、执行、识别、保持、监视、观察、计划、记录、调整、扫视、检验等,然后按认知活动和认知功能对照表(见表 2),确定每项步骤中的认知活动所对应的认知功能。

表 2 认知行为与认知功能关系

行为类型	认知功能				行为类型	认知功能			
	观察	解释	计划	执行		观察	解释	计划	执行
协调			√	√	监视	√	√		
通信				√	观察	√			
比较	√				计划			√	
诊断	√	√			记录		√		√
评估	√	√			调节	√			√
执行				√	扫描	√			
识别	√				检验	√	√		
保持			√	√					

根据机组判断与决策过程,机组成员在遇到问题时首先要询问,同时这也是保证飞行安全的首要前提。而询问这一过程中,最可能的原因是机组成员观察到了一些异常情况并对此进行了初步的诊断,认为可能会对安全带来威胁。当提出问题后,机组成员就要讨论,每一位成员都应该发表自己的看法,而这一过程中最重要的

是机长要比较大家提出的建议。确定方案就是在综合大家意见后,机长做出决断,确定处理问题的方案。然后全体机组成员需要按照机长的分工执行。执行后,机组成员都应积极反馈意见,检验所采取行动的有效性。按照 CREAM 方法提供的认知活动和认知功能对照表,根据以上分析,确定每个认知活动对应的认知功能(见表 3)。

表 3   机组判断与决策过程所对应的认知行为和认知功能

序列	认知活动	认知行为	观察	解释	计划	执行
1	询问	观察	√			
		诊断		√	√	
2	讨论	比较		√		
3	确定方案	计划			√	
4	执行	执行				√
5	反馈	检验	√	√		
		评估		√	√	

2.2.2   评价共同绩效条件(CPC)

根据事件的情景环境条件,对 CREAM 方法所给定的 9 种 CPC 因子的水平进行评价,确定其对绩效可靠性的期望效应。每个 CPC 因子都有多个水平和对应的认知功能权重,如表 4 所示。

表 4   CPC 与绩效可靠性及权重因子[2, 4]

CPC 名称	水平	对绩效可靠性的期望效应	认知功能对应的权重因子			
			观察	解释	计划	执行
组织的完善性	非常有效	改进	1.0	1.0	0.8	0.8
	有效	不显著	1.0	1.0	1.0	1.0
	无效	降低	1.0	1.0	1.2	1.2
	效果差	降低	1.0	1.0	2.0	2.0
工作条件	优越	改进	0.8	0.8	1.0	0.8
	匹配	不显著	1.0	1.0	1.0	1.0
	不匹配	降低	2.0	2.0	1.0	2.0
人机界面与运行支持的完善性	支持	改进	0.5	1.0	1.0	0.5
	充分	不显著	1.0	1.0	1.0	1.0
	可容忍	不显著	1.0	1.0	1.0	1.0
	不适当	降低	5.0	1.0	1.0	5.0
规程/计划的可用性	适当	改进	0.8	1.0	0.5	0.8
	可接受	不显著	1.0	1.0	1.0	1.0
	不适当	降低	2.0	1.0	5.0	2.0
同时出现的目标数量	能力之内的	改进	1.0	1.0	1.0	1.0
	与能力相符	不显著	1.0	1.0	1.0	1.0

（续表）

CPC 名称	水平	对绩效可靠性的期望效应	认知功能对应的权重因子			
			观察	解释	计划	执行
可用时间	超出能力之外	降低	2.0	2.0	5.0	2.0
	充分	改进	0.5	0.5	0.5	0.5
	暂时不充分	不显著	1.0	1.0	1.0	1.0
	连续不充分	降低	5.0	5.0	5.0	5.0
工作时间	白天(可调整)	不显著	1.0	1.0	1.0	1.0
	夜晚(未调整)	降低	1.2	1.2	1.2	1.2
培训和经验的充分性	充分,经验丰富	改进	0.8	0.5	0.5	0.8
	充分,经验有限	不显著	1.0	1.0	1.0	1.0
	不充分	降低	2.0	5.0	5.0	2.0
班组成员的合作质量	非常有效	改进	0.5	0.5	0.5	0.5
	有效	不显著	1.0	1.0	1.0	1.0
	无效	不显著	1.0	1.0	1.0	1.0
	效果差	降低	2.0	2.0	2.0	5.0

　　驾驶舱环境中涉及全部 9 个因子,而这里所讨论的也是驾驶舱中机组成员的普遍认知过程,不涉及具体处理的事件,所以在计算过程中,取全部 9 个因子,并设为最佳环境。也就是说把情景环境设为最佳状态,即对于每一个 CPC 因子权重都取第一行,如表 5 所示。

表 5　驾驶舱内的 CPC 水平与绩效可靠性及权重因子

CPC 名称	水平	对绩效可靠性的期望效应	认知功能对应的权重因子			
			观察	解释	计划	执行
组织的完善性	非常有效	改进	1.0	1.0	0.8	0.8
工作条件	优越	改进	0.8	0.8	1.0	0.8
人机界面与运行支持的完善性	支持	改进	0.5	1.0	1.0	0.5
规程/计划的可用性	适当	改进	0.8	1.0	0.5	0.8
同时出现的目标数量	能力之内的	改进	1.0	1.0	1.0	1.0
可用时间	充分	改进	0.5	0.5	0.5	0.5
工作时间	白天(可调整)	不显著	1.0	1.0	1.0	1.0
培训和经验的充分性	充分,经验丰富	改进	0.8	0.5	0.5	0.8
班组成员的合作质量	非常有效	改进	0.5	0.5	0.5	0.5

2.2.3　识别最可能的认知功能失效

　　CREAM 方法提供了 13 类认知功能失效模式,如表 6 所示。参考 CPC 因子水平,找到每一项认知活动最可能的认知功能失效模式。

表 6　认知功能失效模式与失效概率基本值

认知功能	失效模式	基本值	认知功能	失效模式	基本值
观察	O1 观察目标错误	0.001	计划	P1 优先权错误	0.01
	O2 错误辨识	0.07		P2 不适当的计划	0.01
	O3 观察没有进行	0.07	执行	E1 动作方式错误	0.003
解释	I1 诊断失败	0.2		E2 动作时间错误	0.003
	I2 决策失误	0.01		E3 动作目标错误	0.000,5
	I3 延迟解释	0.01		E4 动作顺序错误	0.003
				E5 动作遗漏	0.03

在询问过程中,涉及"观察"和"诊断"两种认知行为,而从实际情况中可以看出,观察到异常情况比随后的诊断更重要,而在观察的 3 种失效模式中,最容易产生的是"错误辨识"。

讨论过程涉及"解释"中的 3 种失效模式,而通过机组成员的讨论,最容易产生的是"决策失误"。

方案的确定最有可能的是该方案制定的不适当。所以选择"不适当的计划"。

在执行过程中,由于驾驶舱设备的复杂性及多样性,在操作过程中容易在顺序上出现问题,所以选择"动作顺序错误"。

反馈过程中,如果出现了问题,在"观察"、"解释"和"计划"的 9 种失效模式中,最容易产生的是"延迟解释"。也就是说没能及时反馈意见,可能会对后续的操作产生影响。

2.2.4　预测失效概率

CREAM 将认知功能失效概率(Cognitive Failure Probability,CFP)。按任务的操作步骤进行失效概率预测,预测过程如下:

(1) 按表 6 所列认知功能失效模式的基本失效概率值,确定每个认知活动中最可能的认知功能失效模式的失效概率基本值,即可得到该认知活动的标定 CFP 值,记为 $CFP_{标定}$。

(2) 评价 CPC 对 CFP 的影响。CREAM 提供了 CPC 因子对四大认知功能的权重因子表(表 5),进而可得到每个 CPC 因子对每个认知活动的权重因子,再分别求得每个认知活动下所有 CPC 因子的权重因子的乘积,即得到该认知活动的"总权重因子",则修正后的 CFP 值为 $CFP_{修正} = CFP_{标定} \times$ 总权重因子。

(3) 一个操作步骤中的所有认知活动按(1)和(2)得到修正后的 CFP 值之后,即可求得该操作步骤的总的 CFP 值,它需要根据步骤中的所有认知活动的逻辑关系来确定计算方法。

根据预测失效概率的计算步骤,计算结果列于表 7。

表7　认知活动的失误概率

认知活动	最可能失误类型	基本概率($CFP_{标定}$)	总权重	失误概率($CFP_{修正}$)
询问	O2 错误辨识	0.07	0.064	0.004,48
讨论	I2 决策失误	0.01	0.1	0.001
确定方案	P2 不适当的计划	0.01	0.05	0.000,5
执行	E4 动作顺序错误	0.003	0.051,2	0.000,153,6
反馈	I3 延迟解释	0.01	0.1	0.001

由公式 $P = 1 - \prod_{i=1}^{5}(1 - CFP_i)$（其中 $i = 1, 2, 3, 4, 5$，分别对应5种认知功能），可以得到机组判断与决策过程的失误概率为 0.007,12。

3　结论

本文通过对机组判断与决策过程的定量化分析,可得出以下结论:

（1）首先分析了机组的判断与决策过程,然后对其进行了定量计算,实现了对机组的客观评价,为定量分析机组的认知差错提供了一种方法。

（2）在定量计算过程中,本文假设 CPC 因子都处于最高阶水平,即情景环境为最佳状态。而对于具体的一次飞行过程中,CPC 因子水平不可能都处于最高阶,也不一定涉及全部9个因子,而要根据当时的飞行情况而定,但最终的判断是单由某个人进行选择,所以会造成主观性过强,因此在日后对具体事故/事件中飞行员的认知差错进行分析时,需要增加 CPC 因子水平等级评价的客观性方法。

参考文献

［1］　中国民航总局航空安全办公室.中国民航航空安全报告(2009 年)[R].2010.

［2］　中国民用航空杂志社.孟昭蓉,杨春生.世界航空事故汇编[G].2002.

［3］　中国民航局航空安全办公室,中国民航大学民航安全科学研究所.航空安全中人的因素[G].2001.

［4］　Hollnagele. Cognitive reliability and error analysis method [M]. Oxford(UK)：Elsevier Science Ltd. 1998.

［5］　王遥,沈祖培.CREAM——第二代人因可靠性分析方法[J].工业工程与管理,2005,3:17－21.

［6］　王世锦,隋东.空中交通管制员人因可靠性定量分析研究[J].人类工效学,2009,15,4:46－50.

［7］　廖可兵,刘爱群,童节娟等.复杂人机系统班组人误模型与量化分析[J].中国安全科学学报,2007,17,12:42－48.

［8］　高佳,沈祖培,何旭洪.第二代人的可靠性分析方法的进展[J].中国安全科学学报,2004,14,2:15－19.

［9］　高佳,黄祥瑞,沈祖培.第二代人的可靠性分析方法的新进展[J].中南工学院学报,1999,13,2:138－149.

［10］　高文字,张力.人因可靠性分析方法 CREAM 及其应用研究[J].人类工效学,2002,8,4:8－12.

Polymorphic Cumulative Learning in Integrated Cognitive Architectures for Analysis of Pilot-Aircraft Dynamic Environment

Yin Tangwen, Shan Fu

School of Aeronautics and Astronautics, Shanghai JiaoTong University, Shanghai 200240

Abstract   A Polymorphic Cumulative Learning (PCL) conception was proposed in order to make it feasible for Digital Pilots (DP) in trying to have the capability to incorporate various forms of learning mechanisms as abundant as human do. An integrated cognitive architecture for analysis of pilot-aircraft dynamic environment (ICA-APADE) which will facilitate aircraft design and evaluation is devised as a framework to make the analysis and implementation of the PCL conception more concrete and practical.

Keywords   Polymorphic Cumulative Learning (PCL), Digital Pilots (DP), Integrated Cognitive Architecture for Analysis of Pilot-Aircraft Dynamic Environment (ICA-APADE)

1   Introduction

Digital pilots are of great significance in aircraft design and evaluation. Yet few digital pilots have the capability to incorporate various forms of learning mechanisms as abundant as human do, which leads to manifest infidelity and confines their utilization. The purpose of this paper is to figure out more plausible learning mechanisms that could be adopted by digital pilots in integrated cognitive architectures for analysis of pilot-aircraft dynamic environment which will facilitate aircraft design and evaluation.

One of the common characteristics in the shaping of such various forms of learning mechanisms is that human could learn how to recognize and predict erroneous actions at different level, and the achieved learning mechanisms themselves become human's appropriate tools and amplified strength [1] which could be intelligently utilized to improve the working environment by reducing the number of erroneous actions before it becomes unfavorable or even unforgiving [1 - 3].

The fulfillment of the purpose of this paper means that the expected digital pilots could be able to recognize and predict erroneous actions, and be able to reduce the number of erroneous actions [1]. To be more specific, figuring out more plausible learning mechanisms means that the expected digital pilots could be able to establish various forms of learning mechanisms while recognizing and predicting erroneous actions, and be able to accumulate the learning mechanisms established while reducing the number of erroneous actions.

The above establishment and accumulation of various forms of learning mechanisms that could be abundantly incorporate by digital pilots may be characterized as Polymorphic Cumulative Learning (PCL).

2 Overview

2.1 Human's intelligence

Human possess the following essential elements of intelligence [4]:

A. Have definite overall objective.

B. Be able to set specific goals and problems to be solved under certain circumstances.

C. Be able to obtain related information on the premise of some given context of problem-environment-goal.

D. Be able to refine the information and extract relevant knowledge to accomplish cognition.

E. Be able to activate the knowledge to generate intelligent strategies for the confronted context of problem-environment-goal.

F. Be able to convert the intelligent strategies to intelligent behavior, and ultimately solve the problems to achieve the goals.

G. Be able to set new goals and problems to be solved according to the overall objective and the new environment.

Among the above seven essential elements, A, B and G are exclusive to human, while the remaining are desirable for and achievable by human artifacts. Human has the ability to abstract knowledge from information and activate the knowledge to gain intelligence. More importantly, human can convert the intelligence to intelligent tools to make it more intelligent and more capable.

2.2   Digital pilots' intelligence

Digital pilots are a kind of virtual human artifacts, and their intelligence is confined to C, D, E and F. Therefore, digital pilots' intelligence can be defined as the capability to obtain related information purposefully in the context of problem-environment-goal, to properly refine the knowledge obtained to extract relevant knowledge to accomplish cognition, to generate intelligent strategies in combination with the subjective goals, and to successfully solve the problems in the given environment using the intelligent strategies generated.

2.3   Digital pilots' learning

Digital pilots' intelligence highly relies on the accomplishment of the cognition to the environment and the generation of the strategies to tackle the environment. Both the environment cognition process and the strategy generation process should be sufficient enough for the digital pilots to cope with various kinds of tasks in various kinds of scenarios. Therefore, digital pilots' learning mainly lies in the environment cognition process and the strategy generation process. The Polymorphic Cumulative Learning (PCL) conception mentioned above is of indispensable for digital pilots to have the capability to incorporate various forms of learning mechanisms as abundant as human do.

2.4   Integrated Cognitive Architecture

Since digital pilots' intelligence is a subset of that of human's, in order to build intelligent digital pilots, more specifically here to clarify and implement the Polymorphic Cumulative Learning (PCL) conception, the digital pilots should be integrated into a multifunctional architecture. Such a multifunctional architecture should be a quaternity consisting of a Digital Pilot Cognitive Model (DPCM), a Modeling Environment (ME), a Simulation Framework (SF), and a Tool for Digital Pilot Performance Analysis (TDPPA) [5].

The purpose of this paper is an exertion to explore learning mechanisms in Digital Pilot Cognitive Model, with the intermediate goal to analyze the pilot-aircraft dynamic environment and the ultimate objective to facilitate aircraft design and evaluation. For this reason, here the multifunctional architecture mentioned above is regarded as an Integrated Cognitive Architecture (ICA).

2.5   Analysis of Pilot-Aircraft Dynamic Environment

Any preferable integrated cognitive architecture should meet the fundamental and ultimate requirement of supporting intelligent behavior in fulfilling real time tasks, especially in time pressing cases. Cognitive processing is kind of equivalent to deliberation that drives conditioned responses — intelligent behavior. So an integrated cognitive architecture's kernel is cognitive processing. Base on this understanding, an Integrated Cognitive Architecture for Analysis of Pilot-Aircraft Dynamic Environment (ICA-APADE) is proposed in the paper. The integrated cognitive architecture consists of two loosely coupled models, namely, ICA and APADE. The ICA model mainly corresponds to virtual digital pilots along with an interface slot for any virtual external environment to be plugged into, and the APADE model corresponds to the external dynamic environment. The purpose of combining the two models is to provide a context for conceivable learning mechanisms in this paper.

As stated above, the ICA is multifunctional and can be treated as a Modeling Environment (ME), so that it can create and instantiate an APADE model and plug the APADE model into the corresponding interface slot [6].

3 The ICA-APADE and its theory background

The ICA-APADE was motivated by the need for aircraft design and evaluation in earlier stages, and was conceived by inspiration from scientific fields such as System Science, Cognitive Science, Synergetics, Control Theory, Information Theory, Intelligence Science, Modeling and Simulation, and Computer Science.

3.1 System View of the ICA-APADE

The apparent complexity of an ant's behavior over time is largely a reflection of the complexity of the environment in which it finds itself [1]. The complexity of a system's behavior is largely a function of the complexity of the system's local environment and a function of its internal complexity as well.

A system view of the ICA-APADE is draw out with its key components sketched only based on such a reflection rationale (See Fig. 1). For the ICA-

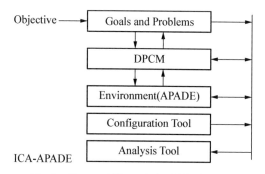

Fig. 1   System View of the ICA-APADE

APADE itself，the Objective module is its external environment. The Goals and Problems module is just a skeleton to be stuffed according to the Objective. The Environment module is a combination of models such as aircraft and its surrounding atmosphere [6]. The DPCM module is a kind of a user to solve the problems specified and to achieve the goals thereafter. The Configuration Tool module and the Analysis Tool module are self-explanatory.

3.2 Control and Information View of the ICA-APADE

An innovative control and information view of the ICA-APADE was conceived above (See Fig. 2). What's unique here is that the DPCM (the virtual pilot) has to handle two tasks concurrently instead of one，and the two tasks are treated equivalently. The first ask is to cope with the environment，that's what the convention control and information systems actually do. The second task is to arrive at the goals and solve the problems. The two tasks are different faces of the same objective. There are two information flows and two control flows as can be seen. All the flows interlock with one another rigidly in the DPCM as they are subject to the same objective.

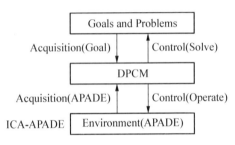

Fig. 2　Control and Information View of the ICA-APADE

Such a conception facilitates the ICA-APADE to utilize theory from Synergetics [7 - 10]，on which the Polymorphic Cumulative Learning (PCL) proposed early was based. Two pieces of information flow represent the acquisition of information from both the goals to achieve and the environment confronted. In terms of Synergetics，the information flows are Macro-to-Micro processes of Information Compression (MMIC) which can be realized by a top-down approach introduced by H. Haken [7, 8]. The two control flows are Micro-Macro processes of Strategy Generation (MMSG) which can be achieved by a bottom-up approach established by H. Haken [9, 10]. What's worth being noticed here is that the acquisition of goal-oriented information is for the control of the environment (to operate the aircraft)，and that the acquisition of environment-oriented information is for the control of the problems (to solve). Each acquisition-control situation is as same as the corresponding task mentioned above. Since the two tasks are just different faces of the same objective，the two situations are bound to be mutually validated to each other. The mutual validation fact is the foundation of Polymorphic Cumulative Learning (PCL).

3.3　Learning View of the ICA-APADE

Based on the insight above, it is natural to reach a learning view of the ICA-APADE as below (See Fig. 3). Prior to the explanation to the learning view, the basics of Synergetics are necessary.

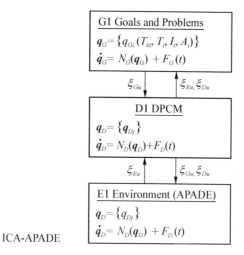

Fig. 3　Learning View of the ICA-APADE

The status of a system could be described by a group of variables $\{q_i\}$; its status vector $q(t) = (q_1(t), \cdots, q_N(t))$ could be expressed by

$$\dot{q} = N(q, \alpha) + F(t) \tag{1}$$

where N representing the determined part, F representing the fluctuation effect, and α representing control parameters [6 - 9].

The top-down approach to realize a Macro-to-Micro process of Information Compression (MMIC) can be summarized as the following [7, 8]:

Suppose a system is in one of its attractor statuses (neutral status) before a message arrives, and the reception of the message means that the initial values of α and q are to be determined. The system may remain in its neutral status or transfer to a new attractor status. If the system enters an attractor k upon a message i, then define a matrix element $M_{ik} = 1$ (otherwize $M_{ik} = 0$). The system may enter several different attractors if the fluctuation effect were taken in to account.

Definition 1: Relative importance of attractor k is P'_k

Definition 2: Relative importance of message i is with fluctuation considered

$$P_j = \sum_k L_{jk} P'_k = \sum_k \frac{M_{jk}}{\sum_{j'} M_{j'k}} P'_k \tag{2}$$

Definition 3: $\qquad S^{(0)} = -\sum P_j \ln P_j, \ S^{(1)} = -\sum P'_k \ln P'_{kj}$ (3)

Theory of maximum information entropy: The probability distribution of a system's statuses could be speculated by maximizing $S^{(0)}$ with the constraint of

$$\sum_j P_j f_j^{(k)} = f_k$$ (4)

where f_k is the mean value of the possible fluctuation.

The message with the maximum value of P_j is the message to be sent. So the theory of maximum information entropy could be utilized to pick out the message to be sent by the G1 module and the E1 module (see Fig. 3).

The bottom-up approach to realize a Micro-Macro process of Strategy Generation (MMSG) can be summarized as the following [9, 10]:

Suppose the solution of the determined equation of formula (2) with a given control parameter $\boldsymbol{\alpha}_0$ is $\boldsymbol{q}_0$: $\dot{\boldsymbol{q}} = N(\boldsymbol{q}, \boldsymbol{\alpha})$, and suppose $\boldsymbol{\alpha}$: $\boldsymbol{q} = \boldsymbol{q}_0 + \boldsymbol{w}$, then formula (3) can be rewritten as $\dot{\boldsymbol{q}}_0 + \boldsymbol{w} = N(\boldsymbol{q}_0 + \boldsymbol{w}, \boldsymbol{\alpha})$. It can to be expanded as an exponential series as $\dot{\boldsymbol{q}}_0 + \boldsymbol{w} = N(\boldsymbol{q}_0, \boldsymbol{\alpha}) + L(\boldsymbol{q}_0)\boldsymbol{w}$, with the previous two items reserved only. Formula (4) can then eventually be converted to

$$\dot{\boldsymbol{w}} = L(\boldsymbol{q}_0)\boldsymbol{w}$$ (5)

The general solution of formulation (5) is $\boldsymbol{w}(t) = e^{\lambda t}\boldsymbol{v}$, with eigen values $\{\lambda_i\}$ and eigen vectors $\{v_i(x)\}$. Then we can get the expression:

$$\boldsymbol{q} = \boldsymbol{q}_0 + \sum_j \xi_j v_j(x) = q_0 + \sum_u \xi_u v_u(x) + \sum_s \xi_s v_s(x), \quad \xi_u > 0, \quad \xi_s < 0$$ (6)

Formula (6) is an order parameter equation, and ξ_u is called order parameter, while ξ_s is called servo parameter which can be expressed by ξ_u. According to Synergetics, ξ_u represents the macro space structure, temporal structure or functional structure of the System. So ξ_u can be used to generate the DPCM's intelligent behavior (see Fig. 3).

To be more explicit, ξ_{Gu} is indicative to identify which statuses of the Goals and Problems module are to be updated, both including the directions and amplitudes; ξ_{Eu} is indicative to identify which kinds of information are to be obtained from the environment; ξ_{Du} is indicative to identify which kinds of intelligent behavior are to be generated to solve the problems and hence to achieve the goals via the manipulation of the environment.

The previous explanation reveals that the top-down approach could be used to determine what the subsystem (Goals and Problems, Environment) needs to be changed (to achieve the goals or to cope with the environment) via information compression, and the bottom-up approach could be used to determine what the subsystem (DPCM) needs to do (to solve the problems or to operate the aircraft) via strategy generation.

Selective, limited attention is an obvious property of the human perceptual system, which has been implicated as a source of 'bounded rationality' in decision-making, has huge effect and notable limitation to intelligent behavior. The meaning for such a property is that attention shift is driven by particular dimension of task information needed. So, the evaluation of each option or preference is confidently based on the currently gathered information. In spite of other use of the currently gathered information for option or preference activation, which forms the course of cognitive processing or deliberation itself, the available information can also be used to validate the internal dynamics of the integrated cognitive architecture, and hence be used to build Polymorphic Cumulative Learning (PCL) mechanisms.

According to the Integrated Cognitive Architecture for Analysis of Pilot-Aircraft Dynamic Environment (ICA-APADE) above, it is clear that Cognitive Processing (Perception, Prediction, Planning, and Learning) or deliberation is at the core of the framework, and two major information loops are established via Cognitive Processing [5]. The essence and effect of ICA-APADE are then made curtain. While implementing the ICA-APADE, the Cognitive Processing module can be built as the kernel, and other ancillary modules can be built and plugged into based on the Cognitive Processing module. Both information loops in the ICA-APADE are Self-Validation Loops (SVL). That's why the available information can be used to build polymorphic cumulative learning mechanisms. During the course of deliberation for decision-making, the Cognitive Processing performs perception, prediction, planning, and learning concurrently.

Since the ICA model and the APADE model are loosely coupled, the ICA model and the APADE model can be independently developed, and other types of compatible APADE can be plugged into the integrated cognitive architecture. Once an initial ICA model and a ready APADE model were at position, the ICA will evolve till sufficiently mature to cope with the scenarios defined by the reference to the integrated cognitive architecture.

4 Results and Conclusion

The ICA-APADE takes consideration of pilots' characteristics and human factors, while traditional nonlinear flight dynamical systems fail to do so. The committed step is the Polymorphic Cumulative Learning mechanisms that draw inferences from cross-attested knowledge structures concealed in cognitive processing.

Acknowledge

This research work was supported by National Basic Research Program of China (973 Program No. 2010CB734103).

References

[1] Erik Hollnagel. The Phenotype of Erroneous Actions: Implications for HCI Design. [M]. London: Academic Press. 1991.

[2] Michael D. Byrne, Alex Kirlik, Using Computational Cognitive Modeling to Diagnose Possible Sources of Aviation Error [J]. The International Journal of Aviation Psychology, 2005,15(2):135 - 155.

[3] Kenneth Leiden, John Keller, Jon French. Context of Human Error in Commercial Aviation [J]. Micro Analysis & Design, Inc. , 2001.

[4] Zhong Y X. An introduction to cognetics: theory of transforms from information to knowledge and further to intelligence [J]. Chinese Engineering Science, 2004,6:1 - 8.

[5] David C Foyle, Becky L Hooey. Human Performance Modeling in Aviation [M]. Taylor & Francis Group, LLC 2008.

[6] Peter H Zipfel. Modeling and Simulation of Aerospace Vehicle Dynamics [M]. The American Institute of Aeronautics and Astronautics, Inc. 2007.

[7] Haken H. Synergetic Computers and Cognition: A Top- Down Approach to Neural Nets [M]. Springer-Verlag Berlin Heidelber, 1991.

[8] Haken H. Information and Self-Organization: A Macroscopic Approach to Complex Systems [M]. Springer-Verlag Berlin Heidelber, 1988,2000.

[9] Haken H. Advanced Synergetics [M], Springer-Verlag Berlin Heidelber, 1983.

[10] Haken H. Principles of Brain Functioning: A Synergetic Approach to Brain Activity, Behavior and Cognition [M]. Springer-Verlag Berlin Heidelber, 1996.

Using NASA-TLX to Evaluate the Flight Deck Design in Design Phase of Aircraft

Zheng Yiyuan, Yin Tangwen, Dong Dayong, Fu Shan

School of Aeronautics and Astronautics, Shanghai Jiao Tong University, Shanghai 200240

Abstract   Safety is the most important consideration in civil aviation, and human factors directly influence the safety of an aircraft. A large part of accidents was induced by human error. This paper focused on implementing the human factors by using improved NASA-TLX to reduce the design-induces error to enhance the safety of aircraft in the design phase.

Keywords   human factors, subjective assessment, NASA-TLX

1   Introduction

Currently, with the development of technique, the rate of civil aircraft accident has been gradually declined. According to NASA's statistics, however, over 75% of accidents related to the performance of human. Therefore, the behavior of pilots in flight deck is the uppermost factor of flight safety.

European Aviation Safety Agency has already published two regulations directly related to human factors in Certification Specifications for Large Aeroplanes CS – 25 [1]: (a) CS – 25 – 1302: Installed System & equipment for use by the flight crew, (b) CS – 25 – 1523: Minimum flight crew. Both of these two regulations concentrate on the safety perform of pilots. And CS – 25 book 2 Acceptable Means of Compliance [2] recommends the workload evaluation could be used as compliance certification to verify the regulations.

The most useful technique to evaluation workload is subjective assessment, including the methods of NASA-TLX, SWAT and CH. Among them, NASA-TLX is the most widely used, and it has achieved some solid goals in human factors research. However, until now this method was only used on the condition of in-fight or aft-flight. Since it is necessary to consider human factors during the design phase, if not, the aircraft would not fit for the pilot to operate, and the safety would decrease dramatically. Therefore whether NASA-TLX could be used in design phase should be discussed. This paper focuses on the possibility of availability of NASA-TLX to implement human factors thoughts in design phase.

2　Description of experiment method

NASA-TLX is a broadly accepted method to evaluate the workload of crew member in-fight or aft-flight phase, which has six dimensions: Mental Demand, Physical Demand, Temporal Demand, Performance, Effort, Frustration Level [3].

Normally, the procedure of implementation this method includes two steps: firstly, grading each dimension based on the different items in questionnaire, the rating scale is from 0 – 20. 0 and 20 represent the lowest and the highest workload respectively, then, comparing each two scores in six parameters to determine the weight of parameters, and drawing a conclusion about the workload [4].

2.1　The usage of nasa-tlxin the research

According to the requirement of research, the implementation differed from traditional NASA-TLX. The rating scale was expended to 0 – 100, and the second step was replaced by measuring the arithmetic average and mean square deviation, with volume normalization.

2.2　Participants

The participants to accomplish the NASA-TLX were six experts. Before scoring, they carefully studied the method of NASA-TLX by themselves. After the calculation, the results were reviewed by two pilots who have now served in Eastern Airlines Corporation.

2.3　The contents of questionnarie

The questionnaire was about the human factors in flight deck design which contained 72 items. These items were classified as the following: Horizon, Display, Control, Systems State, Warning System, Environment, like the extent of the lighting would affect the operation of the pilot, in order to reflect the human factors design philosophy in the design phase.

2.4　Experimental procedure

The experiment was carried out in three stages. Firstly, the eight participants scored the giving 72 items according to their understanding on aviation human factors and NASA-TLX. The method to score the items depended on the way the participants own habits, not only by scoring the row first, but also the column was accepted. Then, the data were analyzed by mean analysis and mean square deviation, with volume normalization analysis.

Secondly, after half week of the first stage, the same items were scored by the same participants. However, before the scoring, the participants were trained on the connotation of the six parameters of NASA-TLX, especially the definition of parameters of performance and frustration level. During the process of scoring, the participants were taught to grade one column firstly, which means to score the same one parameter of the 72 items, then the whole index was finished horizontally, that is to consider the correlation of the six parameters. The new data were also analyzed by mean analysis and mean square deviation, with volume normalization analysis.

Thirdly, the two data were analyzed comparatively. The detail is as following section.

3   Analysis

3.1   The analysis of first stage

In the first stage, the data of weights of mental, physical and time were reasonable, liking the following figures 1, 2, and 3. The horizontal axis is the number of items, and the vertical axis is the means of the items of this parameter among the six. The difference on the items of these three parameters was small, which indicated that about certain item, the thoughts of subjects were roughly

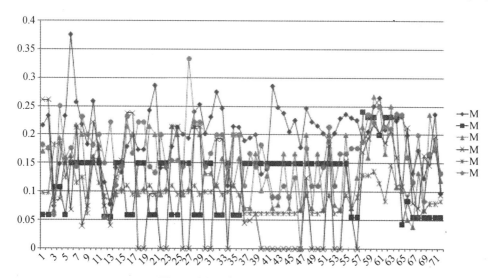

Fig. 1   Mental Weight of first stage

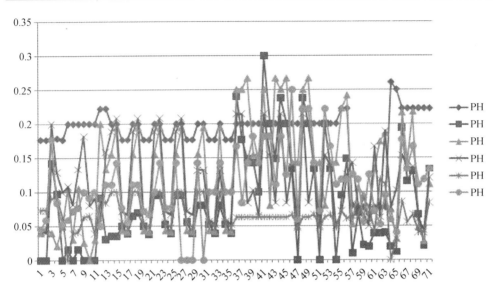

Fig. 2　Physical Weight of first stage

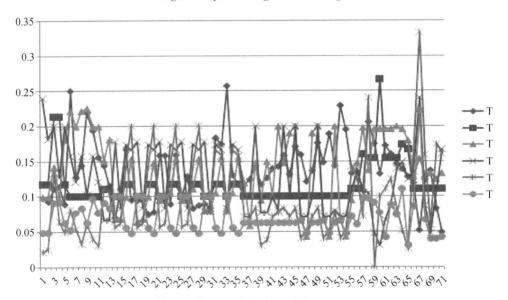

Fig. 3　Temporal Weight of first stage

same. However, the differences in some items were not expected, like item 68 in temporal weight and 5 and 27 in the mental weight, and according to our analysis, the main cause of these differences is the participants' different understandings on these items.

For the weight of effort, performance and frustration level, the differences were quite large. The reason was related on the different understanding on the connotations of the parameters. For example, in performance weight there were many zero, like the magenta curves, which shows that no matter what kind of design on these items, the performance of pilots would not be influenced. This result was unreasonable.

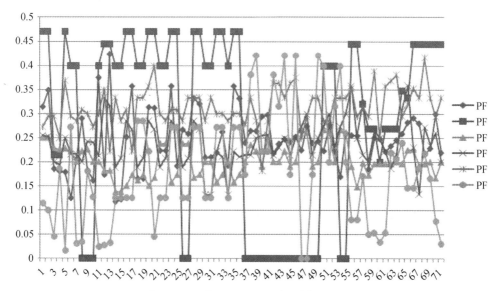

Fig. 4   Performance Weight of first stage

3.2   The analysis of second stage

As described in subsection 2.4, in second stage, the definitions of the six parameters were giving out clearly, and the scores were carried out according to certain rules. Therefore, the results of the weights of effort, performance and frustration level were significantly improved, as figures 5, 6 and 7.

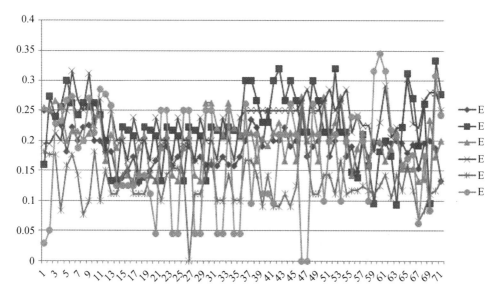

Fig. 5   Effort Weight of second stage

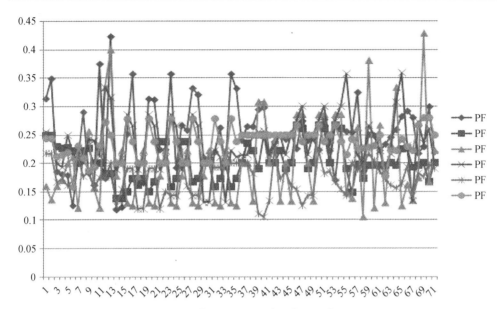

Fig. 6　Performance Weight of second stage

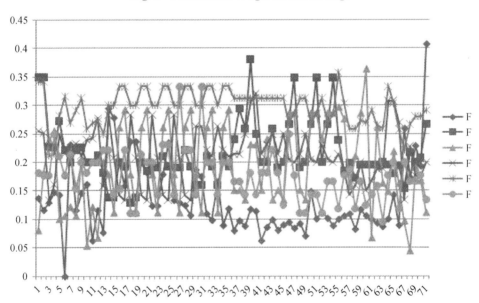

Fig. 7　Frustration Level Weight of second stage

From the above three figures, the distribution of weights of effort were in a reasonable range, no extreme effort existed, which complied with the design philosophy. And there had some large weights in performance and Frustration Level, like item 13 and 37 in performance and items 2, 40, 60 and 72 in frustration level. However, the correlation, which presented by the mean square deviation, of trend of certain weight was much better than the data collected in the first stage, which indicated the fidelity of the result had been improved greatly in the second stage. Nevertheless, the correlations of performance and frustration level still have some unreasonable results, which need further improvement.

3.3 Evaluation

After analyzes, the relatively larger and smaller items in the second stage were listed, like the items mentioned above, to evaluate by the pilots whether the irrational design of the giving item would affect them on certain aspect about the six parameters. And the results were encouraging, with the fidelity of the experiment was confirmed by the pilots.

4 Conclusion

From our analysis, the second set of data outperformed the first one well, especially in some parameters, such as, effort, performance and frustration level. This improvement results from the clear redefinition of the certain parameters, and no different understandings that would degrade the accuracy of the results. Moreover, the factors that would affect the outcome are the configuration of the items and the connotations of the parameters, and the second factor should be minimized in practice.

This experiment verified the implementation of human factors in the design phase by using NASA-TLX. Aiming at the items which were scored larger of smaller, the corresponding parameters should be considered carefully during the design. Statistical data shows that the expense ratio to correct a defect in design, developing and manufacturing phase is 1 : 10 : 100 [5]. Therefore, the earlier consideration of human factors, the less unreasonable cost would spend, and the safer aircraft is.

Acknowledgements

This research work was supported by National Basic Research Program of China (973 Program No. 2010CB734103).

References

[1] Certification Specifications for Large Aeroplanes CS – 25 [S]. EASA. (2009)

[2] CS – 25 BOOK2 Acceptable Means of Complance-AMC [S]. EASA. (2009)

[3] Farmer E. and Brownson A. Revier of workload measurement, analysis and interpretation methods [S]. European Organisation for the Safety of Air Navigation. (2003)

[4] Michael H. Biferno, Willian H. Corwin, Assessment of crew workload measurement

methods, techniques and procedures. Cockpit Integration Directorate Wright Research and Development Center. Air Force Systems Command Wright-Patterson Air Force Base. (1989)

[5]　Stanton N A, Harris D, Salmon P M. Predicting design induced pilot error using HET — A new formal human error identification method for flight decks [G]. Department of Design, Brunel University, Cranfield University. (2006)

A Brief Review on Physiological and Biochemical Evaluations of Human Mental Workload

Ying Lean[1], Fu Shan[2]

1. Dept. of Aviation Medicine and Physiology, School of Aeronautics and
Astronautics, Shanghai JiaoTong University, Shanghai 200240
2. Dept. of Human Factors, School of Aeronautics and Astronautics,
Shanghai JiaoTong University, Shanghai 200240

Abstract　This paper gives a brief review and short summary on physiological and biochemical evaluations of human mental workload. Physiological evaluations consist of peripheral physiological evaluations and central physiological evaluations. Peripheral physiological evaluations mainly include heart rate and heart rate variability (HRV) recorded from ECG, blood pressure, respiration, eye blinks, skin potential, and hemodynamic indices. Central physiological evaluations mainly refer to the cerebral cortex, including EEG and ERP. Biochemical evaluations mainly include catecholamines, cortisol, and immunoglobulin A (IgA) collected from blood, salivary, or urinary samples. They are all objective measurements to evaluate mental workload when conducting particular mental tasks.

Keywords　mental workload, heart rate variability (HRV), eye blink, electroencephalogram (EEG), event-related-potential (ERP), catecholamines, immunoglobulin A (IgA)

1　Introduction

The term 'workload' describes an overall efficiency of human performance when performing a particular task or several tasks under various environmental conditions. While the term 'taskload' refers to the external duties, amount of work, or number of tasks that have to be performed and accomplished, 'workload' refers to the individual effort and subjective experience of a particular person performing the tasks under a certain working environment and in a time interval during which those duties must be accomplished. That is to say, a same task may result in different levels of individual workload due to human differences in capabilities, efforts, attitudes, cognition, skills, limitations, and state of situational awareness, complacency, fatigue, boredom, anxiety, stress, etc. The model of human information processing is not a single-channel model but a multiple-channel model consisted of several separate processing resources, so workload is multi-dimensional.

Workload may be either physical or mental, although they are always connected with each other and can not be separated completely when performing a particular task. This paper focuses mainly on mental workload. Evaluations of mental workload may be either subjective or objective. Subjective evaluations, such as NASA-TLX and some other questionnaires, are inexpensive and easily collected but they are unable to give accurate reports because of individual biases and often require a large number of samples. On the contrary, objective evaluations require a relatively small number of samples and can give more accurate reports than subjective evaluations, but they are more complex, requiring technical skills and operational experience. This paper refers mainly to the physiological and biochemical evaluations of mental workload.

Mental workload leads to the changes of human performance and behaviour. Take driving as an example, Recarte MA et al studied the effects of different mental tasks on visual behavior and driving performance. The increased mental workload was reflected in a significant pupil size increment, a spatial gaze concentration and reduced inspection frequency of mirrors and speedometers (Recarte & Nunes, 2000). Pupil size is an indicator of effort due to the added mental load, spatial gaze variability is an indicator of the alteration of visual search patterns, and the frequency of rearview mirror and speedometer inspection is to measure the degree of situation awareness during driving (Recarte & Nunes, 2003). In their further researches on driving vehicles 4 hours on a highway in real traffic, significant difference was found in pupil diameter for mental tasks compared with the non-task condition, a mean increment about 4% in a range between 2% and 6%. Spatial gaze variability scores showed a significant reduction and spatial gaze concentration effect was created for mental tasks. Mental tasks also caused a reduction in the percentage of glances at the speedometer and rearview mirror inspection. It was also found that regardless of whether correct or not, during non-task conditions, participants detected 77.3% of the targets, but only 64.4% when performing mental tasks. In addition to the significant decrease in the detection probability, correct discrimination responses also decreased. The

percentage of correct responses to the targets was 90. 6% in the non-task
condition, while 83. 8% in the mental task conditions. Response time can be
divided into perception time, inspection time and decision time. With mental
tasks, perception time increased, inspection time decreased, and decision time
remained unaltered. When performing mental tasks, participants glanced at the
targets less frequently, and if looked at, the targets were detected later and
glanced at for less time (Recarte & Nunes, 2003). Horrey WJ et al examined how
driving task and reading information presented on the front LCD screen task
influenced driver performance and visual scanning behaviour with a simulated
traffic environment. A low score of variability in lane keeping represents a good
driving performance. Compared to the simple task, the complex task increased the
score of variability in lane keeping, indicative of a worse driving performance. And
also, when the complexity of the task increased, subjects response time increased,
and percent dwell time to outside world, which represents the proportion of time
drivers spent looking there, decreased. Based on these results, a computational
model of visual attention predicting driver performance and scanning behaviour was
built (Horrey, Wickens, & Consalus, 2006).

Changes of human performance and behaviour resulted from mental workload
are closely related to the physiological and biochemical changes in the body, which
are based on the nervous regulation, humoral regulation and autoregulation. In the
following sections, this paper gives a brief review and summary on this issue.

2 Peripheral Physiological Evaluations

In addition to the changes of pupil diameters, heart rate or Heart Rate
Variability (HRV) recorded from electrocardiogram (ECG) is widely used and has
a long history in the evaluation of mental workload (Jorna, 1992; Roscoe, 1992;
Jorna, 1993; Roscoe, 1993), the sampling of which is non-invasive and safe,
causing no injuries or pains to subjects. As for normal healthy humans, RR
intervals on ECG recordings are not always unchanged but have small differences
since heart rate is controlled dynamically by the autonomic nervous system. HRV
describes beat-to-beat variation in heart rate or small differences in RR intervals,
thus reflecting the function of the autonomic nervous system. In the spectral
analysis of HRV, low frequency band (LF, 0. 04~0. 15 Hz) represents mainly the
cardiac sympathetic nervous activities and high frequency band (HF, 0. 15~0. 40 Hz)
shows the respiratory and cardiac vagal nervous activities. LF increases with an
increased sympathetic tone and HF increases with an increased vagal tone. The

ratio LF/HF therefore indicates the balance of sympathetic tone and vagal tone.

When humans are in the state of heavy workload, acute mental fatigue or stress, cardiac activity is controlled mainly by sympathetic nerves, while in normal cases, cardiac activity is controlled mainly by vagal nerves. This would cause an increase in LF and LF/HF but a decrease in HF during mental workload. Jiao K et al investigated the effects of driving vibration and speeds on drivers mental workload and mental fatigue by means of spectral analysis of HRV (Jiao, 2004; Jiao, Li, Chen, Wang, & Qi, 2004; Jiao, Li, Chen, & Wang, 2005; Li, Jiao, Chen, & Wang, 2003). When the driving speeds were 60, 80, and 100 km/h, while the driving workload increased and mental fatigue appeared, it was found that LF increased, HF decreased and LF/HF also increased, which indicated drivers enhanced sympathetic activities, decreased parasympathetic activities and increased sympathovagal balance. Compared to the low driving speed, this phenomenon was more apparent when the speed was higher (Jiao, 2004; Jiao, Li, Chen, Wang, & Qi, 2004; Jiao, Li, Chen, & Wang, 2005). It was also demonstrated that Chinese traditional magnitopuncture might decrease the drivers LF and increase their HF, which suggested that magnitopuncture might reduce mental workload and fatigue in healthy subjects while performing mental tasks (Li, Jiao, Chen, & Wang, 2003).

Blood pressure, respiration, eye blinks, and some other indices are often used together with heart rate or heart rate variability to get a more comprehensive evaluation of mental workload. Heart rate, blood pressure, respiration and eye blinks were recorded in 14 subjects while performing a complex mental task in a flight simulator environment. These indices were all found to be affected by the mental workload. Slow respiratory activity contributed to HRV, especially after high workload, such as landing. It was found that both systolic and diastolic blood pressure during landing period were higher than rest, flight and after-landing periods. Systolic blood pressure variability and diastolic blood pressure variability in both of the mid-band and high-band were larger during the after-landing period than any other period. Spectral energy of respiration in the mid- and high band were largest during the after-landing period compared with rest, flight and landing periods. The increase in respiratory amplitude after landing was most evident for the mid-band. Subjects tended to breathe deeper and more slowly after landing. Compared with any other period, eye blink interval during landing period was longest and, on the contrary, eye blink duration was shortest during that time. All these physiological facts, combined with subjects subjective feelings and

performance, demonstrated that landing was a period of high mental workload for
pilots (Veltman & Gaillard, 1996).

In another similar research, 12 pilots flew through a tunnel as well as
performing a memory task with different levels of difficulty. Mental workload was
also measured by heart rate, blood pressure, respiration and eye blinks. Heart rate
was sensitive to difficulty levels in the mental tasks. HRV increased when
respiratory activity around 0. 10 Hz increased. The modulus, the gain between
systolic blood pressure and heart period, was hardly influenced by respiration and
therefore appeared to be a better measurement than HRV. The duration of a
respiratory cycle was sensitive to changes in workload. When mental workload
increased and more visual information had to be processed, eye blink intervals,
time between two successive eye blinks, increased and eye blink duration
decreased, with the exception of memory task. All these physiological indices were
sensitive to mental effort (Veltman & Gaillard, 1998). De Rivecourt M et al
examined effects of momentary changes in mental workload on cardiovascular and
eye activity indices. 19 male pilots performed an instrument flight task. Heart rate
changed between rest and tasks. Short term HRV provided more insights into
mental effort. With the increased difficulty of taskload, an increase in heart rate
and a decrease in HRV occurred. Eye activities were also sensitive to mental
effort, which resulted in an increase of mean dwell time and mean fixation duration
(Rivecourt, Kuperus, Post, & Mulder, 2008).

Besides the above measurements, skin potential, other cardiovascular indices
and hemodynamic indices, such as pre-ejection period, left ventricular ejection
time, stroke volume, cardiac output, tissue blood volume, could also be used in
the evaluation of mental workload. In the latest research, these indices were
recorded from 15 male undergraduates when performing the Multi-Attribute Task
Battery (Miyake, Yamada, Shoji, Takae, Kuge, & Yamamura, 2009). Skin
potential level was recorded from both palms. Tissue blood volume was recorded
from the tip of the nose using a laser Doppler blood flow meter. Hemodynamic
indices, including pre-ejection period, left ventricular ejection time, stroke
volume, cardiac output, were recorded using a special admittance cardiograph. It
was demonstrated that compared with other physiological indices, average test/
retest correlation coefficient is highest in skin potential level. Skin potential level
may be most sensitive to mental workload in this experiment. It was also pointed
out that skin conductance, skin temperature difference between nose and forehead,
tissue blood flow may also be reliable indices of mental workload and tissue blood

volume may be a sensitive index of mental workload when the environmental temperature is nearly unchanged (Miyake, Yamada, Shoji, Takae, Kuge, & Yamamura, 2009).

Using blood velocity as an index, Someya N et al researched on the effect of mental task on the coeliac artery and superior mesenteric artery under fasting and postprandial conditions. The mean blood velocities in the coeliac artery and superior mesenteric artery were significantly increased by the mental task under fasting conditions. The vascular conductance in the superior mesenteric artery, but not in the coeliac artery, was significantly decreased by the mental task under fasting conditions (Someya, Endo, Fukuba, Hirooka, & Hayashi, 2010). While researching on the relationship between mental workload (mental arithmetic task) and hot flashes in menopausal women, it was found that regional oxygen saturation, an objective measurement of mental workload, was significantly higher in non-task condition than task condition, but the incidence of hot flashes was greater in task condition than non-task condition. It was suggested that mental workload under time pressure might be a risk factor for menopausal hot flashes (Park, Satoh, & Kumashiro, 2008). Mental workload will cause decrease in visual field area. Using a Goldmann visual perimeter, it was found that the mean area of visual fields decreased to 92.2% under medium workload and 86.4% under heavy workload. The shape of visual field was distorted as well (Rantanen & Goldberg, 1999).

3　Central Physiological Evaluations

Information from outside environment is transmitted through sensory pathways to the thalamus, and then to the cerebral cortex. Human cerebral cortex contains about 10 billion neurons, and which is divided into certain areas to carry out different functions of sensation, cognition and movement (Guillery, 2005; Sur & Rubenstein, 2005; Majewska & Sur, 2006). As electrophysiological recordings of cerebral cortex, Electroencephalogram (EEG) and Event-related-potential (ERP), are widely used to evaluate mental workload in addition to the above mentioned heart rate, HRV, respiration, eye blinks and hemodynamical indices.

Mental workload will lead to the changes of EEG components: alpha band, beta band, theta band, and delta band. While performing a computer-based air traffic control task with three different levels of mental difficulty, EEG recorded from 19 channels, together with eye blink and respiration were recorded from 8 subjects. EEG activity in the theta band was found sensitive to the different

difficulty levels of mental workload. It was obvious at central, parietal, frontal
and temporal site. Percent theta power at these sites increased significantly with
the increase of task difficulty. Percent theta power in the high workload condition
was significantly higher than that of low and medium workload conditions.
Overload condition was associated with significantly increased theta activity.
Besides theta power, higher percent alpha power was revealed in the low workload
condition compared to the high and medium workload conditions. Percent alpha
power was highest for the low workload condition, lower for the high workload
condition, and lowest for the medium workload condition. As for beta 1power, the
overload condition was associated with larger percent beta1 than that of low
workload condition at F7, T4 and high workload condition at T6. However, low
workload condition and medium workload condition were associated with larger
beta 1 power than overload condition at Fz, F3, Pz, F7 and T4. Finally, the
highest percent delta power was found associated with medium workload
condition, lower percent delta power associated with high workload condition, and
low workload condition was associated with the least delta power. In this
experiment, besides the changes of EEG components mentioned above, it was also
found that eye blink rates decreased significantly when the mental task became
more difficult. The blink rates of overload condition were significantly lower than
that of low and medium workload conditions. Respiration rates increased when the
difficulty of mental task increased. Respiration rates of overload condition was
higher than that of low and medium workload conditions (Brookings, Wilson, &
Swain, 1996).

Similarly, during a flight scenario researched by Hankins TC et al, theta band
of EEG increased in mental calculation conditions. Heart rate increased during
take-off and landing period, which required high mental demands. Both EEG and
heart rate were shown sensitive to mental workload (Hankins & Wilson, 1998).
Fournier LR et al examined the effects of single mental task and multiple menal
task on alpha 1 (8 – 10 Hz) event-related desynchronization, alpha 2 (10 – 12 Hz)
event-related desynchronization, and theta (3 – 7 Hz) event-related synchronization,
which were relevant to attentional resource allocation, sensory-motor processing,
information processing or other cognitive activities. It was demonstrated that these
three indices were not sensitive to multiple task workload but were sensitive to
single task workload. Besides EEG, heart rate and breath rate both increased when
the mental workload of single and multiple tasks increased, which were the same
as the above mentioned results (Fournier, Wilson, & Swain, 1999).

A closed-loop biocybernetic system, based on EEG signals which reflected subjects engagement in the mental task, was developed as a method to evaluate automated flight deck compatibilities with human capabilities. Subject performed Multiple-Attribute Task Battery presented on a computer and four feedback indices were used: beta power/alpha power, beta power/(alpha power + theta power), left temporal alpha power/central alpha power, and left occipital alpha power/right occipital alpha power. It was observed that compared with other three indices, beta power/(alpha power + theta power) reflected mental task engagement best (Pope, Bogart, & Bartolome, 1995). A psychophysiological adaptive automation system, which could switch between manual and automatic task modes based on subjects EEG and ERP indices, were built up to evaluate subjects mental workload and engagement and to improve their performance. P300 component of ERP was used as an effective measurement to assess subjects task engagement and performance in an adaptive automation condition (Prinzel, Freeman, Scerbo, Mikulka, & Pope, 2003). A wavelet transform of EEG was used to evaluate the mental workload in a matching task with three levels of task difficulty. The total power of theta, alpha, and beta frequency bands and the time maximum power appeared were extracted, which demonstrated that wavelet transform of EEG was a relatively quick and accurate method for mental workload evaluation (Murata, 2005).

With a wireless sensor headset, EEG was recorded in a vigilance, learning, and memory task. Similar results were found that EEG increased with increasing mental workload. So, it was concluded that EEG reflected the cogintive capabilities of information-gathering, information-processing, problem-solving, analytical reasoning, and allocation of attentions (Berka, Levendowski, Lumicao, Yau, Davis, Zivkovic, et al., 2007). While researching on the relationship among the cerebral cortex, cardiovascular and respiratory functions under mental workload conditions, EEG, ECG, and breath rate were recorded from 29 male subjects. It was found that mental arithmetic task significantly increased breath rate, heart rate and EEG theta power at FC3, FC4 and C4 electrodes. And also, mental task decreased the duration of cardiorespiratory synchronization epochs, which was negatively correlated with the EEG theta power at those electrodes (Zhang, Yu, & Xie, 2010).

Non-linear indices of EEG signals, such as Correlation Dimension, Lyapunov Exponent, and Approximate Entropy, can reflect the cognitive and mental activities of cerebral cortical networks from a dynamical perspective. EEG was recorded in 25 healthy subjects under three conditions: eyes-closed, eyes-open, and mental arithmetic task with eyes-closed. It was found that compared to the eyes-closed condition, Correlation Dimension increased in eyes-open and arithmetic conditions. It was concluded that Correlation Dimension, and some other non-linear indices, may be effective for reflecting information processing during mental tasks (Stam, Woerkom, & Pritchard, 1996).

To research on the relationship between scalp potential and autonomic nervous activity, 43 male subjects participated in a mental arithmetic task. The function of cerebral cortex was evaluated by EEG, and autonomic nervous function was evaluated by HRV recorded from ECG. Approximate Entropy was used as a non-linear index of EEG signals and scalp potential was also determined by the wavelet packet parameters. It was demonstrated that Approximate Entropy significantly increased in mental arithmetic task. And also, relative wavelet packet energy in alpha band of EEG at P3, P4, Pz, O1, O2 and Oz electrodes decreased while beta band at those electrodes increased. As for cardiovascular and autonomic nervous system, mental arithmetic task induced significant increases in heart rate and LF component and a decrease in HF component. Moreover, changes of brain activity were earlier than changes of autonomic nervous activity and there existed significant correlations between HRV and wavelet packet energy. It was concluded that cerebral activities enhanced with the decreased parasympathetic activity and increased sympathetic activity, and the right post-central areas dominated sympathetic activity during stress-induced mental tasks (Yu, Zhang, Xie, Wang, & Zhang, 2009). A latest brief review on the evaluation of mental workload and task engagement by means of EEG was given by Rabbi AF et al (Rabbi, Ivanca, Putnam, Musa, Thaden, & Fazel-Rezai, 2009).

Event-related-potential (ERP) is the potential recorded from the scalp elicited by a visual or (and) auditory stimulus. ERP normally contains the following components: N1, P2, N2, and P3 (P300). N1 (or N2) is the first (or second) negative wave of ERP and P2 (or P3) is the second (or third) positive wave. They are related to attention, information processing, resource allocation and other cognitive activities. Mismatch negativity (MMN), the negative component of ERP usually peaking at about 100 – 200 ms after the onset of stimulus, is often elicited by auditory stimuli, in which the rare deviant tone is different from the repetitive standard tone in its frequency, duration, or intensity (Näätänen, 2000). MMN is widely used in cognitive and neuroscience to measure the function of central auditory processing, allocation of attention, level of workload, and other mental activities.

ERP was used as an objective evaluation to assess pilot performance and mental workload in a simulation of multi-function helicopter. The probe evoked P300 component decreased in amplitude with the increase of communication demands and mental workload (Sirevaag, Kramer, Wickens, Reisweber, Strayer, & Grenell, 1993). In another similar research to examine the effect of mental workload on P300, P300 was elicited by an auditory oddball task which required the detection of deviant tones and a visual oddball task which required the detection of flashes. Increased P300 latency was found with the increase of mental workload (Fowler, 1994). Hohnsbein J et al investigated the effect of time-pressure on two P300 subcomponents: P-SR and P-CR. The latency of P-CR (not P-SR) was shortened with the increase of time-pressure (Hohnsbein, Falkenstein, & Hoormann, 1995).

ERP was recorded from 10 highly trained radar operators when performing simulated radar- monitoring tasks with different difficulty levels. It was observed that the amplitude of N1, N2, and MMN components at the sites of Fz and Cz, and the amplitude of P300 at the site of Pz, decreased with the introduction of mental tasks and decreased further when the difficulty level of tasks increased from low to high. This might be explained as the brain capacities of human mental workload is limited and increases in the difficulty of the primary task resulted in the allocation of the perceptual and cognitive resources and decrease the processing resources available for the secondary task. This allocation of attention and processing resources in the brain caused the changes in amplitudes of ERP components. So, it was concluded that ERP with task-irrelevant auditory probes was effective for the non-intrusive evaluation of mental workload in complex tasks (Kramer, Trejo, & Humphrey, 1995).

When performing gauge monitoring and arithmetic tasks, ERP elicited by irrelevant auditory probes was recorded from 15 healthy adults at the sites of Fz, Cz, Pz and Oz. Similar to the above mentioned results, the amplitude of N1 and P300 decreased significantly during the gauge monitoring and arithmetic tasks (Ullsperger, Freude, & Erdmann, 2001). Effect of visual task on auditory ERP was studied in a zero-back and a one-back visual task. N1 and P2 were created but no MMN elicited by deviant stimuli (Dyson, Alain, & He, 2005). The amplitude of P300 component of ERP as well as short-term HRV metrics were used to measure mental workload in 140s (Henelius, Hirvonen, Holm, Korpela, & Muller, 2009). Mental workload when playing a computer video game was assessed by ERP indices from 14 male subjects. Amplitudes of N1, P2, N2, P300 at Fz, Cz, and Pz electrodes decreased all when the game was more difficult (Allison & Polich, 2008).

　　Simoens V L et al studied the effect of attentional workload and acute mental stress on auditory ERP with the index of MMN. A decrease in MMN amplitude was found at the fronto-central electrodes under stressful and high mental demand condition compared to the non-stressful condition especially at Cz, F3, F4, and Fz sites. The decrease in MMN amplitude during the IQ test, with high attentional workload, was most apparent at right frontal sites, which may indicate that workload of the primary task mainly affected right frontal MMN subcomponents. It was concluded that attentional stress, resulted from high mental workload, may attenuate cortical auditory processing capabilities (Simoens, Istók, Hyttinen, Hirvonen, Näätänen, & Tervaniemi, 2007).

　　Last but not least, cerebral blood flow quantified by Arterial spin labeling perfusion functional magnetic resonance imaging (fMRI) might also be a useful indicator of mental workload and task engagement. While performing a 20-min psychomotor vigilance test for 15 subjects, Arterial spin labeling perfusion fMRI data showed that mental task activated a right lateralized fronto-parietal attentional network in addition to the basal ganglia and sensorimotor cortices. The fronto-parietal network was found less active during the post-task rest compared to the pre-task rest. When the difficulty of mental workload increased and subjects mental fatigue appeared, regional cerebral blood flow in fronto-parietal area decreased, which was happened together with subjects increased reaction time and worse performance. These results demonstrated the persistent effects of mental fatigue in the fronto-parietal network after heavy mental task. In addition, regional cerebral blood flow in the thalamus and frontal gyrus was also associated with subjects mental effort and task engagement (Lim, Wu, Wang, Detre, Dinges, & Rao, 2010).

4　Biochemical Evaluations

　　Catecholamines (including adrenaline and noradrenaline), cortisol, and immunoglobulin A (IgA) collected from blood, salivary, or urinary samples are often used to evaluate mental workload. In earlier studies, effects of mental arithmetic tests with graded difficulty on urinary catecholamines and salivary cortisol levels were researched on 8 young male subjects. A significant increase in adrenaline excretion was observed after all of the tests, but no significant increase was observed for noradrenaline and dopamine. So, it was demonstrated a decreased ratio of noradrenaline/adrenaline and an increased ratio of adrenaline/dopamine. Changes in salivary cortisol concentration during the tests were graded with the difficulty of the tests. Thus, it was concluded that urinary adrenaline and salivary cortisol levels were effective mental workload measurements (Fibiger, Evans, & Singer, 1986). In evaluation of workload in middle-aged steel workers by measuring urinary excretion of catecholamines and cortisol, there existed increases in noradrenaline and adrenaline during working hours, and which was more evident in the middle-aged workers than in the young workers. Changes of dopamine and cortisol were not obvious between these two groups of workers different in ages. On the other hand, it was found that automation caused the age-related differences in urinary noradrenaline and adrenaline not so significant, which meant that automation might reduce the workload in middle-aged workers (Sudo, 1991).

Neuroendocrine responses to a 40 min instrument flight were studied in 35 male volunteers and blood samples were collected. Compared to the control levels, plasma ACTH and cortisol were significantly higher before and 5 min after the flight. Plasma prolactin, noradrenaline and adrenaline increased significantly 5 min after the flight, which represented a mental stress reaction to the flight. Elevated levels of plasma ACTH and cortisol before the flight could be associated with the anticipation of the forthcoming task. High adrenaline levels after the flight correlated significantly with the poor flight performance (Leino, Leppäluoto, Ruokonen, & Kuronen, 1999). Similarly, when performing the Multi-Attribute Task Battery for 16 healthy subjects, it was observed that mental workload significantly increased urinary catecholamines, including adrenaline and noradrenaline (Papadelis, Kourtidou-Papadeli, Vlachogiannis, Skepastianos, Bamidis, Maglaveras, et al, 2003). In another case using urinary excretion rates of catecholamines as indices, it was found that job rotation did not change mental workload as expected (Kuijer, Vries, Beek, Dieën, Visser, & Frings-Dresen, 2004). A research on the effects of flight mental workload on urinary catecholamines was conducted in 54 experienced male military pilots. It was found that adrenaline was significantly higher after flight than the pre-flight period in all pilots, which reflected the responses of autonomic nervous system to the flight workload with the increased sympathetic activity (Otsuka, Onozawa, Kikukawa, & Miyamoto, 2007).

As an indicator of hypothalamic-pituitary-adrenocortical (HPA) axis activation, cortisol awakening response is related to a number of psychosocial factors (Chida & Steptoe, 2009). Chida Y et al gave a systematic review and meta-analysis on the relationship between the cortisol awakening response and psychosocial factors. It was summarized that increased cortisol awakening response was generally related to work stress, job strain, mental overload, overcommitment to work, and other types of life stress. On the other hand, decreased cortisol awakening response was related to fatigue, burnout, exhaustion, and posttraumatic stress syndrome (Chida & Steptoe, 2009).

Biochemical indices collected from blood, salivary, or urinary samples are often used together with heart rate, blood pressure, and other physiological measurements, to get an overall evaluation of mental workload. In a recent research, mental workload of 40 pilots during a new noise-reduced landing approach (Segmented Continuous Descent Approach) was tested compared with the standard Low Drag Low Power procedure in A320 and A330 flight simulators by multiple measurements, including salivary cortisol concentration. In most cases, landing requires a high mental demanding usually accompanies with significant increases in heart rate and decreases in eye blink frequency. However, mean heart rate, blood pressure and salivary cortisol concentration during this new landing approach were not increased significantly. So, it was concluded by the authors that this new landing approach did not require higher workload for pilots compared to the standard approaches (Elmenhorst, Vejvoda, Maass, Wenzel, Plath, Schubert, et al, 2009).

Simoens VL et al also researched on the relationship between cortisol levels and MMN amplitudes in mental workload conditions. It was found that besides the decreases in MMN amplitudes during IQ test, attentional workload and acute mental stress also led to an increase in salivary cortisol levels compared to the non-stressful condition for the majority of subjects. Since cortisol levels during mental workload were inversely related to the changes in MMN amplitudes, it was concluded that cortisol interfered with the preattentive mechanism of MMN and cortical auditory memory-trace formation (Simoens, Istók, Hyttinen, Hirvonen, Näätänen, & Tervaniemi, 2007).

Salivary secretory immunoglobulin A (s-IgA) could also be used as a measurement of mental workload and psychological stress in students. In earlier studies, as an index to study the effect of academic stress on immune function, the secretion rate of salivary s-IgA was collected in 64 first-year dental school students. The s-IgA secretion rate was significantly lower in high stress than low stress periods for the whole subjects. Besides, personality characteristics influenced s-IgA secretion rates. Students with warm personal relationships secreted more s-IgA (Jemmott, Borysenko, Borysenko, McClelland, Chapman, Meyer, et al, 1983). However, other similar researches may have different results and conclusions. The concentration of salivary s-IgA tended to be higher on the day before academic exams and during exams and lower on the days between these exams, which were observed in 10 second-year medical student volunteers. The increase in s-IgA before exams might be due to the anticipation of the forthcoming exams (Otsuki, Sakaguchi, Hatayama, Takata, Hyodoh, Tsujita, et al. , 2004).

According to the researches of Ring C et al, mental arithmetic task led to significant increases in salivary s-IgA concentration and s-IgA secretion rate, as well as modest increase in heart rate. Besides, it was also pointed out that the previous hypothesis that psychological stress of competitive exercise contributes to increased susceptibility to respiratory infections due to reductions in s-IgA required further investigation (Ring, Carroll, Hoving, Ormerod, Harrison, & Drayson, 2005). In addition, salivary s-IgA and cardiovascular activities were measured in 24 males and females conducting mental arithmetic tasks. It was shown that mental arithmetic tasks resulted in increased s-IgA concentration and s-IgA secretion rate, as well as increased blood pressure and pulse rate, compared to the rest periods. But there were no significant correlations between s-IgA and cardiovascular activities (Ring, Drayson, Walkey, Dale, & Carroll, 2002).

Tasks which conducted in a short period of time and required subjective efforts, such as memory task and mental arithmetic task, usually lead to increases in s-IgA levels. The effects of four mental tasks on s-IgA, which was also sampled non-invasively from saliva, was studied by Wetherell MA et al. These four tasks were used simultaneously: mental arithmetic task, memory search, auditory monitoring, and visual monitoring. It was observed that these tasks resulted in increased s-IgA secretion. Furthermore, subjects differences in their perceptions of mental workload could be classified according to their high or low s-IgA reactivities: Subjects who perceived the tasks to be too difficult demonstrated low s-IgA reactivities, while subjects who engaged in the tasks but did not feel too demanding demonstrated high s-IgA reactivities. It was concluded that s-IgA reaction to acute mental stress depended not only on the external tasks, but also on the individual perceptions of the tasks (Wetherell, Hyland, & Harris, 2004).

Mechanisms of mental workload associated with stress, anxiety, and other negative mental factors are always researched at the molecular and genetic levels. According to the medical ethics, these researches are often conducted in animals, such as rats and mice, instead of humans, by means of foot shock, maze, repeated forced swimming, etc (Passerin, Cano, Rabin, Delano, Napier, & Sved, 2000; Korte & Boer, 2003; Funk, Li, & Lê, 2006; Banerjee, Shen, Ma, Bathgate, & Gundlach, 2010). However, they are beyond the topic of this paper and would be discussed in the near future, not in this paper.

Acknowledgement

This research is supported by 'Chinese National Key Basic Research Program (973 Program)' with the number of : 2010CB734103.

References

[1] Allison B Z, Polich J. Workload assessment of computer gaming using a single-stimulus event-related potential paradigm [J]. Biological Psychology, 2008,77(3),277 – 283.

[2] Banerjee A, Shen P J, Ma S, et al. Swim stress excitation of nucleus incertus and rapid induction of relaxin – 3 expression via CRF1 activation [J]. Neuropharmacology, 2010,58 (1),145 – 155.

[3] Berka C, Levendowski D J, Lumicao M N, et al. EEG correlates of task engagement and mental workload in vigilance, learning, and memory tasks [J]. Aviation, Space, and Environmental Medicine, 2007,78(5),231 – 244.

[4] Brookings J B, Wilson G F, Swain C R. Psychophysiological responses to changes in workload during simulated air traffic control [J]. Biological Psychology, 1996,42(3),361 – 377.

[5] Chida Y, Steptoe A. Cortisol awakening response and psychosocial factors: a systematic review and meta-analysis [J]. Biological Psychology, 2009,80(3),265 – 278.

[6] Dyson B J, Alain C, He Y. Effects of visual attentional load on low-level auditory scene analysis [J]. Cognitive, Affective, and Behavioral Neuroscience, 2005,5(3),319 – 338.

[7] Elmenhorst E M, Vejvoda M, Maass H, et al. Pilot workload during approaches: comparison of simulated standard and noise-abatement profiles [J]. Aviation, Space, and Environmental Medicine, 2009,80(4),364 – 370.

[8] Fibiger W, Evans O, Singer G. Hormonal responses to a graded mental workload [J]. European Journal of Applied Physiology and Occupational Physiology, 1986,55(4),339 – 343.

[9] Fournier L R, Wilson G F, Swain C R. Electrophysiological, behavioral, and subjective indexes of workload when performing multiple tasks: manipulations of task difficulty and training [J]. International Journal of Psychophysiology, 1999,31(2),129 – 145.

[10] Fowler B. P300 as a measure of workload during a simulated aircraft landing task [J]. Human Factors, 1994,36(4),670 – 683.

[11] Funk D, Li Z, Lê A D. Effects of environmental and pharmacological stressors on c-fos and corticotropin-releasing factor mRNA in rat brain: Relationship to the reinstatement of

alcohol seeking [J]. Neuroscience, 2006,138(1),235 - 243.

[12] Guillery R W. Anatomical pathways that link perception and action [J]. Progress in Brain Research, 2005,149,235 - 256.

[13] Hankins T C, Wilson G F. A comparison of heart rate, eye activity, EEG and subjective measures of pilot mental workload during flight [J]. Aviation, Space, and Environmental Medicine, 1998,69(4),360 - 367.

[14] Henelius A, Hirvonen K, Holm A, et al. Mental workload classification using heart rate metrics [R]. Conference Proceedings — IEEE Engineering in Medicine and Biology Society, 1,1836 - 1839.

[15] Hohnsbein J, Falkenstein M, Hoormann J. Effects of attention and time-pressure on P300 subcomponents and implications for mental workload research [J]. Biological Psychology, 1995,40(1 - 2),73 - 81.

[16] Horrey W J, Wickens C D, Consalus K P. Modeling drivers' visual attention allocation while interacting with in-vehicle technologies [J]. Journal of Experimental Psychology: Applied, 2006,12(2),67 - 78.

[17] Jemmott J B. 3rd, Borysenko J Z, Borysenko M, et al. Academic stress, power motivation, and decrease in secretion rate of salivary secretory immunoglobulin A [J]. Lancet. 1(8339): 1400 - 1402.

[18] Jiao K. Study on automobile driving fatigue and remission methods [D]. Shanghai Jiao Tong University, PhD thesis, 2004.

[19] Jiao K, Li Z, Chen M, et al. Effect of different vibration frequencies on heart rate variability and driving fatigue in healthy drivers [J]. International Archives of Occupational and Environmental Health, 2004,77(3),205 - 212.

[20] Jiao K, Li Z, Chen M, et al. Synthetic effect analysis of heart rate variability and blood pressure variability on driving mental fatigue [J]. Sheng Wu Yi Xue Za Zhi, 2005,22(2): 343 - 346.

[21] Jorna P G. Spectral analysis of heart rate and psychological state: a review of its validity as a workload index [J]. Biological Psychology, 1992,34(2 - 3),237 - 257.

[22] Jorna P G. Heart rate and workload variations in actual and simulated flight [J]. Ergonomics, 1993,36(9),1043 - 1054.

[23] Korte S M, Boer S F. A robust animal model of state anxiety: fear-potentiated behaviour in the elevated plus-maze [J]. European Journal of Pharmacology, 2003,463(1 - 3):163 - 175.

[24] Kramer A F, Trejo L J, Humphrey D. Assessment of mental workload with task-irrelevant auditory probes [J]. Biological Psychology, 1995,40(1 - 2),83 - 100.

[25] Kuijer P P, Vries W H, Beek A J, et al. Effect of job rotation on work demands, workload, and recovery of refuse truck drivers and collectors [J]. Human Factors. 2004,46(3):437 - 448.

[26] Leino T K, Leppäluoto J, Ruokonen A, et al. Neuroendocrine responses to psychological workload of instrument flying in student pilots [J]. Aviation, Space, and Environmental Medicine, 1999,70(6):565 - 570.

[27] Li Z, Jiao K, Chen M, et al. Effect of magnitopuncture on sympathetic and parasympathetic nerve activities in healthy drivers — assessment by power spectrum analysis of heart rate variability [J]. European Journal of Applied Physiology, 2003,88(4 - 5),404 - 410.

[28] Lim J, Wu W C, Wang J, et al. Imaging brain fatigue from sustained mental workload: An ASL perfusion study of the time-on-task effect [J]. Neuroimage, 2010,49(4),3426 – 3435.

[29] Majewska A K, Sur M. Plasticity and specificity of cortical processing networks [J]. Trends in Neurosciences, 2006,29(6),323 – 329.

[30] Miyake S, Yamada S, Shoji T, et al. Physiological responses to workload change. A test/retest examination [J]. Applied Ergonomics, 2009,40(6),987 – 996.

[31] Murata A. An attempt to evaluate mental workload using wavelet transform of EEG [J]. Human Factors, 2005,47(3),498 – 508.

[32] Näätänen R. Mismatch negativity (MMN): perspectives for application [J]. International Journal of Psychophysiology, 2000,37(1),3 – 10.

[33] Otsuka Y, Onozawa A, Kikukawa A, et al. Effects of flight workload on urinary catecholamine responses in experienced military pilots [J]. Perceptual and Motor Skills, 2007,105(2),563 – 571.

[34] Otsuki T, Sakaguchi H, Hatayama T, et al. Secretory IgA in saliva and academic stress [J]. International Journal of Immunopathology and Pharmacology, 2004,17(2),45 – 48.

[35] Papadelis C, Kourtidou-Papadeli C, Vlachogiannis E, et al. Effects of mental workload and caffeine on catecholamines and blood pressure compared to performance variations [J]. Brain and Cognition, 2003,51(1),143 – 154.

[36] Park M K, Satoh N, Kumashiro M. Mental workload under time pressure can trigger frequent hot flashes in menopausal women [J]. Industrial Health, 2008,46(3),261 – 268.

[37] Passerin A M, Cano G, Rabin B S, et al. Role of locus coeruleus in foot shock-evoked Fos expression in rat brain [J]. Neuroscience, 2000,101(4),1071 – 1082.

[38] Pope A T, Bogart E H, Bartolome D S. Biocybernetic system evaluates indices of operator engagement in automated task [J]. Biological Psychology, 1995,40(1 – 2),187 – 195.

[39] Prinzel L J, Freeman F G, Scerbo M W, et al. Effects of a psychophysiological system for adaptive automation on performance, workload, and the event-related potential P300 component [J]. Human Factors, 2003,45(4),601 – 613.

[40] Rabbi A F, Ivanca K, Putnam A V, et al. Human performance evaluation based on EEG signal analysis: A prospective review [C]. Conference Proceedings — IEEE Engineering in Medicine and Biology Society, 2009,1,1879 – 1882.

[41] Rantanen E M, Goldberg J H. The effect of mental workload on the visual field size and shape [J]. Ergonomics, 1999,42(6),816 – 834.

[42] Recarte M A, Nunes L M. Effects of verbal and spatial-imagery tasks on eye fixations while driving [J]. Journal of Experimental Psychology: Applied, 2000,6(1),31 – 43.

[43] Recarte M A, Nunes L M. Mental workload while driving: effects on visual search, discrimination, and decision making [J]. Journal of Experimental Psychology: Applied, 2003,9(2),119 – 137.

[44] Ring C, Drayson M, Walkey D G, et al. Secretory immunoglobulin A reactions to prolonged mental arithmetic stress: inter-session and intra-session reliability [J]. Biological Psychology, 2002,59(1),1 – 13.

[45] Ring C, Carroll D, Hoving J, et al. Effects of competition, exercise, and mental stress on secretory immunity [J]. Journal of Sports Sciences, 2005,23(5),501 – 508.

[46] Rivecourt M, Kuperus M N, Post W J, et al. Cardiovascular and eye activity measures as

indices for momentary changes in mental effort during simulated flight [J]. Ergonomics, 2008,51(9),1295 - 1319.

[47] Roscoe A H. Assessing pilot workload: why measure heart rate, HRV and respiration [J]? Biological Psychology, 1992,34(2 - 3),259 - 287.

[48] Roscoe A H. Heart rate as a psychophysiological measure for in-flight workload assessment [J]. Ergonomics, 1993,36(9),1055 - 1062.

[49] Simoens V L, Istók E, Hyttinen S, et al. Psychosocial stress attenuates general sound processing and duration change detection [J]. Psychophysiology, 2007,44(1),30 - 38.

[50] Sirevaag E J, Kramer A F, Wickens C D, et al. Assessment of pilot performance and mental workload in rotary wing aircraft [J]. Ergonomics, 1993,36(9),1121 - 1140.

[51] Someya N, Endo M Y, Fukuba Y, et al. Effects of a mental task on splanchnic blood flow in fasting and postprandial conditions [J]. European Journal of Applied Physiology, 2010,108 (6),1107 - 1113.

[52] Stam C J, Woerkom T C, Pritchard W S. Use of non-linear EEG measures to characterize EEG changes during mental activity [J]. Electroencephalography and Clinical Neurophysiology, 1996,99(3),214 - 224.

[53] Sudo A. Evaluation of workload in middle-aged steel workers by measuring urinary excretion of catecholamines and cortisol [J]. Sangyo Igaku-Japanese Journal of Occupational Health, 1991,33(6),475 - 484.

[54] Sur M, Rubenstein J L. Patterning and plasticity of the cerebral cortex [J]. Science, 2005, 310(5749),805 - 810.

[55] Ullsperger P, Freude G, Erdmann U. Auditory probe sensitivity to mental workload changes — an event-related potential study [J]. International Journal of Psychophysiology, 2001,40(3),201 - 209.

[56] Veltman J A, Gaillard A W. Physiological indices of workload in a simulated flight task [J]. Biological Psychology, 1996,42(3),323 - 342.

[57] Veltman J A, Gaillard A W. Physiological workload reactions to increasing levels of task difficulty [J]. Ergonomics, 1998,41(5),656 - 669.

[58] Wetherell M A, Hyland M E, Harris J E. Secretory immunoglobulin A reactivity to acute and cumulative acute multi-tasking stress: relationships between reactivity and perceived workload [J]. Biological Psychology, 2004,66(3),257 - 270.

[59] Yu X, Zhang J, Xie D, Wang J, et al. Relationship between scalp potential and autonomic nervous activity during a mental arithmetic task [J]. Autonomic Neuroscience, 2009,146 (1 - 2):81 - 86.

[60] Zhang J, Yu X, Xie D. Effects of mental tasks on the cardiorespiratory synchronization [J]. Respiratory Physiology & Neurobiology, 2010,170(1),91 - 95.

Civil Cockpit Human-machine Interface Evaluation

Xu Haiyu, Zhang An, Tang Zhili, Zhang Chao

School of Electronic Information, Northwestern Polytechnical University, Xi'an 710072

Abstract   According to Fuzzy Comprehensive Evaluation based on combination weighting, civil cockpit Human-Machine interface was evaluated. Evaluation index system described by uncertain language was set up, and combination weighting was obtained by linear combination of subjective weight, which was available from G1 method and information entropy method, then index system was comprehensive evaluated using fuzzy comprehensive evaluation based on combination weighting. The validity and efficiency of this method proposed in this paper is proved, and research results provide reference for comprehensive evaluation and designing on civil cockpit Human-Machine interface.

Keywords   civil cockpit, Human-Machine interface, combination weighting, fuzzy comprehensive evaluation

1   Introduction

With the application of large display control computer, automation and intelligent equipments, based on the design concept of psychology, aesthetics and ergonomics, besides the high efficiency, designers of civil cockpit are paying more and more attention to comfortability, security and economy, which calls for the corresponding development of human-machine interface evaluation model. Literature [1 - 5] introduced the study of flight interface and terrain avoidance system, the qualitative evaluation of plane human-machine interface based on pilot's recognition simulator, the qualitative and quantitative evaluation of plane human-machine interface, multi-characteristics decision, "cards categorization", human-machine interface design based on state recognition and human-machine interface study based on virtual simulation evaluation. Researches at home and abroad are mainly concentrated on qualitative evaluation and virtual simulation evaluation, while the study of quantitative evaluation is relatively less. Literature [6] introduced several useful civil cockpit human-machine interface qualitative evaluation methodologies, such as Checklist, Observation, Questionnaires, and so on. Virtual simulation evaluation is carried out in virtual environment, where human's operation is modeled by digital human body model provided by analysis software, with the experimental data of physiology, dynamics and ergonomics.

2　Civil cockpit human-machine interface evaluation index system

Based on the conclusion of human-machine interface researches at home and abroad, a civil cockpit human-machine interface comprehensive evaluation index system described with uncertain language is proposed below.

Civil cockpit human-machine interface evaluation index system (shown as Fig. 1) includes: level of display, economic effectiveness, comfortableness, task ability, environment, airworthiness, ability of control. ① Task ability describes the level of aircrafts' achieving expected target in flight. Much high and new technology has been used to improve aircrafts' task ability and insure the fight task's successful finish in any complicated situation or emergency. The evaluation of task ability considers factors such as efficiency, work load, security, and so on. ② The world civil market is developing very fast now. The COTS is used as the manufacture materials, which reduce the manufacture cost and enhance the economy efficiency to a large degree. ③ Owing to the improvement of electronic equipments and the application of large control computers, from hand operation to semi automation operation, till today's full automation operation, the conformability of flying an aircraft has been raised a lot. ④ Since Europe strengthened aircrafts' landing prerequisite, the environmental demand has been raised. Environment factors in human-machine interface mainly concerned are electromagnetic radiation, noise, energy consumption, and so on. ⑤ Airworthiness is the most basic demand in aircraft design. ⑥ The characteristic of cockpit display and control is the main object of human-machine interface evaluation. Former evaluation index systems of display and control ability were built based on the design of human-machine interface, the level of display is comprised of 7 factors, display accuracy, display timeliness, workload, display definition, reliability, controllability and security.

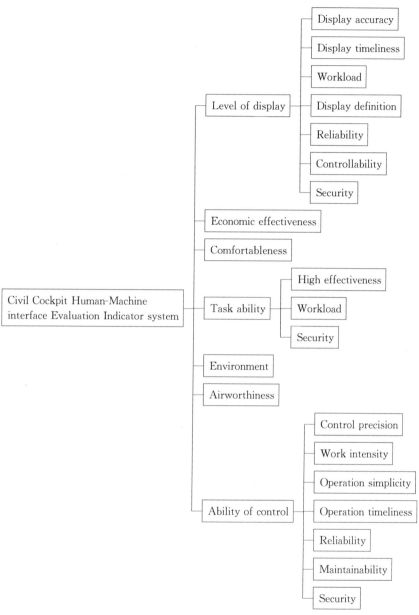

Fig. 1

3   The fuzzy comprehensive evaluation

3. 1   The determination of the weights

Faced with fuzzy or uncertain object, the fuzzy comprehensive evaluation shows some flexibility and practicability, so it is adopted as the evaluation means of civil cockpit human-machine interface in this paper. To fill in the gap of subjectivity of qualitative analysis, the idea of combination weighting is taken, and G1 method is used to determine the weight.

3.2   The determination of subjective weight with G1 method

G1 method's computing is relatively simple, the operation is convenient, and its result is accuracy, which proved to be a fine subjective weighting method.

The steps of G1 are illustrated below:

(1) The determination of the orders of indexes. According to certain principles, each evaluation index's relative importance to others is determined, $u_i^* > u_j^*$, then the order of all indexes based on their relative importance is arranged $u_1 > u_2 > \cdots > u_n$.

(2) Compare the relative importance of two adjacent indexes. The importance level of Index u_{k-1} and index u_k is compared by calculating $\omega_{k-1}'/\omega_k' = r_k (k = 2\cdots n)$, and the value of r_k can be checked out from the reference table below.

Tab. 1

r_k	Explanation
1.0	the same importance
1.1	between little more importance and same importance
1.2	little more importance
1.3	between more importance and little more importance
1.4	more importance
1.5	between much more importance and more importance
1.6	much more importance
1.7	between extremely more importance and much more
1.8	extremely more importance

(3) calculate the weight ω_i' of index u_i.

Substitute the value gotten from Tab. 1 for r_i in eq. (1) and (2) ω_i' is gotten.

$$\omega_n' = \left(1 + \sum_{k=2}^{n} \prod_{i=k}^{n} r_i\right)^{-1} \tag{1}$$

$$\omega_{k-1}' = r_k \omega_k' \ (k = n\cdots 2) \tag{2}$$

3.3   The determination of the objective weight by comentropy method

Methods commonly used to determine the objective weight include: comentropy method, deviation maximization and relation judge matrix method. Literature [8, 9] illustrated that with the use of comentropy method, the determination of objective weight is convenient and the result is objective, so it is adopted in this paper. Comentropy method is based on the definition of entropy in informationism, in which the information included in each index is concerned. With this method, the larger one index's information content is, the larger its influence to the final evaluation or decision is, and the bigger its weight is. Fuzzy evaluation matrix $\boldsymbol{D} = (d_{ij})_{m \times n}$ is calculated with formulas below:

$$r_{ij} = d_{ij} \bigg/ \sum_{i=1}^{n} d_{ij} \tag{3}$$

$$E_j = -k \sum_{i=1}^{n} r_{ij} \ln r_{ij} \, (k = 1/\ln n) \tag{4}$$

$$\omega_j'' = (1 - E_j) \bigg/ \sum_{j=1}^{m} (1 - E_j) \, (j = 1 \cdots m) \tag{5}$$

Index weight ω_j'' is gotten through the calculating of eq. (3), (4), (5).

3.4 Combination weighting

Subjective weighting has great subjective haphazardry in the ordering of scheme, and is affected by experts' lack of knowledge or experience, besides these, in some situation the inner relation of indexes is hard to be recognized by human's subjective consciousness. Objective weighting can overcome the shortcoming above. The evaluation result of objective weighting may be inconsistent with the actual situation, which can be made up by subjective weighting. Combination weighting combines the advantages of subjective and objective weighting and give the result which is objective and propitious for comprehensive evaluation. The calculate method is illustrated below:

The subjective weight and objective weight gotten from eq. (1) and (2) are $\boldsymbol{\Omega}' = (\omega_1', \cdots, \omega_n')$, $\boldsymbol{\Omega}'' = (\omega_1'', \cdots, \omega_n'')$. Set $\alpha + \beta = 1$, $\omega_i = \alpha \omega_i' + \beta \omega_i''$ (α is the influence coefficient of subjective weight and β is that of objective weight, the value of α and β should be determined regarding the practical situation.), and get the index weight $\boldsymbol{\Omega} = (\omega_1, \cdots, \omega_n)$. The value of α and β shows the relative importance between subjective weight and objective weight, whose determination doesn't belong to the topic of this paper. With the use of combination weighting, we can get relatively objective index weight and finally get more accuracy result.

3.5 Fuzzy comprehensive evaluation

Fuzzy theory is proposed by L. A. Zaden in 1965. It is based on fuzzy set and studies uncertain objects, its main idea is to qualify fuzzy factors to information that can be processed by computer. By this theory, build fuzzy map from evaluation index set to comment set, build fuzzy evaluation matrix and finally get the evaluation result.

Set factors $U = \{u_1 \cdots u_m\}$ and comments set $S = \{s_1 \cdots s_n\}$. U represents the index set of m groups of comprehensive evaluation, S represents comments set. Ask several experts give marks for the evaluation object, and get the evaluation matrix of all indexes, $\boldsymbol{D} = (d_{ij})_{m \times n}$. Then, calculate $r_{ij} = d_{ij} \Big/ \sum\limits_{j=1}^{n} d_{ij}$, get the membership degree matrix $\boldsymbol{R} = (r_{ij})_{m \times n}$. The final comprehensive evaluation result is gotten by calculating $E = \boldsymbol{\Omega} \cdot \boldsymbol{R} = (e_1 \cdots e_n)$, in which $e_j = \sum\limits_{i=1}^{m} \omega_i \cdot r_{ij}$, $\boldsymbol{\Omega}$ is the index weight matrix gotten above. $e_k = \max\{e_j\}$ $(k = 1, \cdots, n)$, according to the principle of maximum membership degree, check out s_k in the comments set corresponding to e_k and s_k determine the level of the evaluation result.

4　Case analysis

Select 10 experts to give marks to the evaluation index system shown in Fig. 1 and get the indexes evaluation matrixes $\boldsymbol{D}$, $\boldsymbol{D}_1$, $\boldsymbol{D}_2$, $\boldsymbol{D}_3$. Then, set 5 levels of human-machine interface evaluation: high, relative high, medium, relative low, low, and set the comments system of index system comprehensive evaluation: $S = \{s_1, s_2, s_3, s_4, s_5\}$.

$$\boldsymbol{D} = \begin{bmatrix} 7 & 3 & 0 & 0 & 0 \\ 1 & 3 & 4 & 2 & 0 \\ 0 & 4 & 5 & 1 & 0 \\ 5 & 4 & 1 & 0 & 0 \\ 0 & 0 & 3 & 3 & 4 \\ 0 & 1 & 2 & 3 & 4 \\ 6 & 4 & 0 & 0 & 0 \end{bmatrix} \qquad \boldsymbol{D}_1 = \begin{bmatrix} 4 & 5 & 1 & 0 & 0 \\ 5 & 4 & 1 & 0 & 0 \\ 6 & 4 & 0 & 0 & 0 \\ 0 & 0 & 4 & 3 & 3 \\ 0 & 5 & 4 & 1 & 0 \\ 1 & 3 & 3 & 3 & 0 \\ 0 & 1 & 2 & 5 & 2 \end{bmatrix}$$

$$\boldsymbol{D}_2 = \begin{bmatrix} 6 & 4 & 0 & 0 & 0 \\ 7 & 3 & 0 & 0 & 0 \\ 0 & 4 & 5 & 1 & 0 \end{bmatrix} \qquad \boldsymbol{D}_3 = \begin{bmatrix} 5 & 4 & 1 & 0 & 0 \\ 0 & 2 & 4 & 4 & 0 \\ 4 & 3 & 3 & 0 & 0 \\ 4 & 5 & 1 & 0 & 0 \\ 0 & 0 & 3 & 4 & 3 \\ 0 & 0 & 5 & 4 & 1 \\ 0 & 2 & 4 & 4 & 0 \end{bmatrix}$$

With the calculation above, we get the subjective and objective weight of 7 indexes, then we set the combination weight: $\alpha = 0.5$, $\beta = 0.5$, meaning that the subjective weight is as important as the objective. Fig. 2 shows the difference between the subjective weight of G1, the objective weight of comentropy, and the weight of combination weight.

We can get some conclusions from Fig. 2: ① The subjective weight and objective weight of task ability, display level, control ability and airworthiness reach unanimity. The weight of control ability, display level and control ability are bigger, while that of worthiness is smaller. ② The difference of subjective weight and objective weight of economic effectiveness, environment and security is significant. Subjective weighting

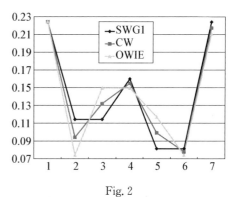

Fig. 2

arranges the importance orders of the three indexes based on human's experience, that is, economic effectiveness, environment and security, from the lowest to the highest. Objective weighting recognizes the connotative information from the given data, and give the importance orders, which is, comfortableness, economic effectiveness and security, from low to high. ③ The results of subjective weighting and objective weighting show obvious difference, so the combination weighing which combined them by α and β can improve the comprehensive evaluation result.

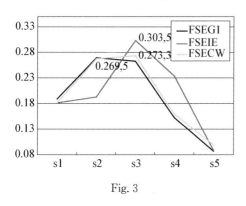

Fig. 3

Calculate the results of FSEG1 (the fuzzy synthetic evaluation based on G1 method), FSEIE (the fuzzy synthetic evaluation based on comentropy) and FSECW (the fuzzy synthetic evaluation based on combination weighting), which is shown is Fig. 3.

As shown in the figure, ① the evaluation result of this type of civil cockpit of FSEG1 is "relative high", while that of FSEIE and FSECW is medium. The largest and the smallest value of the result of FSEG1 is 0. 2695 and 0. 2626, with little difference; FSEIE and FSECW show significant difference on the value of maximum value and second maximum value. It can be concluded that, the subjectivity of subjective evaluation and fuzzy comprehensive evaluation lead to the difference. FSECW combines the advantages of subjective evaluation and objective evaluation, turns to be a nice evaluation method.

5   Conclusion

This paper carries out the evaluation of civil cockpit human-machine interface with fuzzy comprehensive evaluation based on combination weighting. By case analysis, this method proved to be validity, objective and effective. ① The evaluation index build in this paper is based on uncertain language, which is with certain engineering significance, but for the complexity of civil cockpit human-machine interface, the index system needs to studied and improved further and evaluation model that is more effective, convenient, and objective is in demand. ② The fuzzy comprehensive evaluation based on combination weighting can be applied in civil cockpit evaluation, as well as other fields' human-machine interface evaluation. ③ The original data used in this paper is gotten from experts' marks, which is with strong subjectivity, so the objectivity of evaluation data is the direction of future study. ④ The change of the influence coefficient in combination weighting can improve the evaluation result. The optimization of math formulas and the proof based on objective data can bring vital and effective modified methods.

Acknowledgement

The authors of this paper would like to thank for the funding support of the national 973 project.

References

[1]   Coppenbarger R A, Cheng V H L. Simulation Evaluation of a Pilot Interface with an Automated Rotorcraft Obstacle Avoidance System. Control Applications [C]. Second IEEE Conference on, 1993, 649 - 657.

[2]   Daisuke, Karikawal, Makoto, et al. Human-Ma-chine System Simulation for Supporting the Design and Evaluation of Reliable Aircraft Cockpit Interface. SICE - ICASE [C], International Joint Conference, 2006, 55 - 60.

[3]   Senol M B, Dagdeviren M. Evaluation of Cockpit Design by Using Quantitative and Qualitative Tools. Industrial Engineering and Engineering Management [C]. IEEM 2009, IEEE International Conference on 2009, 847 - 851.

[4]   Jin H B, Wang L. Simulation Applying Situation Awareness to Human-Machine Interface Design of Aviation. Wuhan: Knowledge Acquisition and Modeling [C], KAM '09. Second International Symposium on 2009, 387 - 390.

[5]   Wang L J, Xiang W. The Virtual Evaluation of the Ergonomics Layout in Aircraft Cockpit. Computer-Aided Industrial Design & Conceptual Design [C]. CAID & CD 2009, IEEE 10th

International Conference on 2009,1438 – 1442.

[6] Neville S. Human Factors in Consumer Products. Taylor & Francis. Inc, 1997,12.

[7] Xie J X, Qin Y. The fuzzy comprehensive evaluation of the highway emergency plan based on G1 method. Computer Science and Information Technology (ICCSIT) [C]. 2010 3rd IEEE International Conference on 2010,313 – 316.

[8] Wei G F, Zhang X. Research on Risk Assessment and Optimization Control of Information Systems Development: Based on Improved Fuzzy Comprehensive Evaluation and Dynamic Programming Algorithm [C]. Wireless Communications Networking and Mobile Computing (WiCOM), 2010 6th International Conference on 2010,1 – 5.

[9] Gao H J, Song Y F. Fuzzy Comprehensive Evaluation Model on University Teaching Quality [C]. Business Intelligence and Financial Engineering, 2009. BIFE 09 International Conference on 2009,190 – 193.

Man-machine Function Allocation Based on Uncertain Linguistic Multiple Attribute Decision Making

Zhang An, Tang Zhili, Zhang Chao*

School of Electronics and Information, Northwestern Polytechnical University, Xi'an 710129

Abstract   Function allocation is one of the necessary stages in the design course of human-machine systems since appropriate function allocation makes the whole system more effective, reliable and inexpensive. Therefore, our research mainly focuses on the problems of function allocation between human and machine in human-machine systems, analyses each capability advantage of man and machine according to their respective inherent characteristics and makes a comparison between them. In view of highly uncertain characteristics of decision attribute value in the practical process, we introduce the ULMADM (uncertain linguistic multiple attribute decision making) method in the function allocation process. Meanwhile, we also use the UEWAA (uncertain extended weighted arithmetic averaging) method to determine the automation level range of the operator functions, then, we eventually establish the automation level of man-machine function allocation by using the multi-attribute decision making algorithm, which is combined by UEWAA and ULHA (uncertain linguistic hybrid aggregation) operators. Finally, an example about function allocation is given, that is, fault diagnosis in the cockpit of civil aircrafts. The final result of the example demonstrates that the proposed method about function allocation is feasible and effective.

Keywords   human-machine systems, function allocation, uncertain linguistic multiple attribute decision making (ULMADM), levels of automation, fault detection, cockpits, reliability

1   Introduction

Currently, there is no a unified theory about function allocation, even though many scholars have proposed some theories and research methods since Fitts [1] firstly put forward the concept of function distribution. For example, Parasuraman [2], etc, put forward an automation classification and level design method; Endsely [3 – 5], etc, Sheridan and Verplank [6] have come up with a famous classification idea. Although these methodologies can solve some man-machine function allocation problems in their related fields, however, the ways used to establish the appropriate automation level of a function are primarily based on the qualitative and lack effective quantitative analysis. Therefore, making part of functions allocated reasonably between man and machine — with the appropriate automation level — can not only complete people's advantages, like the intuition, experience, initiative, etc, but also realize the advantages of an automatic system, such as speed, accuracy more than the two working alone when both complementary advantages are combined. Our research mainly focuses on the function allocation in man-machine systems, according to the comparisons of each capability advantage between man and machine, the automation level range of the function can be established due to the method uncertain extended weighted arithmetic averaging (UEWAA) [7] of the uncertain linguistic multiple attribute decision making (ULMADM). After that, we can use the multi-attribute decision making algorithm, which is based on the combination of the UEWAA and ULHA [7] operators, to determine the final automation level of the function allocation in man-machine system.

2   Comparisons between the capability advantages of man and machine

At first, the study on the capability advantages of man and machine is needed before the functions in man-machine systems are allocated. One of the most famous researches on the comparison of human capabilities and machine capabilities is Fitts List [1], which was presented by P. M. Fitts in 1951, one of the founders in the man-machine system fields. Fitts List has pointed out the inherent advantages and deficiencies of man and machine. According to the comparisons between them, it seems easier to find which functions can be suitable for man and which functions should be allocated to machine. Normally, the advantages of one side are just the shortcomings of the other. So if the two worked together, the whole performance of the man-machine systems would be improved greatly, which is far better than each of them working alone. In other words, the system can harvest both advantages of the two. Therefore, various methods of function allocation are all just based on Fitts List. That's why the method was widely applied in the early years of the simple industrial automation surveillance and control systems.

Along with the computer and the rapid development of automatics and the deepening of understanding ourselves, Fitts List gradually shows some limitations in analyzing the advantages of man and machine and some updates on the basis of it are also appeared. So far, although most of the man-machine abilities in the Fitts List are still used, however, they cannot fully summarize the characteristics of both. For example, people have advantages in complex information transmission, symbol reasoning, pattern recognition, conceptualization, intuition, learning of experience, memory and other application fields. In addition, people can also perceive complex situations in overall. In fact, it is the advantages that pay a significant role in man-machine systems.

We will give more detailed capability advantages of man and machine during the process of function allocation in man-machine systems in Tab. 1 [8] and Tab. 2 [8].

Tab. 1　Advantages of machine's abilities

The advantages of machine's abilities
The ability of data and information visualization
Strong abilities in managing and storing data
Memory of large quantities of data, knowledge and rules
The logical reasoning ability based on the rules
Simple and repeat decision-making ability
The ability of complex mathematical operations for a long time
The ability of dealing with several variable combinatorial problems
The ability to work for a long time
Parallel processing capacity of knowledge
Fast computing speed. The calculation speed is essential for time-sensitive problems
Highly accurate calculation
Higher predictability because of the procedure set before running
The ability in treating benign structural problems
In the long run, own high cost-effectiveness ratio
The ability of working in harsh environment

Tab. 2　Advantages of human's abilities

The advantages of human's abilities
Own the ability of flexibility and adaptability and can change to adapt the environment
Has creativity in solving problems

(continued)

The advantages of human's abilities
Has a good visual perception ability
Has the emotion and can consider the abstract concept
Own strong ability of accumulating experience and learning
The ability of predicting and processing accidents
Communicate complex information in a variety of ways
Own the ability of conceptualization and have advantages in dealing with fuzzy and inaccurate information
Good symbol or spatial reasoning ability
Greatly shorten the time of making decision because of the experience and intuition of instinct
Pattern recognition. Be good at describing and classifying observed things
Predict the development of situation according to common sense knowledge
Can well control error and uncertain events
Effectively narrow the search space of the optimal decisions
In the process of solving decision-making results, when a satisfactory solution can occur, calculation can be timely ended and thus computing resources will be saved
Use more wide strategic sets to solve problems
The ability in treating adverse structural problems
The ability of assessing and perceiving external environment as a whole

3   The automation level of the function allocation in man-machine systems

In the process of function allocation in man-machine systems, through the allocation function between man and machine, it can be argued that man and machine are a cooperative relationship. A function may not be completed by man or machine alone, instead, it is needed to finish it by cooperation. The relationship represents in the different levels of automation (LOAs) of man-machine functions.

Currently, there are a variety of divisions about LOAs of man-machine interaction systems. In the references [2 - 6, 9, 10], Sheridan, Verplank and Parasuraman etc have put forward a method about automation level of man-machine interaction systems, as shown in Tab. 3.

Tab. 3   Automation level

LOAs	Description
1	System does not provide any assistance, one must complete all the decisions and manipulation
2	System provides decision-making or action plan
3	System narrows scheme selection

（continued）

LOAs	Description
4	System provides a proposal
5	Execute the plan if people agree
6	The scheme is allowed to be vetoed in the limited time before the implementation
7	Implement Automatically unless it is necessary to notice human being
8	Told them if people want to know
9	Whether or not to inform people is decided by computer
10	System decides all the work and refuses to man's intervention

In Tab. 3, automation level is divided into 10 levels, ranging from fully manual to completely automation and the level of automation is gradually improved. The automation level partitioning raised by Sheridan etc has become a widely quoted principle. Nowadays, many other schemes are just based on some certain merged or refined level.

4　The method about man-machine function allocation based on the ULMADM

4.1　The definition of ULMADM

Man-machine function allocation is a typical multiple attribute decision making (MADM) problem. Under the incomplete information, in order to solve the highly uncertain decision problems, we introduce a kind of the ULMADM method and use UEWAA and ULHA operators to solve the function allocation problems in man-machine systems. Firstly, we will introduce some basic concepts related [11].

Assume $M = \{1, 2, \cdots, m\}$, $N = \{1, 2, \cdots, n\}$, $L = \{1, 2, \cdots, l\}$.

Definition 1 [12]: Set WAA: $R^n \to R$, if

$$\mathrm{WAA}_\omega(\alpha_1, \alpha_2, \cdots, \alpha_n) = \sum_{j=1}^{n} \omega_j \alpha_j \tag{1}$$

$\omega = (\omega_1, \omega_2, \cdots, \omega_n)$ is a weighted vector of a group data $(\alpha_1, \alpha_2, \cdots, \alpha_n)$, $\omega_j \in [0, 1]$ $(j = 1, 2, \cdots, n)$, $\omega_1 + \omega_2 + \cdots + \omega_n = 1$, so the WAA is called a weighted arithmetic averaging operator, for short, WAA operator.

Consider that decision makers generally need an appropriate linguistic assessment scale when they are in qualitative measure. For this reason, beforehand, we can set a linguistic assessment scale: $S = (s_\alpha \mid \alpha = -l', \cdots, l')$, l' is a natural number.

Notice that the number of terms in S is usually odd. Meanwhile, the scale must meet the following conditions: ① if $\alpha > \beta$, so $s_\alpha > s_\beta$; ② Negative operator is existed, $\mathrm{neg}(s_\alpha) = s_{-\alpha}$.

In order to facilitate calculation and avoid missing information, an expanded scale $\bar{S} = \{s_\alpha \mid \alpha \in [-q, q]\}$ is defined on the basis of original scale $S = \{s_\alpha \mid \alpha = -l', \cdots, l'\}$ and $q(q > l')$ is a sufficiently large natural. If $\alpha \in \{-l', \cdots, l'\}$, s_α is a nature term. If $\alpha \notin \{-l', \cdots, l'\}$, s_α is an expanded term. The expanded scale still meets conditions ① and ②.

The algorithms of linguistic assessment scale will be defined as follows [13]:

Definition 2: Set s_α, $s_\beta \in \bar{S}$, y, y_1, $y_2 \in [0, 1]$, therefore,

(1) $s_\alpha \oplus s_\beta = s_{\alpha+\beta}$;

(2) $s_\alpha \oplus s_\beta = s_\beta \oplus s_\alpha$;

(3) $ys_\alpha = s_{y\alpha}$;

(4) $y(s_\alpha \oplus s_\beta) = ys_\alpha \oplus ys_\beta$;

(5) $(y_1 + y_2)s_\alpha = y_1 s_\alpha \oplus y_2 s_\alpha$.

Definition 3 [13]: Set EWAA: $\bar{S}^n \rightarrow \bar{S}$, if

$$\text{EWAA}_\omega(s_{\alpha_1}, s_{\alpha_2}, \cdots, s_{\alpha_n}) = \omega_1 s_{\alpha_1} \oplus \omega_2 s_{\alpha_2} \oplus \cdots \oplus \omega_n s_{\alpha_n} = s_{\bar{\alpha}} \qquad (2)$$

Notice that $\bar{\alpha} = \omega_1 \alpha_1 + \omega_2 \alpha_2 + \cdots + \omega_n \alpha_n$, $\omega = (\omega_1, \omega_2, \cdots, \omega_n)$ is a weighted vector of the language data s_{α_j} $(j \in N)$, and $s_{\alpha_j} \in \bar{S}$, $\omega_j \in [0, 1]$ $(j \in N)$, $\omega_1 + \omega_2 + \cdots + \omega_n = 1$, therefore, the function EWAA is called an extended weighted arithmetic averaging (EWAA) operator.

Definition 4 [14]: Set $\tilde{\mu} = [s_a, s_b]$, s_a, $s_b \in \bar{S}$, s_a and s_b are the lower and upper limit of $\tilde{\mu}$ respectively, so $\tilde{\mu}$ is called an uncertain linguistic variable.

Assume that $\tilde{S}$ is a set of all uncertain linguistic variables. Consider any two linguistic variables $\tilde{\mu}$ and $\tilde{\upsilon}$, $\tilde{\mu} = [s_a, s_b]$, $\tilde{\upsilon} = [s_c, s_d] \in \tilde{S}$, β, β_1, $\beta_2 \in [0, 1]$. The algorithms of them are defined as follows:

(1) $\tilde{\mu} \oplus \tilde{\upsilon} = [s_a, s_b] \oplus [s_c, s_d] = [s_a \oplus s_c, s_b \oplus s_d] = [s_{a+c}, s_{b+d}]$;

(2) $\beta\tilde{\mu} = \beta[s_a, s_b] = [\beta s_a, \beta s_b] = [s_{\beta a}, s_{\beta b}]$;

(3) $\tilde{\mu} \oplus \tilde{\upsilon} = \tilde{\upsilon} \oplus \tilde{\mu}$;

(4) $\beta(\tilde{\mu} \oplus \tilde{\upsilon}) = \beta\tilde{\mu} \oplus \beta\tilde{\upsilon}$;

(5) $(\beta_1 + \beta_2)\tilde{\mu} = \beta_1 \tilde{\mu} \oplus \beta_2 \tilde{\mu}$.

Definition 5 [14]: Set $\tilde{\mu} = [s_a, s_b]$, $\tilde{\upsilon} = [s_c, s_d] \in \tilde{S}$, meanwhile, set $l_{ab} = b - a$, $l_{cd} = d - c$, so the possible degree when $\tilde{\mu} \geqslant \tilde{\upsilon}$ is defined as follows:

$$p(\tilde{\mu} \geqslant \tilde{\upsilon}) = \max\left\{1 - \max\left(\frac{d-a}{l_{ab} + l_{cd}}, 0\right), 0\right\} \qquad (3)$$

Similarly, the possible degree when $\tilde{\upsilon} \geqslant \tilde{\mu}$ is defined as follows:

$$p(\tilde{\upsilon} \geqslant \tilde{\mu}) = \max\left\{1 - \max\left(\frac{b-c}{l_{ab} + l_{cd}}, 0\right), 0\right\} \qquad (4)$$

Definition 6 [14]：Set UEWAA：$\tilde{S}^n \to \tilde{S}$, if

$$\text{UEWAA}_\omega(\tilde{\mu}_1, \tilde{\mu}_2, \cdots, \tilde{\mu}_n) = \omega_1\tilde{\mu}_1 \oplus \omega_2\tilde{\mu}_2 \oplus \cdots \oplus \omega_n\tilde{\mu}_n \tag{5}$$

Notice that $\boldsymbol{\omega} = \omega_1, \omega_2, \cdots, \omega_n$ is a weighted vector of the uncertain linguistic variable $\tilde{\mu}_i (i \in N)$, and $\omega_j \in [0, 1]$ $(j \in N)$, $\omega_1 + \omega_2 + \cdots + \omega_n = 1$, so the function UEWAA is called an uncertain EWAA operator (UEWAA).

Definition 7 [14]：Assume ULHA：$\tilde{S}^n \to \tilde{S}$, if

$$\text{ULHA}_{\omega, w}(\tilde{\mu}_1, \tilde{\mu}_2, \cdots, \tilde{\mu}_n) = w_1\tilde{v}_1 \oplus w_2\tilde{v}_2 \oplus \cdots \oplus w_n\tilde{v}_n \tag{6}$$

Notice that $\boldsymbol{w} = (w_1, w_2, \cdots, w_n)$ is a weighted vector related with ULHA, $w_j \in [0, 1]$ $(j \in N)$, $w_1 + w_2 + \cdots + w_n = 1$. $\tilde{v}_j$ is the $j-th$ larger element of the weighted uncertain linguistic variable group $(\tilde{\mu}'_1, \tilde{\mu}'_2, \cdots, \tilde{\mu}'_n)$ $(\tilde{\mu}'_i = n\omega_i\tilde{\mu}_i, i \in N)$, here, $\boldsymbol{\omega} = (\omega_1, \omega_2, \cdots, \omega_n)$ is a weighted vector of an uncertain linguistic variable group, $(\tilde{\mu}_1, \tilde{\mu}_2, \cdots, \tilde{\mu}_n)$ $(i \in N)$, $\omega_j \in [0, 1]$ $(j \in N)$, $\omega_1 + \omega_2 + \cdots + \omega_n = 1$, and n is a balance factor, so the function ULHA is called an uncertain linguistic hybrid aggregation (ULHA) operator.

4.2 Establish the automation level range of function allocation

The automation level range can be determined according to a comprehensive comparison about each of capability advantages of man and machine. The method about how to determine the automation level range based on UEWAA operator will be given in this section. The concrete steps are as follows：

Step 1：Assume that X is a function set waiting for being allocated, H and G are the respective sets of the capability advantages of man and machine, the weighted vectors of them are $\boldsymbol{\omega} = (\omega_1, \omega_2, \cdots, \omega_m)$ and $\boldsymbol{\xi} = (\xi_1, \xi_2, \cdots, \xi_l)$ respectively and $\omega_j \in [0, 1]$ $(j \in M)$, $\omega_1 + \omega_2 + \cdots + \omega_m = 1$, $\xi_j \in [0, 1]$ $(j \in L)$, $\xi_1 + \xi_2 + \cdots + \xi_l = 1$.

Decision makers point out the evaluation values of the uncertain linguistic influence degree $\tilde{r}_{ij}$ and $\tilde{q}_{ij}$ respectively, which the capability advantages of man and machine, $h_j \in H$ and $g_j \in G$, treat undistributed function $x_i \in X$, and acquire the evaluation matrixes $\tilde{R} = (\tilde{r}_{ij})_{n \times m}$, $\tilde{Q} = (\tilde{q}_{ij})_{n \times l}$, $\tilde{r}_{ij}$, $\tilde{q}_{ij} \in \tilde{S}$.

Step 2：The linguistic assessment information in the $i-th$ line of evaluation matrix $\tilde{R}$ and $\tilde{Q}$ is respectively aggregated by UEWAA operator, and obtain the comprehensive evaluation results $\tilde{y}_i(\omega)$ and $\tilde{z}_i(\xi)$ $(i \in N)$ which the capabilities of man and machine treat undistributed function $x_i \in X$.

$$\tilde{y}_i(\omega) = \text{UEWAA}_\omega(\tilde{r}_{i1}, \tilde{r}_{i2}, \cdots, \tilde{r}_{im})$$
$$= \omega_1\tilde{r}_{i1} \oplus \omega_2\tilde{r}_{i2} \oplus \cdots \oplus \omega_m\tilde{r}_{im}, i \in N \tag{7}$$

$$\tilde{z}_i(\xi) = \text{UEWAA}_\xi(\tilde{q}_{i1}, \tilde{q}_{i2}, \cdots, \tilde{q}_{il})$$
$$= \xi_1\tilde{q}_{i1} \oplus \xi_2\tilde{q}_{i2} \oplus \cdots \oplus \xi_l\tilde{q}_{il}, \ i \in N \tag{8}$$

Step 3: According to the equation (3), calculate the possible degree $p_i = p(\tilde{y}_i(\omega) \geqslant \tilde{z}_i(\xi))$ ($i \in N$) between comprehensive evaluation results $\tilde{y}_i(\omega)$ and $\tilde{z}_i(\xi)$ ($i \in N$), and then obtain a possible degree vector $P = \{p_1, p_2, \cdots, p_n\}$, $0 \leqslant p_i \leqslant 1$. Notice that $\tilde{y}_i(\omega)$ and $\tilde{z}_i(\xi)$ ($i \in N$) are comprehensive evaluation results when the capabilities of man and machine treat undistributed function $x_i \in X$.

Step 4: According to the possibility p_i, automation level range A of function $x_i \in X$ can be established. The concrete rules are described as follows:

$$\begin{cases} floor((1-p_i) \times 10) - 1 \leqslant A \leqslant floor((1-p_i) \times 10) + 1 \\ A \in \{1, 2, \cdots, 10\} \end{cases} \tag{9}$$

Notice that $floor(x)$ is a Gauss integral function.

4.3 Establish the automation level of function allocation

Automation level range of function allocation has been established, namely, several different solutions to function allocation are given. However, selecting the optimal scheme from the solutions is still needed according to function allocation assessment criteria. Eventually, the final automation level of man-machine function allocation can be determined. In the practical evaluation process, in order to reduce experts' subjective deviation, different schemes are usually graded by several evaluation experts in accordance with the assessment criteria.

The multi-attribute decision-making method based on UEWAA and ULHA operators is used to determine the automation level of function allocation. The concrete steps are described as follows [7]:

Step 1: As for a multiple attribute decision making problem, assume that X, U and D respectively are scheme set, attribute (assessment criteria) set and decision maker (expert) set. The weighted attribute vector is $\omega = (\omega_1, \omega_2, \cdots, \omega_m)$, $\omega_j \in [0, 1]$ ($j \in M$), $\omega_1 + \omega_2 + \cdots + \omega_m = 1$. The weighted vector for decision makers is $\lambda = (\lambda_1, \lambda_2, \cdots, \lambda_t)$, $\lambda_k \geqslant 0$ ($k = 1, 2, \cdots, t$), $\lambda_1 + \lambda_2 + \cdots + \lambda_t = 1$. Decision maker $d_k \in D$ points out the value $\tilde{r}_{ij}^{(k)}$ of the uncertain linguistic assessment under the attribute $u_j \in U$ of the scheme $x_i \in X$ and then obtain the evaluation matrix $\tilde{R}_k = (\tilde{r}_{ij}^{(k)})_{n \times m}$, $\tilde{r}_{ij}^{(k)} \in \tilde{S}$

Step 2: The uncertain evaluation information in the $i - th$ line of evaluation matrix $\tilde{R}_k$ is aggregated by using UEWAA operator, and then we will get the comprehensive property appraisal value $\tilde{z}_i^{(k)}(\omega)$ ($i \in N$, $k = 1, 2, \cdots, t$) of the allocation scheme x_i, given by decision makers d_k. Notice that:

$$\tilde{z}_i^{(k)}(\omega) = \mathrm{UEWAA}_\omega(\tilde{r}_{i1}^{(k)}, \tilde{r}_{i2}^{(k)}, \cdots, \tilde{r}_{im}^{(k)}) = \omega_1 \tilde{r}_{i1}^{(k)} \oplus \omega_2 \tilde{r}_{i2}^{(k)} \oplus \cdots \oplus \omega_m \tilde{r}_{im}^{(k)}$$

$$(10)$$

Step 3: The comprehensive property appraisal values $\tilde{z}_i^{(k)}(\omega)$ $(k=1, 2, \cdots, t)$ given by t decision makers are aggregated by using ULHA operator, and then we will acquire the group comprehensive property appraisal value $\tilde{z}_i(\lambda, w')$ $(i \in N)$ of the allocation scheme x_i,

$$\tilde{z}_i(\lambda, w') = \mathrm{ULHA}_{\lambda, w'}(\tilde{r}_i^{(1)}, \tilde{r}_i^{(2)}, \cdots, \tilde{r}_i^{(t)})$$
$$= w_1' \tilde{v}_i^{(1)} \oplus w_2' \tilde{v}_i^{(2)} \oplus \cdots \oplus w_t' \tilde{v}_i^{(t)} (i \in N) \tag{11}$$

Notice that $w' = (w_1', w_2', \cdots, w_t')$ is a weighted vector of ULHA operator, $w_k' \in [0, 1]$ $(k=1, 2, \cdots, t)$, $w_1'+w_2'+\cdots+w_t'=1$, $\tilde{v}_i^{(k)}$ is the k-th larger element in a group of weighted uncertain linguistic variables $(t\lambda_1\tilde{r}_i^{(1)}(\omega), t\lambda_2\tilde{r}_i^{(2)}(\omega), \cdots, t\lambda_t\tilde{r}_i^{(t)}(\omega))$, t is a balance factor.

Step 4: According to the equation (3), calculate the possible degree $p_{ij} = p(\tilde{z}_i(\lambda, w') \geqslant \tilde{z}_j(\lambda, w'))$ $(i, j \in N)$ of the comprehensive attribute values $\tilde{z}_i(\lambda, w')$ $(i \in N)$ between each plan, and possible degree matrix $\boldsymbol{P} = (p_{ij})_{n \times n}$.

Step 5: Calculate the priority vector $\boldsymbol{v} = (v_1, v_2, \cdots, v_n)$ of the possible degree matrix $\boldsymbol{P}$, and rank the schemes according to the component size of v, namely, to harvest a optimum solution. Notice that (see [15])

$$v_i = \frac{1}{n(n-1)}\left(\sum_{j=1}^n p_{ij} + \frac{n}{2} - 1\right), i \in N \tag{12}$$

5 Example and analysis

In this section, we take the fault diagnosis (FD), one of the man-machine function allocations in the cockpit, for example, and our research adopts the method mentioned above to determine the optimal automation level.

5.1 Automation level range of FD

In order to determine the scope of the automation level of the FD function, at first, we need to make a comparison about capability advantages between machine and man. Here, we select the projects from Tab. 1 and 2, which are comparatively related with the FD function, and then the projects compose the sets of capability advantages of man and machine, therefore, the sets can be defined as $\boldsymbol{H} = \{h_1, h_2, h_3, h_4, h_5, h_6, h_7, h_8\}$ and $\boldsymbol{G} = \{g_1, g_2, g_3, g_4, g_5, g_6, g_7, g_8\}$.

Analytic Hierarchy Process (AHP) is used to determine the weight coefficient of each element in the two sets. First of all, we use the five-scale method to make a comparison between the elements of the capability advantage sets; secondly, we obtain the comparison matrix, and then calculate the eigenvector of the judging matrix and the consistency; finally, establish the weighted vector of the elements respectively. The weighted vectors of elements in the set H and set G are:

$$\omega = (0.287, 0.106, 0.081, 0.142, 0.119, 0.019, 0.193, 0.053)$$
$$\xi = (0.163, 0.097, 0.228, 0.126, 0.154, 0.106, 0.062, 0.064).$$

Establish language assessment scale:
$$S = \{s_a \mid a = -5, \cdots, -5\} = \{\text{minimum, very small, small, comparative}$$
small, little small, normal, little large, comparatively large, large, very large, maximum}.

The degree (contribution) on FD function influenced by each element of the capability advantages of man and machine is evaluated by experts. The result of the evaluation is

$$\tilde{R} = ([s_2, s_3], [s_0, s_2], [s_3, s_4], [s_1, s_3], [s_0, s_2], [s_2, s_3], [s_{-2}, s_0], [s_0, s_2])$$
$$\tilde{Q} = ([s_0, s_2], [s_2, s_4], [s_0, s_1], [s_2, s_3], [s_2, s_4], [s_3, s_4], [s_1, s_3], [s_{-3}, s_{-1}])$$

According to the equations (7) and (8), we use UEWAA operator to aggregate the results of the assessment and then obtain comprehensive evaluation results which the capabilities of man and machine influence the FD function. The results are

$$\tilde{y}(\omega) = [s_{1.138}, s_{2.224}]$$
$$\tilde{z}(\xi) = [s_{0.942}, s_{2.482}]$$

Using the expression(3), we calculate the possible degree $p(\tilde{y}(\omega) \geqslant \tilde{z}(\xi)) = 0.538$ while $\tilde{y}(\omega) \geqslant \tilde{z}(\xi)$. According to expression(9), we can easily know that the automation level range of FD function is $3 \leqslant A \leqslant 5$.

5.2 Automation level of FD

After the scope of the automation level of FD function has been determined, the established result is equivalent to point out three different given allocation schemes. Therefore, we can adopt the multiple attribute decision making method based on UEWAA and ULHA operators, and establish the final automation level of FD function.

Assume is a set of the allocation schemes about FD function, $X = \{x_3, x_4, x_5\}$, x_i means that the automation level of the scheme x_i is i ($i = 3, 4, 5$). Allocation assessment criteria set $U = \{u_1, u_2, u_3, u_4, u_5\}$, all the elements of u

are corresponding to five main evaluation criteria during the process of the function allocating in the cockpit, u_1—Mental workload [16 - 17]; u_2—Situation awareness [18]; u_3—Reliability; u_4—Decision-making risk; u_5—System cost. Also, using AHP, we can obtain the attribute weighted vector $\boldsymbol{\omega}$ = {0.351, 0.227, 0.284, 0.037, 0.074}. Decision maker set $D = \{d_1, d_2, d_3\}$, d_i is the $i-th$ decision maker, $i =$ 1, 2, 3. Suppose the weighted vector $\boldsymbol{\lambda}$ of $\boldsymbol{D}$, $\boldsymbol{\lambda} = (0.33, 0.33, 0.34)$.

The three decision-makers point out uncertainty language evaluation matrices in Tab. 4 - 6 according to the language assessment scale. $S = \{s_\alpha \mid \alpha = -5, \cdots, 5\} =$ {Worst, very bad, bad, comparative bad, little bad, normal, little good, comparative good, good, very good, best}.

The ULHA weighted (position) vector $\boldsymbol{w}$ is needed before comprehensive property appraisal values $\tilde{z}_i^{(k)}(\omega)$ ($i = 3, 4, 5$; $k = 1, 2, 3$) of the scheme x_i given by three decision-makers are aggregated by using ULHA operator. We use the discrete normal distribution method to calculate the position weighted vector [19 - 20], we can obtain $\boldsymbol{w} = (0.243, 0.514, 0.243)$. Then, according to steps in the section 2.3, we can draw the priority vector $\boldsymbol{v}$ of probability matrix $\boldsymbol{P}$:

$$\boldsymbol{v} = (0.295, 0.409, 0.296).$$

Rank the order according to the component size of v, so the order is

$$x_4 > x_5 > x_3$$

Thus the optimum scheme is x_4, that is to say, the automation level of FD function takes 4 most appropriately.

Tab. 4　Decision-making matrix $\widetilde{R}_1$

	u_1	u_2	u_3	u_4	u_5
x_3	$[s_0, s_2]$	$[s_3, s_4]$	$[s_{-2}, s_0]$	$[s_2, s_4]$	$[s_2, s_3]$
x_4	$[s_0, s_3]$	$[s_0, s_2]$	$[s_3, s_5]$	$[s_1, s_4]$	$[s_2, s_4]$
x_5	$[s_1, s_3]$	$[s_2, s_4]$	$[s_2, s_4]$	$[s_1, s_3]$	$[s_0, s_1]$

Tab. 5　Decision-making matrix $\widetilde{R}_2$

	u_1	u_2	u_3	u_4	u_5
x_3	$[s_{-1}, s_0]$	$[s_2, s_3]$	$[s_{-1}, s_1]$	$[s_2, s_3]$	$[s_1, s_2]$
x_4	$[s_0, s_1]$	$[s_0, s_2]$	$[s_2, s_3]$	$[s_1, s_2]$	$[s_2, s_3]$
x_5	$[s_0, s_2]$	$[s_0, s_1]$	$[s_1, s_2]$	$[s_1, s_3]$	$[s_0, s_1]$

Tab. 6 Decision-making matrix $\widetilde{R}_3$

	u_1	u_2	u_3	u_4	u_5
x_3	$[s_1, s_2]$	$[s_3, s_4]$	$[s_2, s_3]$	$[s_1, s_2]$	$[s_2, s_3]$
x_4	$[s_2, s_4]$	$[s_0, s_1]$	$[s_3, s_4]$	$[s_2, s_4]$	$[s_1, s_2]$
x_5	$[s_{-1}, s_1]$	$[s_2, s_3]$	$[s_{-1}, s_1]$	$[s_3, s_4]$	$[s_1, s_3]$

6 Conclusions

At the beginning of our research, the concept of man-machine system function allocation is presented on the basis of the traditional concept of function allocation. According to contrast results based on the capability of man and machine of Fitts List etc, we make comparisons between the capability advantages of man and machine. In view of highly uncertain characteristics of decision attribute value in the practical processes, the method ULMADM is introduced during the process of function allocation. Meanwhile, use UEWAA operator to determine the automation level range of functions and the multi-attribute decision-making algorithm based on the combination of UEWAA and ULHA operators, eventually establishes the automation level of man-machine function allocation in the cockpit. Compared with the methodologies [3 – 6] mentioned above, the method ULMADM provides us effective qualitative and quantitative analysis of some man-machine function allocation problems so that it is an effective means of establishing the highly uncertain decision attribute values in the practical function allocation process. In this respect, the method ULMADM makes man-machine function allocation become much easier and more efficient. Finally, we take fault diagnosis as an example of function allocation, after the calculation, the result demonstrates that the proposed method about function allocation is feasible and effective.

Acknowledgements

This work was funded by National Basic Research Program of China (973Program: 2010CB734104).

References

[1] Fitts P M. Human Engineer for an Effective Air Navigation and Traffic Control System [R]. Washington D C: National Research Council, 341.
[2] Parasuraman R, Sheriden T B, Wickens C D. A Model for Types and Levels of Human Interaction with Automation [J]. IEEE Transactons on Systems, Man, and Cybernetics-

Part A: Systems and Humans. 2000,30(3):286 - 297.

[3] Endsley M R, Kaber D B. Level of automation effects on performance, Situation awareness and workload in a dynamic control task [J]. Ergonomics, 1999,42:462 - 492.

[4] Endaley M R. Toward a theory of situation awareness in dynamic systems [J]. Human Factors, 1995,37:32 - 64.

[5] Endsley M R. The application of human factors to the development of expert systems for advanced cockpits [C], in Proc. 31st Annu. Meeting Human Factors Soc, 1987,1388 - 1392.

[6] Sheriden T B, Verplank W L. Human and computer control of undersea teleoperator. // In Man-machine systems laboratory report [R]. Cambridge M A: MIT, 1978.

[7] Xu Z S. Uncertain linguistic aggregation operators based approach to multiple attribute group decision making under uncertain linguistic environment [J]. Information Sciences, 2004,168:171 - 184.

[8] US Department of Defense. Human engineering procedures guide [M]. Washington D C: DoD - HDBK - 763, 1987.

[9] Wickens C D, Gordon S E, Liu Y. An introduction to human factors engineering [M]. New York: longman, 1998.

[10] Arciszewski H F R, Tjerk E de Greef, Jan H van Delft. Adaptive Automation in a Naval Combat Management System [J]. IEEE Transactions on Systems, Man, and Cybernetics-Part A: Systems and Humans, 2009,39,6:1188 - 1199.

[11] Xu Z S. Uncertain multi-attribute decision-making method and application [M]. Beijing: Tsinghua University Press, 2004 [in Chinese].

[12] Xu Z S, Da Q L. An overview of operators for aggregating information [J]. International Journal of Intelligent Systems, 2003,18:953 - 969.

[13] Xu Z S. A method based on linguistic aggregation operators for group decision making with linguistic preference relations [J]. Information Sciences, 2004,166:19 - 30.

[14] Xu Z S. A direct approach to group decision making with uncertain additive linguistic preference relations [J]. Fuzzy Optimization and decision making, 2006,5:23 - 35.

[15] Xu Z S. Algorithm for priority of fuzzy complementary judgement matrix [J]. Journal of Systems Engineering, 2001,16(4):311 - 314. [In Chinese].

[16] Price H E. The Allocation of Functions in Systems [J]. Human Factors, 1985,27(1):33 - 45.

[17] Older M, Clegg C, Waterson. Report on the Revised Method of Function Allocation and Its Preliminary Evaluation [R]. Institute of Work Psychology, University of Sheffield, 1996.

[18] Dearden A, Harrison M, Wright P. Allocation of Function: Scenarios, Context and the Economics of Effort [J]. International Journal of Human-Computer Studies, 2000,52(2): 289 - 318.

[19] Xu Z S. An Overview of methods for determining OWA weight [J]. International Journal of intelligent Systems, 2005,20:843 - 865.

[20] Wang Y, Xu Z S. A new method of OWA operator empowerment [J]. Mathematics Knowledge and Practice, 2008,38(3):51 - 61. [in Chinese].

Pilot Attention Allocation Model Based on Fuzzy Theory

Wanyan Xiaoru, Zhuang Damin, Wei Hengyang, Song Jianshuang

School of Aeronautic Science and Engineering, Beijing University
of Aeronautics and Astronautics Beijing 100191

Abstract　Quantitative research of pilot's attention allocation mechanism is required in the optimization design of aircraft human-machine interface and system evaluation. After making a comprehensive consideration of several factors, including the importance of information, information detective efficiency and human errors, pilot attention allocation model was built on the basis of hybrid entropy. In order to make verification of pilot attention allocation model, a simulation model of head-up display (HUD) used to present flight indicators was developed. After setting the membership degrees of the importance for different indicators according to their priorities, the experiments of key-press response and eye-movement tracking were designed and carried out under the cruise and hold modes. As the experiment results are in good agreement with the theoretical model, the effectiveness of pilot attention allocation model based on fuzzy theory is confirmed.

Keywords　attention allocation, hybrid entropy, information importance, eye-movement tracking, cognitive engineering

1　Introduction

In the human-machine interaction system of the modern aviation, the role of pilot is transforming from the operator to monitor due to the improved aircraft performance and automation [1]. The pilot needs to keep monitoring various indicators simultaneously when carrying out flight missions. Therefore, getting visual information effectively highly depends on reasonable allocation of pilot's limited attention resource. It was found that the factor of attention allocation always ranks in the top twenty after ordering 114 human factors which relates to flight deck automation according to different criteria [2]. Consequently, researching the attention allocation behavior of the pilot is of significance to offer scientific reference for the human-machine interface design of aircraft cockpit and thus it is helpful to improve the flight performance and safety.

As the mechanism of attention allocation is manipulated by the human brain information processing system, and is affected by the physiological and psychological factors to a great extent, it is difficult to definitely interpret human attention allocation mechanism using biomedical method at present stage. Therefore, there has been a substantial interest in making quantitative solution to this problem from the perspective of engineering psychology.

Human behavior models have been used effectively in the process of analysis, design and evaluation for various human-machine interaction systems. However, because of the non-linearity, randomicity, discretization and time-variation of the human behavior, many difficulties exist and great attention has been paid to the in-depth study of this field in cognitive engineering [3]. In the prior studies, many valuable models have been proposed to describe and predict the attention allocation strategy in the monitoring behavior of human-machine system. In 1964, Sender provided one of the first quantitative models of monitoring behavior by introducing the concept of bandwidth [4]. Kleinman built the attention allocation model using gradient algorithm within the framework of optimal control model (OCM) of human response and verified it according to the hover control task for a CH - 46 helicopter [5]. Bohnen and colleagues distinguished the factor of alarm rate from bandwidth and make optimization of Sender's model [6]. Wickens and colleagues researched the effect of value, effort, salience, habit, expectancy and context on pilot's scanning behavior and built the SEEV model [7 - 9]. In order to express the fuzziness of human behavior, fuzzy control models have been applied well to explain the uncertain phenomenon relates to human psychological state and thinking activity. Based on it, Nobuyuki Matsui and Yan Lou built the fuzzy control models of human attention allocation behavior [10 - 11]. However, some influencing factors which may affect human attention allocation behavior have not been considered in these models, besides, the usability of the models have not been verified.

For the monitoring task of indicators in the aviation field, the acquisition of flight information is mainly based on pilot's previous knowledge and experience. It belongs to voluntary attention driven by the top-down information processing mechanism [12]. Therefore, the attention resource allocated to a certain indicator depends on its significance evaluated by the pilot. The prior studies also indicated that the value which represents the importance of indicator is the most important influencing factors of pilot attention allocation behavior [13]. Besides, the attention intensity for a certain indicator is also affected by pilot's physiological and psychological states. Insufficient understanding of the indicators or unreasonable design of the information display interface usually results in human errors, such as omitting, misreading or misjudgment of the information. For such a reason, even the most important indicator may be ignored and cannot activate the attention mechanism. It means that the attention mechanism has randomness and whether the importance of a certain indicator can be correctly evaluated depends on some incidence probability. Therefore, human errors as an influencing factor should be also taken into consideration when the model is built. Moreover, in the practical application, due to different visual coding (such as color, shape, size, etc.) and processing depth (such as identification, memory, calculation, etc.) of the information, the detective efficiencies of different indicators are not the same. The indicator with low detective efficiency consumes more attention resource than the one with high detective efficiency, even if they are evaluated with the same importance by the pilot. Thus, we consider the detective efficiency as another influencing factor when building pilot attention allocation model.

The present study built pilot attention allocation model on the study basis of Kleinman and Nobuyuki Matsui. Using the theory of hybrid entropy, this model take several influencing factors, including information importance, human errors and detective efficiency into consideration synthetically. In order to apply the pilot attention allocation model to the aeronautic human-machine interface and make validation of it, a HUD simulation model used to display indicators was developed. According to different importance of every indicator under two flight modes, the membership degree of the importance for each indicator was set by fuzzy membership function. Through measuring the behavioral performances of the participants, meanwhile combining with the real-time eye-movement data derived from the eye tracker, the actual attention allocation situation of the participants were obtained and compared with the theoretical model.

2 Pilot Attention Allocation Model

2.1 Pilot Attention Allocation Model Based on Hybrid Theory

In human-machine interaction systems, if the human brain is considered as an information receptor, then a vector can be used to express n indicators which are monitored by the pilot simultaneously when they present on the information display interface.

$$Y = (y_1, y_2, \cdots, y_i, \cdots, y_n) \tag{1}$$

We use ω_i to represent the significance of a certain indicator y_i which was evaluated by the pilot.

$$U = (\omega_1, \omega_2, \cdots, \omega_i, \cdots, \omega_n) \tag{2}$$

A fuzzy vector X is given to express the vagueness of such a evaluation, where μ_i is the membership degree of the importance for a certain evaluation ω_i.

$$X = (\mu_1, \mu_2, \cdots, \mu_i, \cdots, \mu_n) \tag{3}$$

The concept of fractional attention f_i was introduced by Kleinman in OCM, where f_i is the fractional number of sensory channels that carry the information when the human is considered as a multi-channel processor. It is an equivalent representation to consider the human as a time-shared, single processing channel, and then f_i is the fraction of time devoted to y_i, and f_{tot} is defined as the total fractional attention or capacity devoted to the monitoring task [5]. The constraints which f_i and f_{tot} should satisfy are

$$\sum_{i=1}^{n} f_i = f_{tot} = 1, \quad f_i \geqslant 0 \tag{4}$$

If the pilot is regarded as an ideal monitor, then the pilot should allocate his attention resource according to the importance of each indicator, so that to make optimization of the resource utilization. Combined with the definition of fractional attention in OCM, f_i can be expressed with equation (5).

$$f_i = \frac{\mu_i}{\sum\limits_{i=1}^{n} \mu_i}, \quad (i = 1, 2, \cdots, n) \tag{5}$$

However, as the existence of the randomness of attention allocation mechanism, p_i can be assumed as the incidence probability that the pilot can correctly evaluate the importance of a certain indicator y_i.

$$P = (p_1, p_2, \cdots, p_i, \cdots, p_n) \tag{6}$$

Then, the fractional attention which is allocated to a certain indicator y_i is modified as f'_i.

$$f_i' = \frac{p_i \mu_i}{\sum\limits_{i=1}^{n} p_i \mu_i}, \quad (i = 1, 2, \cdots, n) \tag{7}$$

Considering that the fuzziness and randomness of human attention mechanism are two kinds of uncertainty. They can complement but cannot substitute each other. Thus, the uncertainty results from the fuzziness and randomness together should be measured by the hybrid entropy H_{tot} [14]. As the consciousness for the importance of the indicators originates from the uncertain subjective evaluation of the pilot, in addition, with the increase of such uncertainty, the desire for obtaining information and the anxiety resulting from information insufficiency will be strengthened, thus it is helpful to enhance the attention level. Such a phenomenon is in line with the common cognitive law. Therefore, the hybrid entropy H_{tot} can be defined as the psychological entropy of the pilot [10]. Assuming that A is a fuzzy subset of domain U, then the H_{tot} is described as follows [15]

$$H_{tot}(A, P) = m(A, P) + H(P) \tag{8}$$

where $m(A, P)$ is the fuzzy entropy given by equation(9), and $H(P)$ is the probability entropy given by eq. (10) [15].

$$m(A, P) = \sum_{i=1}^{n} p_i S(\mu_i) \tag{9}$$

$$H(P) = -\sum_{i=1}^{n} p_i \ln p_i \tag{10}$$

In equation (9), $S(\mu_i)$ is the binary fuzzy entropy of μ_i [15], then

$$S(\mu_i) = -\mu_i \ln \mu_i - (1 - \mu_i) \ln(1 - \mu_i) \tag{11}$$

Thus, According to the Shannon additivity rule [16], the average hybrid entropy $H_{avg}(A, P)$ of n indicators can be obtained by eq. (12).

$$H_{avg}(A, P) = \frac{1}{n} \sum_{i=1}^{n} H_{tot}(A, P) = \frac{1}{n} \sum_{i=1}^{n} (p_i S(\mu_i) - p_i \ln p_i) \tag{12}$$

The value of p_i can be estimated reasonably by introducing the maximum entropy principle [17]. It is easy to see that the constraints which p_i should satisfy are

$$p_i \geqslant 0, \quad \sum_{i=1}^{n} p_i = 1 \tag{13}$$

According to the maximum entropy principle, the value of p_i should make the average hybrid entropy $H_{\text{avg}}(A, P)$ reach its maximum. After calculating the extreme value of Lagrangian function L under the constraints eq. (13), the value of p_i^* which makes $H_{\text{avg}}(A, P)$ reach its maximum is given by (15).

$$L = \frac{1}{n}\sum_{i=1}^{n}(p_i S(\mu_i) - p_i \ln p_i) - \lambda\left(\sum_{i=1}^{n} p_i - 1\right) \tag{14}$$

$$p_i^* = \frac{\exp S(\mu_i)}{\sum_{i=1}^{n}\exp S(\mu_i)}, \quad (i = 1, 2, \cdots, n) \tag{15}$$

Substituting p_i^* into eq. (12), then the maximum average hybrid entropy $H_{\text{avg}}^*(A, P)$ is expressed as eq. (16). In such a case, the pilot has the highest attention level.

$$H_{\text{avg}}^*(A, P) = \frac{1}{n}\ln\sum_{i=1}^{n}\exp S(\mu_i) \tag{16}$$

The detective efficiency ψ_i of a certain indicator y_i can be defined by the reciprocal of its mean response time t_i, as shown in eq. (17). Detective efficiency ψ_i decreases with the increase of mean response time t_i. Taking the influencing factor of the detective efficiency into consideration, the fractional attention is rewritten as F_i, as shown in eq. (18).

$$t_i = \frac{1}{\psi_i} \quad (i = 1, 2, \cdots, n) \tag{17}$$

$$F_i = \frac{p_i^* \mu_i t_i}{\sum_{i=1}^{n} p_i^* \mu_i t_i}, \quad (i = 1, 2, \cdots, n) \tag{18}$$

2.2　Optimal Number of Indicators for Attention Allocation

In order to determine the optimal number of the indicators for the pilot to make attention allocation, it is necessary to research the relationship between the indicators quantity n and their maximum average hybrid entropy $H_{\text{avg}}^*(A, P)$. The membership function is supposed to satisfy the following conditions.

$$\begin{cases} \mu_i < \mu_{i+1}, & (i = 1, 2, \cdots, n) \\ \mu_1 = 0, \ \mu_n = 1 \end{cases} \tag{19}$$

A simple form of the membership function which satisfies the conditions above can be expressed by eq. (20). Aiming at different change rate k, three typical cases are selected, including $k = 1(\Delta^2\mu/\Delta i^2 = 0)$, $k = \dfrac{1}{2}(\Delta^2\mu/\Delta i^2 < 0)$ and $k = 2(\Delta^2\mu/\Delta i^2 > 0)$.

$$\mu_i = \left[\dfrac{i-1}{n-1}\right]^k, \quad (i = 1, 2, \cdots, n) \tag{20}$$

After calculating, the relationship of the indicators quantity n and their maximum average hybrid entropy $H^*_{\text{avg}}(A, P)$ is shown in Fig. 1. It can be seen that $H^*_{\text{avg}}(A, P)$ present the same trend although the change rates are different. When the quantity of the indicators are 3 or 4, $H^*_{\text{avg}}(A, P)$ exhibit the higher values. $H^*_{\text{avg}}(A, P)$ decrease monotonically with the increase of the quantity of the indicators after then. Therefore, the optimal number of the indicators to which pilot can make attention allocation effectively are 3 or 4.

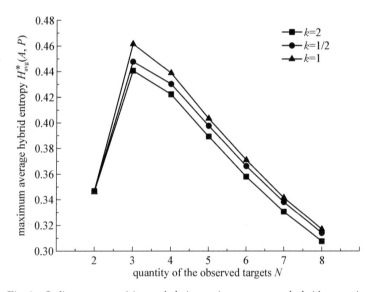

Fig. 1   Indicators quantities and their maximum average hybrid entropies

3   Method

3.1   Experiment Interface Design

The experiment interface was designed with reference to the typical layout of HUD. According to the optimal number of indicators for pilot's attention allocation, four indicators of pilot's normal work were selected, including indicated airspeed, barometric altitude, pitching angle and heading. The colors of all the four indicators were commonly used green in HUD. The indicators presented on a 19-inch liquid crystal display and the resolution was $1,280 \times 1,024$.

Disturbances which made the indicator display unusual were set for each indicator by programming. The disturbance of each indicator randomly appeared with equal probability and disappeared after presenting transitorily. According to the military standard MIL – STD – 1787B that the reaction time of the unusual attitude recovery should be less than 1 s, we set the duration of the disturbance to 0. 8 s, and the average inter-stimulus interval between disturbances was 1 s. Each indicator appeared 20 disturbances during once experiment, and no more than one disturbance appeared simultaneously. Participants used the standard keyboard and mouse in the process of human-machine interaction.

The situation of human attention allocation can be mirrored by the eye-movement to a great extent. In order to record the eye-movement data of the participants objectively, Smart Eye system, which is a non-contact eye tracker, was introduced in the experiment. It can track the eye-movements in the complete natural state with two infrared cameras.

3. 2　Experiment Task

The task of key-press response was performed in the experiment. When the experiment began, the participants needed to monitor the four indicators simultaneously, and allocated their attention resource according to different flight modes. When the disturbance was detected, the participants were asked to make a response to eliminate the disturbance by pressing the corresponding key within the given time. No response, mistaken response or delayed response of the disturbance was all considered as the noneffective attention, and the probable attention resource input was neglected. The accuracy rates and reaction times of the participants were recorded as the evaluation indexes of the behavioral performance.

3. 3　Participants

12 students from Beijing University of Aeronautics and Astronautics participated in the study. All the participants (8 males, 4 females; ranging from 22 to 28 years old, mean age 24. 8 years) are familiar with the basic operation of computer and have the background knowledge of aeronautics. All participants are right-handed with normal or corrected to normal vision.

3. 4　Procedure

The membership degrees of the importance for the four indicators under the cruise mode and hold mode were set based on their relative priorities [18]. In order to simulate pilot's potential experience of the importance of each indicator in real flight environment, the membership degrees were transformed into scores by a certain ratio so that the importance of each indicator was easier to be understood by the participants. For a certain indicator, correct response to its disturbance of each time would make the participant win the corresponding score. The membership degrees of the importance and the scores for each indicator under two flight modes are shown in Tab. 1 and Tab. 2.

Tab. 1 Membership degrees of the importance and scores under the cruise mode

Indicator	Indicated airspeed	Barometric altitude	Pitching angle	Heading
Score	9	8	5	1
Membership degree	0. 9	0. 8	0. 5	0. 1

Tab. 2 Membership degrees of the importance and scores under the hold mode

Indicator	Indicated airspeed	Barometric altitude	Pitching angle	Heading
Score	6	7	9	3
Membership degree	0. 6	0. 7	0. 9	0. 3

Each participant participated in the experiments both under the cruise and hold mode. The order of modes was counterbalanced across participants. Participants practiced enough times before the formal experiment to get familiar with the process of the experiment and memorize the scores of indicators. During the experiment, the participants were required to allocate their attention reasonably according to the importance of the indicators, and try to reach the highest total score. At the same time, the Smart Eye system kept real-time tracing.

4 Results

4. 1 Theoretical Results of the Mathematical Model

Assuming that the importance of the four indicators were the same, thus the mean response time t_i of each indicator can be measured respectively. Then the theoretical values of the incidence probability p_i and the fractional attention $F_i(\%)$ calculated by pilot attention allocation model are shown in Tab. 3 and Tab. 4.

Tab. 3 Theoretical values under the cruise mode

Indicator	Indicated airspeed	Barometric altitude	Pitching angle	Heading
detective efficiency	2. 04	2. 08	2. 33	2. 02
incidence probability	0. 22	0. 26	0. 31	0. 21
fractional attention	35. 01	36. 40	24. 65	3. 94

Tab. 4 Theoretical values under the hold mode

Indicator	Indicated airspeed	Barometric altitude	Pitching angle	Heading
detective efficiency	2.04	2.08	2.33	2.02
incidence probability	0.28	0.26	0.20	0.26
fractional attention	28.72	30.92	26.70	13.66

4.2 Experiment Results of the Key-press Response

In the experiment of key-press response, the experiment value of the fractional attention F'_i for a certain indicator y_i in once experiment can be defined as (21), where κ_i is the correct response times of y_i, t_i is the mean response time of y_i, and n is the quantity of indicators.

$$F'_i = \frac{\kappa_i t_i}{\sum_{i=1}^{n} \kappa_i t_i}, \quad (i = 1, 2, \cdots, n) \tag{21}$$

According to the recorded experiment data and combined with (21), the experiment values of the fractional attention F'_i under the cruise and hold modes were obtained after statistical analysis, as shown in Tab. 5.

Tab. 5 Mean percentages of the fractional attention F'_i under the cruise and hold modes

Flight mode	Fractional attention $F'_i/\%$			
	Indicated airspeed	Barometric altitude	Pitching angle	Heading
Cruise	37.62	35.85	20.95	5.57
Hold	29.24	33.26	26.25	11.25

4.3 Experiment Results of the Eye-movement Tracking

In the experiment of eye-movement tracking, the infrared images were transformed into digital images by PCI frame grabber with the sampling rate of 60 Hz. Therefore, in once experiment, the ratio of the fixation points m_i for a certain indicator y_i to the fixation points for all the four indicators can be defined as the experiment value of the fractional attention F''_i. n is still the quantity of indicators.

$$F''_i = \frac{m_i}{\sum_{i=1}^{n} m_i}, \quad (i = 1, 2, \cdots, n) \tag{22}$$

As shown in Fig. 2, a certain participant's fixation points were recorded by the Smart Eye system under the cruise mode in once experiment. The distribution of the fixation points for each indicator can be seen intuitively. In Fig. 2, the indicated airspeed, barometric altitude, pitching angle and heading have 33.03%, 33.49%, 30.89% and 2.60% fixation points, respectively. Fig. 3 presents the fixation points under the hold mode in once experiment, where the indicated airspeed, barometric altitude, pitching angle and heading obtain 26.95%, 31.15%, 26.63% and 15.26% fixation points, respectively.

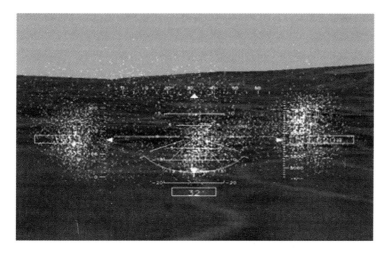

Fig. 2   Fixation points under the cruise mode

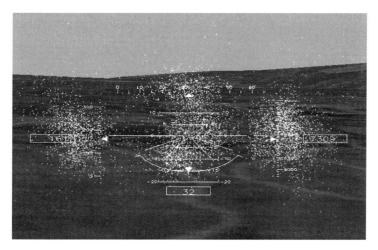

Fig. 3   Fixation points under the hold mode

Statistically analyzing the recorded fixation points, the experiment values of the fractional attention F_i'' under the cruise and hold modes are shown in Tab. 6.

Tab. 6　Mean percentages of the fractional attention F_i'' under the cruise and hold modes

Flight mode	Fractional attention $F_i'/\%$			
	Indicated airspeed	Barometric altitude	Pitching angle	Heading
Cruise	30. 14	40. 33	26. 08	3. 46
Hold	27. 754	35. 88	21. 59	14. 78

4. 4　Comparison of the Theoretical and Experiment Results

Under the cruise and hold mode, the fractional attention values of the key-press response experiment F_i' and the eye-movement tracking experiment F_i'' as well as the theoretical value F_i are compared, as shown in Fig. 4 and Fig. 5.

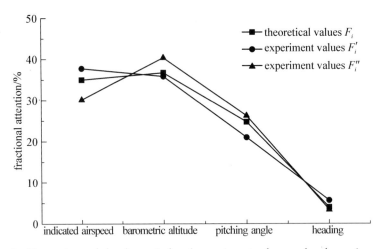

Fig. 4　Comparison of the theoretical and experiment values under the cruise mode

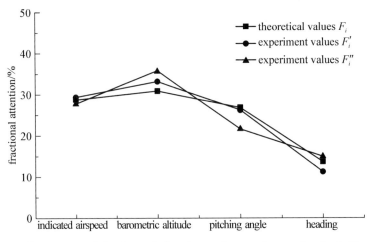

Fig. 5　Comparison of the theoretical and experiment values under the hold mode

5   Discussion

It can be seen from Fig. 4 and Fig. 5 that the data gathered from experiments are consistent with the theoretical values calculated from pilot attention allocation model, thus, the effectiveness of the model is confirmed. According to pilot attention allocation model, the indicator with the highest priority may not obtain the most attention resource due to the existences of human errors and detective efficiency. This conclusion has also been embodied in the experiments. For example, the priority of pitching angle was the highest under the hold mode, however, the actual attention resource allocated to it was less than the ones allocated to indicated airspeed and barometric altitude in the key-press response and eye-movement tracking experiments, as shown in Fig. 5.

In our study, two different experiment methods have been adopted for analyzing the actual situations of participants' attention allocation. According to Fig. 4 and Fig. 5, it can be seen that the results of key-press response experiment were in better agreement with the theoretical values. This is may because that the fixation points recorded by eye tracker can not always mirror participant's actual observation positions. For a instance, without moving the positions of eyes, the participant can monitor other indicator by split vision. Comparatively, the action of key-press proves effectively that the participant has allocated his attention to the indicator responded.

According to pilot attention allocation model, the optimal number of indicators to which pilot can make attention allocation effectively is 3 or 4. At this time, the pilot owns the highest attention level and can allocate his attention resource reasonably according to the information priority.

Although pilot's attention allocation behavior is mainly driven by the top-down mode, several bottom-up factors which influencing the involuntary attention also exist. However, related research manifested that the bottom-up factors hardly disturb the scanning strategies for the well-trained pilot [9]. Therefore, the effects of bottom-up factors on pilot's attention allocation behavior were ignored when the model was built.

In addition, the factor of indicator location which relates to scanning habit should also affect pilot's attention allocation behavior. At present, this study has not taken this factor into consideration yet. In our experiments, it is found that different indicator locations did not have significant effect on the actual attention allocation of the participants. As shown in Fig. 4 and Fig. 5, the pitching angle which located in the good visual field did not earn more attention resource than the prediction of the theoretical model, and the heading which located in the poor visual field did not lose attention resource than the prediction. This may because that the distribution of the four indicators was not dispersed enough. This part of study will be further performed in the next step.

6　Conclusion

After making comprehensive analysis of several influencing factors, such as the importance of information evaluated by the pilot, information detective efficiency and human errors, pilot attention allocation model was built from the perspective of fuzzy mathematics and cognitive psychology. According to the simulation experiments performed in our study, the model is suggested to be used for predicting pilot's attention allocation to a group of indicators. Therefore, it can provide reference for the humanity evaluation of cockpit interface design.

Acknowledgments

This study is supported by National Basic Research Program of China (Program Grant No. 2010CB734104) and the National High Technology Research and Development Program of China (Program Grant No. 2009AA012101). The authors thank the anonymous reviewers for their helpful comments to improve the manuscript.

References

[1] Wickens C D, Lee J D, Yili Liu, An Introduction to Human Factors Engineering [M], second ed., Prentice Hall, New Jersey, 2007.

[2] Funk K, Suroteguh C, Wilson J, et al. Fight deck automation and task management [J], Systems, Man, and Cybernetics 1998,1:863 – 868.

[3] Liu W, Yuan X G. Ergonomics interaction design and evaluation [M], Beijing: Science and Technology Publishing House, 2008.

[4] Senders J W. The Human Operator as a Monitor and Controller of Multidegree of Freedom Systems [J], Human Factors in Electronics 1964,5:2 – 5.

[5] Kleinman D L. Solving the optimal attention allocation problem in manual control [J], Automatic Control 1976,21:813 – 821.

[6] Bohnen H G M, Leermakers M A M, Venemans P J. Sampling behavior in a four instrument monitoring task: effects of signal bandwidth and number of events per signal [J], Systems Man and Cybernetics 1996,21:413 – 422.

[7] Wickens C D, Helleberg J, Goh J, et al. Horrey, Pilot Task Management: Testing an

Attentional Expected Value Model of Visual Scanning [R], Savoy, University of Illinois Institute of Aviation, ARL - 01 - 14/NASA - 01 - 7,2001.

[8] Wickens C D, Helleberg J, Xu X. Pilot maneuver choice and workload in free flight [J], Human Factors 2002,44:171 - 188.

[9] Wickens C D, Goh J, Helleberg J, et al. Attentional models of multitask pilot performance using advanced display technology [J], Human Factors 2003,45:360 - 380.

[10] Matsui N, Bamba E. Evaluative cognition and attention allocation in human interface [J], Association Symposium of Measurement and Automatic Control, 1987,J70 - D:2321 - 2326.

[11] Lou Y, Hanwu He H W, Lu Y M. Attention allocation behavior modeling of virtual driver [J]. Micro-computer Information, 2008,24:274 - 276.

[12] Yuan X G, Zhuang D M, Zhang X G, Simulation of human machine engineering [M], Beijing: Press of Beijing University of Aeronautics and Astronautics, 2005.

[13] Miller S M, Kirlik A, Kosorukoff A, et al. Ecological Validity as a Mediator of Visual Attention Allocation in Human-Machine Systems, Savoy [R], University of Illinois Institute of Aviation, AHFD - 04 - 17/NASA - 04 - 6,2004.

[14] Zadeh L A. Probability theory and fuzzy logic are complementary rather than competitive [J], Technometrics, 1995,37:271 - 276.

[15] Pal N R, Pal S K. Higher order fuzzy entropy and hybrid entropy of a set [J], Information Sciences, 1992,61:211 - 231.

[16] Bezzi M. Quantifying the information transmitted in a single stimulus [J], Biosystems, 2007,89:4 - 9.

[17] Jaynes E T. Clearing up Mysteries — the Original Goal [C], in: J. Skilling (Eds.), Proceedings of the 8'th International Workshop in Maximum Entropy and Bayesian Methods, Kluwer Academic Publishers, Holland (1989) 1 - 27.

[18] MIL - STD - 1787B (USAF), Aircraft Display Symbology [S], The Department of Defense of United States of America, 1996.

Study on the Information Coding Design Humanity of Man-machine Display Interface

Yingwei Zhou, Damin Zhuang, Guoling Song, and Xiaoru Wanyan

School of Aeronautic Science and Engineering, Beijing University
of Aeronautics and Astronautics, Beijing 100191

Abstract   The humanity of coding design for human-machine display interface in the product design is studied by experimental method. During the experiment, three character types, three eye distances from the display interface, five character sizes are adopted. The correct rate and reaction time are chosen as the evaluation indexes. Analyzed by ANOVA, mathematical models of the information coding are established. Mathematical models proposed in this paper can be widely applied for man-machine display interface coding design, and provide theoretical basis for different kinds of products design.

Keywords   man-machine display interface, information coding, product design, correct rate, reaction time

1   Introduction

Man-machine display interface design is one of the most important aspects during the process of product design and development. Good display interface design can greatly improve the performance of product and reduce the burden of operator [1]. At present, man-machine display interface has become more and more important for many kinds of electromechanical products, electronic products, and some large man-machine systems, including airplane, nuclear power station, etc. Over 70% of the information obtained by man-machine interaction comes from man-machine display interface, where characters, number and identifier are the most common information forms. Therefore, making reasonable man-machine display interface design is important for operator to identify information rapidly and reduce human errors [2 - 3].

In the non-computer era, Peters and Adams have proposed the following equation for calculating the reasonable English character size [4]:

$$H = 0.002,2D + 25.4 \times (K_1 + K_2) \tag{1}$$

Where, H represents the height (mm) of the English character, D represents the distance (mm) from the eye to the characters observed; K_1 is the coefficient associates with illumination condition, the common values of K_1 are 0.06, 0.16 and 0.26; K_2 is the coefficient associates with the importance of contents, the common value of K_2 is 0 or 0.075 in the important case.

The usability of the above equation has been proved in Ref [2] to some extent. However, in the current computer age, especially in the field of man-machine display interface coding design of products, high intensity and fast changes of information flow make it necessary to carry out humanity research on character coding design.

2   Study Method and Experimental Design

$3 \times 3 \times 5$ mixed design is adopted in this experiment. Taken as the between-subject factor, factor 1 is the eye distance from the display interface, which includes three levels: 500, 700 and 900 mm. Taken as the within-subject factors, factor 2 is the character type which includes three levels: English character, numerical character and Chinese character, as well as factor 3 is the character size which includes five levels: 5 points, 8 points, 13 points, 20.5 points and 33 points. When the experiment begins, the subjects enter the experiment by choosing different character sizes. Half of the subjects are required to perform the experiments from small character size to large one, and that is from 5 points to 33 points. Others are required to perform the experiments from large character size to small one, and that is from 33 points to 5 points. The characters types are presented randomly by computer.

The experiment interface is implemented by Visual Basic programming, and is presented on a 17-inch widescreen LCD monitor. Referring to the relatively mature form of Rapid Serial Visual Presentation (RSVP), one or multi-characters is shown in the fixed region of the screen every time, then after a certain time interval, the subsequent characters cover the previous ones [5]. One complete stimulus sequence is presented on the experiment interface during each experiment. In the current study, a pair of characters or a numerical character is shown in the center of the screen at each time. The subjects are asked to press certain buttons to make a response during the period of character presentation, and the system records the right and wrong responses as well as the reaction time automatically.

There are 100 high frequency common used Chinese characters chosen from "Modern Chinese Character Set for Common Use", including 50 near-synonyms and 50 antonyms, which are all bold black characters. English characters include the 26 capital letters and 26 lowercase letters, which are all Times New Romar characters. Numerical characters of three-figure number are generated randomly, which are all bold black characters.

28 graduate students from Beijing University of Aeronautics and Astronautics participate in this study, including 11 males and 17 females, between the ages of 20 and 24. All the subjects do not suffer from color blindness, and their visual acuity is level 1. 0 or above. The subjects are divided into three groups according to the eye distances from the display interface. Five character sizes under fixed eye distance are tested for each subject, and each character size is tested for 60 times.

The subjects are required to sit on the positions where 500, 700 or 900 mm far from the experiment display interface, and keep the upper body, especially the head, stable during the experiment. After filling the personal information and experimental parameters (information presentation time; screen contrast; eye distance from the display interface; character size) into the blanks of initial experimental interface, the subjects press the "start" button to enter the formal experimental interface. One group of characters is presented in the middle of the experiment interface each time, and the presentation time is set as 2 s. The subjects are asked to press the buttons as rapidly as possible to make a response after the characters are shown.

Firstly, a random combination of one capital letter and one lowercase letter are presented on the experiment interface, such as "a - A", "d - H". No. 7 button is required to be pressed in case of the capital letter is corresponding to the lowercase letter, such as "a - A", "b - B", and No. 8 button is required to be pressed in case of the capital letter is not corresponding to the lowercase letter, such as "f - A", "b - R". After the English letters are shown for 20 times, a three-figure numerical character from 100 to 999 is shown. If the numerical character is larger than 500, No. 7 button is required to be pressed. If not, No. 8 button is required to be pressed. Finally, a group of Chinese characters is shown, such as "Far-Near". No. 7 button is required to be pressed if the two characters are near-synonyms, and No. 8 button is required to be pressed if they are antonyms.

Dynamic data record and save are achieved by programming, including the subjects' personal information, the times of no response, error response, correct response, the total times of the displayed information as well as the reaction time of the correct response.

3   Experimental Results and Analysis

3. 1   Experimental Results and Data Processing

The data of the correct rates and reaction times obtained from the experiment are shown in Fig. 1 and Fig. 2, and statistically analyzed by ANOVA analysis in SPSS for Windows 13. 0.

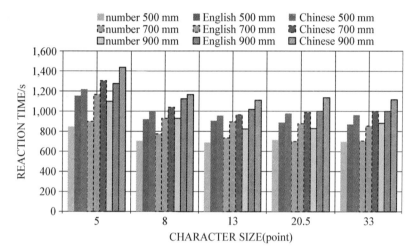

Fig. 1   Dada of the Correct Rate

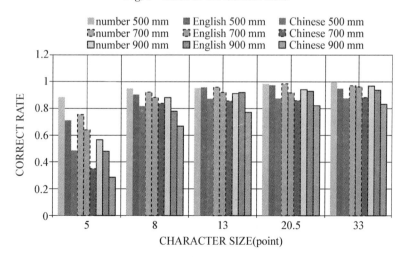

Fig. 2   Dada of the Reaction Time

For the English character, ANOVA analysis of the correct rate shows that, at 0. 05 significance level, the main effect of the eye distance from the display interface is not significant ($P = 0.076$), the main effect of the character size is significant ($P = 0.007$), and the interaction effects between the eye distance from the display interface and character size are also significant ($P = 0.022$). ANOVA analysis of the reaction time shows that, the main effect of the eye distance from the display interface ($P = 0.029$) and the character size ($P = 0.001$) are significant, and the interaction effects between the eye distance from the display interface and character size are not significant ($P = 0.289$).

For the numerical character, ANOVA analysis of the correct rate shows that, the main effect of the eye distance from the display interface ($P = 0.009$) and the character size ($P = 0.014$) are significant, and the interaction effects between the eye distance from the display interface and character size are not significant ($P = 0.479$). ANOVA analysis of the reaction time shows that, the main effect of the eye distance from the display interface ($P = 0.003$) and the character size ($P = 0.001$) are significant, and the interaction effects between the eye distance from the display interface and character size are not significant ($P = 0.350$).

For the Chinese character, ANOVA variance analysis of the correct rate shows that, the main effect of the eye distance from the display interface is not significant ($P = 0.078$), the main effect of the character size is significant ($P < 0.001$), and the interaction effects between the eye distance from the display interface and character size are not significant ($P = 0.125$). ANOVA analysis of the reaction time shows that, the main effect of the eye distance from the display interface ($P = 0.013$) and the character size ($P < 0.001$) are significant, and the interaction effects between the eye distance from the display interface and character size are not significant ($P = 0.713$).

3.2　Experimental Results Analysis

The critical character size is obtained by analyzing the relationship between the character size and correct rate as well as response time, using the Least Square Fitting Method in the Matlab.

For the correct rate, the critical character sizes of English under three different eye distances from the display interface (500, 700, 900 mm) are 9.04, 11.66 and 12.39. For the reaction time, the critical character sizes of English under three different eye distances from the display (500, 700, 900 mm) are 8.11, 11.98 and 12.73.

Similarly, for the correct rate, the critical character sizes of number under three different eye distances from the display interface conditions (500, 700, 900 mm) are 8.09, 8.97 and 11.85. For the reaction time, the critical character sizes of number under three different eye distances from the display interface (500, 700, 900 mm) are 8. 17, 8. 92 and 10. 29. For the correct rate, the critical character sizes of Chinese under three different eye distances from the display interface (500, 700, 900) are 11. 95, 16. 78 and 17. 88. For the reaction time, the critical character sizes of the Chinese under three different eye distances from the display interface (500, 700, 900 mm) are 11. 79, 16. 08 and 17. 26.

According to equation (1) which is proposed by Peters and Adams, there are linear relationships between the eye distance from the display interface and character size. Therefore, without considering the effects of illumination conditions, the relationships between the eye distance from the display interface and character size can be described as $H = a \times D + b$. Where, H is the height of the English character, which is converted from the character size, D is the eye distance from the display interface. A and b are the unknown coefficients, which can be determined by taking mean value through multiple solution. The values of a and b can be ascertained by choosing two groups of data from the three ones. After calculating three group values of a and b, their final values can be fixed by taking their mean values.

Finally, the equations between the eye distance from the display interface and character size are shown as equation (2), (3) and (4).

$$H_{\text{English}} = 0.003,5 \times D + 1.545 \qquad (2)$$

$$H_{\text{number}} = 0.002,6 \times D + 1.407 \qquad (3)$$

$$H_{\text{Chinese}} = 0.005,0 \times D + 2.060 \qquad (4)$$

Comparing the equation (1) and (2) which are both English character equations, it can be seen that when K_1 takes 0. 06 under the relatively good illumination conditions, the characters height equation advised by Peters and Adams is $H = 0.002,2D + 1.524$. As the values of D are 500, 700, 900 mm, the differences of the English character sizes which calculated by equation (1) and equation (2) are only 0. 67, 0. 93 and 1. 19 mm, respectively. It is can be seen from equation (2) to equation (4) that under the same eye distance from the display interface, the order of the required character sizes is numerical character $<$ English character $<$ Chinese character.

4　Conclusions

In this paper, the relationships between the character size of the English, number, Chinese and eye distance from the display interface are discussed through human cognition. The experimental date including the correct rate and reaction time are analyzed by ANOVA. Mathematical models of the relationships between character sizes of different types and eye distance from the display interface are established. Compared with the conclusions obtained by foreign scholars, the usability of mathematical models and feasibility of experimental methods are verified. The experimental results can provide theoretical basis for man-machine display interface coding design for different kinds of products.

Acknowledgements

This work was financially supported by the National Basic Research Program of China (Program Grant No. 2010CB734104), and the National High Technology Research and Development Program of China (Program Grant No. 2009AA012101).

References

[1] Minyan Xia, Xuehua Tang: Human-machine interface design rules of electromechanical product based on knowledge of cognitive psychology [J]. Machinery Design & Manufacture, 2010. 1:183, in Chinese.

[2] Lei Zhang: The study on encoding of aircraft cockpit display interface [D]. University of Aeronautics and Astronautics (2010), in Chinese.

[3] Damin Zhuang, Rui Wang. Research of target identification based on cognitive characteristic [J], Journal of Beijing University of Aeronautics and Astronautics, 2003:1051, in Chinese.

[4] Gavriel Salvendy. Handbook of Human Factors [M]. edited by Wile-Interscience Publication, 1984.

[5] Kazuhiro Fujikake, Satoshi Hasegawa: Readability of Character Size for Car Navigation Systems [J]. Human Interface, 2007:503.

Applying Situation Awareness to Human-machine Interface Design of Aviation

Huibin Jin[1], Lei Wang[2]

1. Research institute of Civil Aviation Safety, Civil Aviation University of China, Tianjin 300300
2. Research institute of Civil Aviation Safety, Civil Aviation University of China, Tianjin 300300

Abstract   This paper discusses relationship between Situation Awareness and aviation accident, and proposed SA-oriented Design process for human-machine interface design. After some common designing principles, we discuss emphatically on the 10 levels of automation, and dynamic functional allocation that combines the human and machine generation is preferred.

Keywords   Situation Awareness, human-machine interface, SA-oriented Design, Level of Automation

1   Introduction

In complex and dynamic environments, people often have trouble in arriving at effective performance because he must process the vast amount of data around them to know what's on and what will come, i. e. he should have good situation awareness (SA).

The most popular definition is proposed by Dr. Endsley: the perception of the elements in the environment within a volume of time and space, the comprehension of their meaning and the projection of their status in the near future [1, 2]. She divides the process of SA into 3 levels, Level 1 are perceiving critical factors in the environment, Level 2 is understanding what those factors mean, particularly when integrated together in relation to the operator's goals, and at the highest level, an understanding of what will happen with the system in the near future. Endsley describes a theoretical framework model of SA as Fig. 1.

Since SA lies at the heart of all human decision making and performance, a large portion of errors attributed to human operators actually accrue from errors in

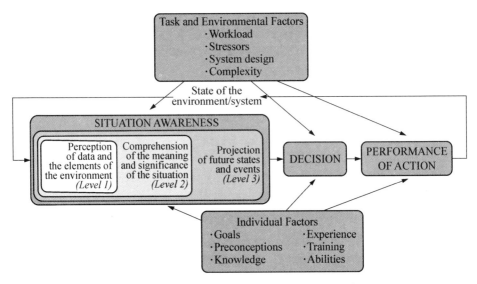

Fig. 1 SA framework model proposed by Endsley

their situation awareness. So we will study the relationship between SA and human errors, and find helpful cues in human-machine interface design of aircraft.

2 Decision-making and SA Error

The failures of human decision-making are frequently cited in investigations of error in a wide variety of systems. In aviation mishaps, failures in decision-making are cited as a causal factor in approximately 51.6% of all fatal accidents and 35.1% of non-fatal accidents, of the 80%–85% of accidents which are generally attributed to human error. While some of these incidents may represent failures in actual decision-making (action selection), a high percentage are actually errors in situation awareness [3]. The accident data of Civil Aviation of China from 1999 to 2008 led to a similar conclusion and more than 75% (See Fig. 2) accidents are related to human error [4].

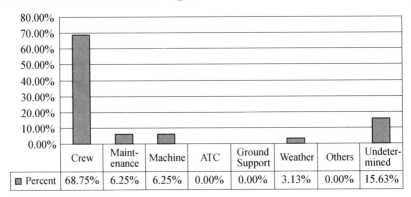

	Crew	Maint-enance	Machine	ATC	Ground Support	Weather	Others	Undeter-mined
▣ Percent	68.75%	6.25%	6.25%	0.00%	0.00%	3.13%	0.00%	15.63%

Fig. 2 Accident Factors of Civil Aviation of China from 1999 to 2008

According to the 3 levels of the SA definition, SA errors are also described in 3 levels. At the most basic level, people may fail to perceive the situation correctly. The reason may be: ① data not available; ② hard to discriminate or detect data; ③ failure to monitor or observe data; ④ misperception of data; ⑤ memory loss. The second level of SA level is improper integration or comprehension of information. The reason may be: ① lack of or incomplete mental model; ② use of incorrect mental model; ③ over-reliance on default values; ④ other. The highest level of SA error is incorrect projection of future actions of the system. The reason may be: ① lack or incomplete mental model; ② over-projection of current trends; ③ other. According to D. G. John's investigation, 72% SA errors belongs to level 1, and failure to monitor or observe data is the major one of all, which holds 35% of total SA errors.

3   SA-oriented Design

By creating designs that enhance an operator's awareness of what is happening in a given situation, decision making and performance can improve dramatically. The SA-oriented Design process provides a means for that purpose, which including a determination of SA requirements, design principles for SA enhancement, and measurement of SA in design evaluation [5].

3.1   SA Requirements Analysis

Before summarizing the SA-oriented Design principles, we must know what aspects of the situation are important for a particular operator's SA. A form of cognitive task analysis called a goal-directed task analysis is used to analysis. As shown in Tab. 1.

Tab. 1   Format of Goal-Directed Task Analysis

Goal
  —Sub-goal
    —Decision
      —Projection (SA Level 3)
      —Comprehension (SA Level 2)
      —Data (SA Level 1)

In such analysis, first task is to identify the major goals, then the major sub-goals necessary for meeting each of these goals, and then the major decisions for sub-goal, at last the SA factors.

3.2   Common Design Principles for SA Enhancement

After SA requirement analysis, we know the important factors of and should put more emphasis on them in human-machine interface design. Endsley and John summarized some principles in SA-oriented Design. Examples including [6]:

(1) Direct presentation of higher level SA needs (comprehension and projection) is recommended, rather than supplying only low level data that operators must integrate and interpret manually.

(2) Goal-oriented information displays should be provided, organized so that the information needed for a particular goal is co-located and directly answers the major decisions associated with the goal.

(3) Support for global SA is critical, providing an overview of the situation across the operator's goals at all times and enabling efficient and timely goal switching and projection.

(4) Critical cues related to key features of schemata need to be determined and made salient in the interface design.

(5) Extraneous information not related to SA needs should be removed.

(6) Support for parallel processing, such as multi-modal displays should be provided in data rich environments.

3.3 Measurement of SA

Measurement of SA provides a measure of operator's ability to dynamically integrate multiple pieces of information into a coherent picture under operational challenges. It not only will check the result of SA-oriented Design, but also can feedback to enhance the design. The different method is adopted according to different stages, as shown in Fig. 3 [7, 8].

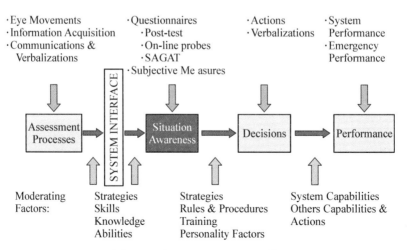

Fig. 3　different SA measurements in different stage

4   Level of Automation

Nowadays, many engineers inclined to design the human-machine interface to the most possibility of technology. Yet too much automation can push the user out-of-the-loop, causing them to lose SA in regards to the status of the elements under systems control [9, 10]. With the discussion of part 2, we know that failure to monitor or observe data is the major one of all SA errors, so we will cast more light on the function allocation between human and machine.

In terms of performance, human operators benefit most from automation of the implementation portion of the task, but only under normal operating conditions; in contrast, removal of the operator from task implementation is detrimental to performance recovery if the automated system fails [11].

T. B. Sheridan divided Automation into 10 levels, Endsley and Kaber improved the taxonomy on his basis. As shown in Tab. 2 [12, 13].

Tab. 2   Endsley and Kaber's LOA Taxonomy

LEVEL OF AUTOMATION	FUNCTIONS			
	MONITORING	GENERATING	SELECTING	IMPLEMENTING
1. Manual Control	Human	Human	Human	Human
2. Action Support	Human/Computer	Human	Human	Human/Computer
3. Batch Processing	Human/Computer	Human	Human	Computer
4. Shared Control	Human/Computer	Human/Computer	Human	Human/Computer
5. Decision Support	Human/Computer	Human/Computer	Human	Computer
6. Blended Decision Making	Human/Computer	Human/Computer	Human/Computer	Computer
7. Rigid System	Human/Computer	Computer	Human	Computer
8. Automated Decision Making	Human/Computer	Human/Computer	Computer	Computer
9. Supervisory Control	Human/Computer	Computer	Computer	Computer
10. Full Automation	Computer	Computer	Computer	Computer

According to Endsley and Kaber's research, the level of automation has a significant impact on automated system performance. Specifically, the level of automation that combine human generation of options with computer implementation (such as level 4, level 5, level 6 and level 8) produce superior overall performance during normal operations, as compared to purely manual control (level 7, level 9 and level 10) and to higher level of automation (level 1, level 2, level 3 and level 4) involving computer generation of options.

5　Conclusion

From discussion above, we can see that SA error accounts a great proportion of all the reasons of aviation accidents. To improve the decision making and performance, operators should have better awareness of the situation, thus SA-oriented Design is proposed. Besides common design principles, the level of automation is a key problem disturbs the human-machine interface designers. In this paper dynamic functional allocation that combines the human and machine generation is preferred.

Acknowledgement

This research is supported by the National Basic Research Program of China (No. 2010CB734105) and the Doctor Degree Program' Building Fun of Civil Aviation University of China.

References

[1]　Endsley M R. Theoretical underpinnings of situation awareness: a critical review, in Endsley, M. R., Garland, D. J. (Eds), Situation Awareness Analysis and Measurement [M], Lawrence Erlbaum Associates, Mahwah, NJ, 2000.

[2]　Endsley, Mica R. (1988): Situation Awareness in Aircraft Systems [C]. In: Proceedings of the Human Factors Society 32nd Annual Meeting 1988,207 – 230.

[3]　Jones, D. G., and Endsley, M. R. Sources of situation awareness errors in aviation [J]. Aviation, Space and Environmental Medicine, 1996,67(6):507 – 512.

[4]　Safety report of Civil Aviation of China [M], 2008.

[5]　Mica R. Endsley, Cheryl A. Bolstad, Debra G. Jones, et al. Situation awareness oriented design: from user's cognitive requirements to creating effective supporting technologies [C]. Human Factors and Ergonomics 47th Annual Meeting, 2003,47(3):268 – 272.

[6]　M. J. Adams, Y. J. Tenney and R. W. Pew, Situation Awareness and the Cognitive Management of Complex Systems [J]. Human Factors, 1995,37(1):85 – 104.

[7]　Kaemf, G. L., Klein, G., Thordsen, M. L. and Wolf, S. Decision making in complex naval command-and-control environments [J]. Human Factors, 1996,38(2):220 – 231.

[8]　Fracker, M. L. A theory of situation assessment: Implications for measuring situation awareness [C]. In proceeding of the human factors society 32nd annual meeting. Santa Monica. 1988.

[9]　B. M. Muir. Trust in Automation: Part Ⅱ. Experimental Studies of Trust and Human Intervention in a Process Control Simulation [J], Ergonomics,1996,39,3:429 – 460.

[10]　Endsley, M. R., Bolte, B., and Jones, D. G. Designing for situation awareness: An approach to human-centered design [M]. London: Taylor & Francis. 2003.

[11]　M. R. Endsley; R. Mogford. Level of automation effects on performance, situation

awareness and workload in a dynamic control task [J]. Ergonomics, 1999,42,3:462 - 492.

[12]   Endsley, M. R. (1995). A taxonomy of situation awareness errors [C]. in Fuller, R. , Johnston, N. , McDonald, N. (Eds), Human Factors in Aviation Operations, Avebury Aviation, Ashgate Publishing, Aldershot.

[13]   T. B. Sheridan. Human supervisory control. Handbook of Systems Engineering and Management [M], Chap. 16,1999.

Enhance Team Situation Awareness by Sharing Information to Avoid Human Errors in Aviation

Jin Huibin, Sun Ruishan, Kong Xiangfen

Department of Industry Engineer, Civil Aviation University of China, Tianjin 300300

Abstract　Firstly, this paper discussed the relationship between team situation awareness (SA), and information sharing, and proposed the opinion of enhancing team SA by information sharing to avoid human errors in aviation; Then compared the SA requirement of pilot and ATC; At last, discussed the necessary and effect of information sharing between pilot and controller, weather and traffic in detailed, and suggested human-centered and goal-directed design to maximize the advantages of information sharing and minimize the disadvantages.

Keywords　information sharing, team SA, human error, human-centered

1　Introduction

Human error is defined as inappropriate human behavior that lowers levels of system effectiveness or safety, which may or may not result in an accident or injury [1].

The accident data of civil aviation of China from 1999 to 2008 led to a similar conclusion and more than 75% accidents are related to human error, as shown in Fig. 1 [2].

In the process of aviation, most actions occur in teams or crews, where individuals interact dynamically, interdependently, and adaptively toward a common and valued goal/objective/mission. Each one in the team have been assigned specific roles or functions to perform and have a limited life span of membership [3, 4].

In the team, each crew member has a sub-goal pertinent to his/her specific role that feeds into the overall team goal.

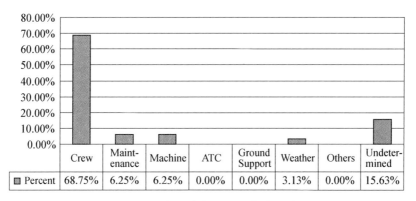

	Crew	Maint-enance	Machine	ATC	Ground Support	Weather	Others	Undeter-mined
Percent	68.75%	6.25%	6.25%	0.00%	0.00%	3.13%	0.00%	15.63%

Fig. 1   Accident Factors of Civil Aviation of China from 1999 to 2008

Associated with each crew member's sub-goal are a set of situation awareness
(SA) elements about which he/she is concerned. As the members of a team are
essentially interdependent in meeting the overall team goal, some overlap between
each member's sub-goal and their SA requirements will be present. It is this subset
of information that constitutes much of team coordination [5].

2   Team SA for Flight

2.1   Pilot Communication Model

From the view of pilot, a safety flight depends on a effectively coordination
with ATC, airline, steward, vice pilot and other ones, as shown in Fig. 2.

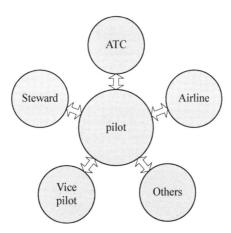

Fig. 2   Communication Model of Pilot

Each one in the team has his/her own SA requirement for his/her responsibilities.
In 5/2009, an Air Force sergeant traveling on a United Airlines nonstop flight

between Chicago and Tokyo noticed a fuel leak on the Boeing 747 information sharing is critical in enhancing the team SA. The pilot's communication model with information shared is shown in Fig. 3.

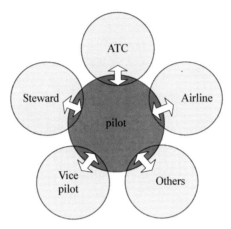

Fig. 3　Information Shared Communication Model of Pilot

During in-flight operations, the most important and frequently communication is the one between pilot and ATC, and this paper will cast more light on this.

2.2　Pilot/ATC SA Requirement

In order to understand how pilots and controllers might act on share information, it was necessary first to identify their roles, their motives and their informational needs. A goal-directed task analysis was performed for commercial airline pilots and ATC by Endsley and each task analysis constructed a comprehensive goal hierarchy from which the specific situation awareness information requirements were derived [6, 7]. Here we take the sub-goal about perturbation and flight path as an example to analysis, as shown in Tab. 1.

Tab. 1　Goal-directed task analysis of pilot/ATC

Agent	Main goal	Sub-goal	Decision
Pilot	Assure flight safety	Handle perturbations	Minimize impact of hazardous weather
			Respond to emergencies
			Minimize impact of abnormal situation
			Minimize impact of abnormal ATC situations
		Select best flight path	Manage current flight plan
			Develop alternate routes
			Determine best option

(continued)

Agent	Main goal	Sub-goal	Decision
ATC	Assure flight safety	Handle perturbations	Minimize impact of hazardous weather
			Respond to emergencies
			Resolve non-conformance
			Handle special operations
		Manage traffic flows	Manage arrival flows
			Manage departure flows
			Develop alternate routes

It is indicated that pilot and ATC have parallel main goals of "assure flight safety", and there is considerable overlap between their sub-goals and decisions. From the comparison of SA requirement, we can identify areas of common or competing interest between pilot and ATC. Information sharing helps them avoid misunderstanding and have better team SA to achieve the common interest.

2.3 Team SA Device

The process used for achieving shared SA across a team can take place through direct communication, shared displays or a shared environment.

Direct communication includes a simple verbal exchange either directly or transmitted by radio or phone, or it may rely on non-verbal communications such as finger pointing or facial expression.

Shared display means different crew members may directly view much of the information through displays that are available to them, including visual displays, audio displays and displays that use other senses (e. g. tactile device).

Environment is another important way to achieve team SA. When sharing the same environment, team members receive common information through cues around.

3 Information Sharing of Pilot and Controller

The interaction between Pilot and ATC is focused on the management of the specific flight for which they have common responsibility. The principal interaction element is the assigned clearance which constitutes the contractual agreement between the Pilot and Controller for the airspace, runway or airport surface resources assigned to the flight. The possibility of human errors increased when the clearance must be amended due to some conflict (e. g. weather, traffic, airspace), and need for shared information increases accordingly. The most important factors in pilot/ATC information sharing are weather and traffic [8].

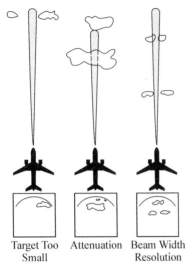

Target Too   Attenuation   Beam Width
Small   Resolution

Fig. 4　Sketch Map of Airborne
Weather Radar

3.1　Shared Weather Information

Weather is a primary causal factor for Controller/ATC interactions. In route operation, convective weather, turbulence, winds and icing are the major weather factors which must be considered. In terminal and surface operations visibility, ceiling, surface winds and braking action are also key weather factors. Bad weather can make planned trajectories unacceptable for both safety and ride quality reasons [9].

With airborne weather radar, pilot can visually observe the en route weather, as shown in Fig. 4 [10]. Since ATC have relative fewer information about the weather in particular area, he rely on pilot reports to build a mental representation of the spatial extent of weather within their sector.

3.2　Shared Traffic Information

Traffic is another key factor of pilot/ATC interaction. It is indicated by the Tab. 1 that pilot and ATC have related but differing goals with regard to traffic. The Pilot is observed to have an aircraft-centric view and is primarily concerned with traffic which will impact current or planned trajectories. Conversely the ATC has to takes the overall situation into consider [11, 12].

In the current radar controlled environment, Controllers have access to current state information (position, altitude, velocity) of all radar-observedaircraft as well as flight plan information. As shown in Fig. 5.

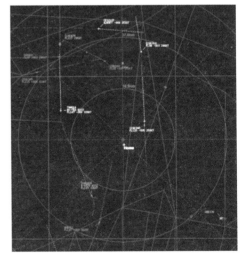

Fig. 5　Radar Screen of ATC

Due to the clear information superiority of the ATC, Pilot will generally defers to ATC's requests with regard to traffic. The traffic information shared will both increase the Pilot's team SA and distribute responsibility for traffic separation.

3.3 Effect on Team SA of Shared Information

Nowadays, the technology of datalink communication is well developed. Current datalink applications include pre-departure clearance delivery via the Tower Data Link System, global voice and data communications via satellite, and weather uplinks via Terminal Weather Information for Pilots. With the development of datalink, information which is not uniformly accessible today will be more easily to be shared between pilots, controllers and other users.

In general, information sharing is expected to have advantages for enhancing team SA for all the agents. From the experiment conducted by Todd [6], it is concluded that, with information shared, the controller's SA about weather increased from 40% to 93%, and pilot's SA about traffic increased from 0% to 60%.

However, the sharing of information may effect a less desirable outcome, one characterized by increased voice communications, increased workload, and increased contention between agents. To maximize the advantages of information sharing and minimize the disadvantages, human-centered designation idea should be introduced and only the necessary and abstracted information be shared in goal-directed way.

4 Conclusion

From the discussion above, we know that human error is a primary cause of accident of aviation and it is necessary to find way to avoid human errors. Information sharing makes each agent more informed and have better SA, which result in improved overall negotiations and decisions.

During in-flight operations, the most important and frequently communication is the one between pilot and ATC and the information sharing, whose primary factors of are weather and traffic, improves their SA remarkably. Of course, information sharing also has side-effect, such as increased workload, and human-centered and goal-directed designation should be introduced to overcome such problems.

Acknowledgements

This research is supported by the National Basic Research Program of China (No. 2010CB734105) and the Doctor Degree Program's Building Fund of Civil Aviation University of China.

References

[1] Wickens, Christopher D. , Gordon, et al. An Introduction to Human Factors Engineering [M]. Addison-Wesley Educational Publishers Inc. , New York. 1998.

[2] Safety report of Civil Aviation of China of 2008 [R].

[3] Endsley, M. R. Design and evaluation for situation awareness enhancement [C]. In Proceedings of the Human Factors Society 32nd Annual Meeting. Santa Monica, CA: Human Factors Society. 1998.

[4] Salas, E. , Dickinson, T. L. , Converse, S. , & Tannenbaum, S. I. Toward an understanding of team performance and training [M]. In R. W. Swezey & E. Salas(Eds.), Teams: their training and performance. Norwood, NJ: Ablex. 1992.

[5] Mica R. Endsley, William M. Jones. Situation awareness information dominance & information warfare [R]. Tech Report 97 – 01,1997.

[6] M. R. Endsley, T. C. Farley, W. M. Jones, et al. Situation Awareness Information Requirements for Commercial Airline Pilots [R], MIT International Center for Air Transportation Report ICAT – 98 – 1, Cambridge, MA. 1998.

[7] M. R. Endsley & M. D. Rodgers. Situation Awareness Information Requirements for En Route Air Traffic Control (DOT/FAA/AM – 94/27) [S], Federal Aviation Administration Office of Aviation Medicine, Washington, D. C. 1994.

[8] Todd C. Farley, R. John Hansman, Mica R. Endsley. The Effect of Shared Information on Pilot/Controller Situation Awareness and Re-Route Negotiation.

[9] R. John Hansman and Hayley J. Davison. The Effect of Shared Information on Pilot/ Controller and Controller/Controller Interactions [C]. 3rd USA/Europe Air Traffic Management R&D Seminar Napoli, 2000.

[10] John Dutcher. Applying Human Factors and Systems Theory to Weather: Weather Risk Management [C]. International Seminar on Aviation Human Factors . Tianjin, China, 2009.

[11] Farley, T. , and Hansman, R. J. , An Experimental Study of the Effect of Shared Information on Pilot/Controller Re-Route Negotiation [R], MIT International Center for Air Transportation Report, ICAT – 99 – 1, January 1999.

[12] Hansman, R. J. , Endsley, M. , Farley, T. , Vigeant-Langlois, L. , and Amonlirdviman, K. , The Effect of Shared Information on Pilot/Controller Situation Awareness and Re-Route Negotiation [C], FAA/Eurocontrol 2nd International Air Traffic Management R&D Seminar (ATM 98), Orlando, FL, December 1998.

第二篇　建模仿真

飞行员信息处理时间计算模型

张晓燕，薛红军

西北工业大学航空学院，西安 710072

摘要 飞行员在驾驶舱中的特性是驾驶舱设计达到人机有效匹配的基础。文章建立了飞行员信息处理模型对执行任务过程中飞行员的特性进行研究。飞行员信息处理模型以信息加工阶段理论为基础，将飞行员信息处理过程简化为感觉、知觉和执行三个阶段，每个阶段基于飞行员的职业特殊性作了一定的假设。通过对中央控制面板信息处理过程的实验研究对飞行员信息处理模型进行了正确性验证，结果表明模型能够对飞行员信息处理过程进行预测。飞行员信息处理模型对驾驶舱设计、飞行安全评估以及操作程序的合理性验证等具有指导意义。

关键词 飞行员模型，信息处理模型，视觉搜索，Fitts 定律，最佳搜索策略，人机匹配

Modeling of Pilot Information Processing time

Zhang Xiaoyan，Xue Hongjun

School of Aeronautics，Northwestern Polytechnical University，Xi'an 710072

Abstract Perfect match of man-machine in the cockpit can be arrived by the research of pilot performance. The paper built a pilot information process model to study the pilot performance. The model built based on the information process phased theory simplified the whole process as perceive, cognition and execution. Each phase has its own hypothesis based on the pilots' specified job. The verification of the model is achieved by the experiment of pilot dealing with information on the central control panel, and the results suggested that the model can predict the information proceed process by pilot. The pilot information process model can instruct cockpit design, flight safety evaluation and rationality evaluation of pilot operation procedures.

Keywords　pilot model，information process model，visual search，Fitts'Law，optimized Search strategy，man-machine match

　　现有的航空事故有70%是由人为失误引起的,而人为失误中超过2/3是飞行员失误,失误的根源在于人机匹配不合理。为了达到高效的人机匹配性能,需要对飞行员在驾驶舱中发挥的定性和定量的"性能"进行研究。飞行员的"性能"主要是飞行员在驾驶舱各种环境中的信息处理能力,是认知能力和操纵能力的综合。国内外学者在人的认知特性研究方面建立了EMMA(Eye Movement and Movement of Attention)模型[1-2],ACT-R(Adaptive Control of Thought-Rational)模型[3-5]等;在操纵规律研究方面建立了Fitts定律[6-8],控制模型[9-10]等;在人的失误研究方面建立了各种层次的人为失误分类模型,试图从组织管理等方面找出诱发人为失误的原因[11]。飞行员特性的研究主要是利用人的特性研究成果结合飞行员的职业特殊性进行了一些特定的研究,NASA组织了五个团队针对飞行员进近和着陆阶段的认知特性以及飞行员认知失误开展研究[12-13];也有学者针对工作负荷、情景感知等可能诱发飞行员失误的因素进行研究[14]。目前对飞行员能力研究工作或是针对飞行员信息处理的某个阶段如认知模型,或是将飞行员看作一个笼统的响应函数如与飞行品质相关的控制模型,并没有全面考虑飞行员的智能性、随机性和自适应等特性,也没有针对飞行员信息处理过程的完整模型,因此尽管这些研究成果具有较好的解决特定问题的能力,但不具有很好的适应性。本文针对飞行员信息处理的整个过程,结合信息加工阶段理论建立了完整的飞行员信息处理模型,模型可以对驾驶舱设计、飞行安全评估以及操作程序的合理性验证等进行指导。

1　飞行员信息处理模型

　　依据人的信息加工阶段模型(见图1[15])将飞行员信息处理过程(见图2)简化为三步,即感觉、知觉以及执行。感觉为飞行员将注意力转移到处理对象上,主要参与系统为工作记忆;知觉是飞行员对处理对象的识别、判断和定位,主要参与系统为长

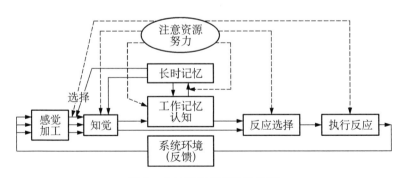

图1　人的信息加工阶段模型

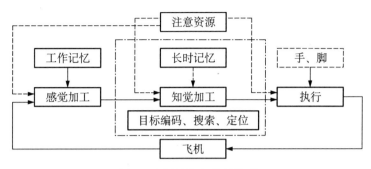

图 2 飞行员信息处理模型

时记忆系统;执行是对处理对象的最终反应,主要是由手、脚、口等器官执行;注意它们参与整个信息处理过程。

1.1 模型假设

针对飞行员的信息处理过程对模型做以下假设:

(1)飞行员通过视觉获取信息。飞行员执行任务时80%以上的信息获取依赖于视觉。

(2)飞行员的搜索策略为"最佳搜索策略",如图3所示。飞行自左至右搜索至中点位置,再由中点位置自上而下搜索至中心位置,再由中心位置向四边搜索。"最佳搜索策略"假设飞行员认为目标出现在中心或中心周围的几率最大。

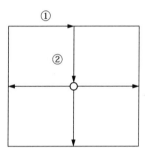

(3)信息在飞行员长时记忆中以最小特征单元存储,如颜色、形状等。

图 3 飞行员视觉搜索策略

(4)飞行员的目标编码与目标搜索定位是并行的,即飞行员在看到目标之后在长时记忆中对目标进行特征匹配,同时在视野范围内搜索定位目标。

(5)飞行员对控制面板的操纵为触点运动,运动规律服从 Fitts 定律。此模型考虑飞行员对面板的点击输入,暂时不考虑旋转、推拉等动作输入。

1.2 飞行员信息处理模型

飞行员的信息处理过程如图2所示。整个信息处理过程为:注意转移,编码目标,搜索定位目标,执行。飞行员信息处理过程不仅与信息的自身特性如目标大小、目标距离、目标频率等相关,而且与人自身的能力特性如人的视觉搜索策略、操纵类型等相关。作者综合两者的影响建立飞行员信息处理模型如下:

$$T = T_1 + T_2 + T_3 + T_4 \tag{1}$$

式中:T 为信息处理时间;

T_1 为注意力转移到现有目标上的时间,采用普遍值 85 ms[1];

T_2 为编码目标时间,与目标在飞行员长时记忆里出现的频率以及与飞行员视网膜中央窝的距离有关。由下式表示:

$$T_2 = C(-\log_2 f_i)e^{d_i} \tag{2}$$

式中:C 为常数,取 0.01;f_i 为目标 i 的出现频率;d_i 为目标 i 距离飞行员中央窝的距离,以视角表示;

T_3 为目标的搜索定位时间,搜索到面板中心的时间与目标的行数 m 和列数 n 有关,其中的 63.2 为目标搜索时的扫视单个目标的时间(ms)。

$$T_3 = 63.2 \times ([(m+1)/2] - 1 + [(n+1)/2] - 1) \tag{3}$$

式中:"$[k]$"表示对 k 取最大整数;

T_4 为对目标的执行时间:

$$T_4 = a + b\log_2(A/W + 1) \tag{4}$$

式(4)中:a, b 为常数,根据作者前期对触点运动规律的研究,取 $a = 95.1, b = 90.9$;A 为目标距离手部初始位置的距离;W 为目标的有效宽度。

2 实验验证

通过设计中央控制面板的信息输入实验对飞行员信息处理模型的准确性进行验证。

2.1 实验设计

(1) 被试:10 人,右手为优势手。

(2) 仪器:计算机,触摸屏。

(3) 实验变量水平:目标数目(1, 9, 12, 25, 30, 36, 42)×变量位置(中心位置和四个随机位置),按钮大小为边长 10 mm 的正方形。每种水平实验 20 次,实验次数为 $20 \times 5 \times 7 = 700$ 次。

(4) 每种变量水平按钮的排列依次为:1×1, 3×3, 4×3, 5×5, 6×5, 6×6, 7×6($i \times j$ 表示有 i 行,每行 j 个元素)。虚拟面板的尺寸为 10 cm×10 cm 的虚拟正方形,面板中心位置不变,其他目标围绕中心平均分布。

2.2 实验结果与分析

选择中心位置的目标作为验证对象,中心位置距离被试手部的距离恒为 15 cm。图 4 为不同按钮密度面板目标的反应时间和失误率。

由实验结果可知,随着控制面板上目标密度的增加,信息处理时间逐渐增加,增加比较平缓。被试的失误率与面板上按钮的密度并不是简单的递增或递减的关系,在目标数量为 9 的时候达到最大值。在目标数量为 9 时,面板上的按钮矩阵首次由单行增加到 3 行 3 列,被试的信息处理模式并没有及时转换,信息处理数量增加,视觉搜索模式首次由自左至右转换为两个过程:自左至右和自上而下,认为失误主要

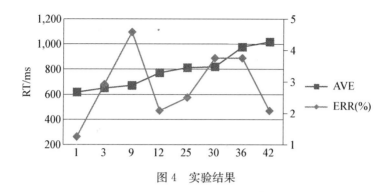

图 4　实验结果

出现在对目标的定位阶段；按钮矩阵由 9 个转换到 12 个时，由于数量增加较少，并且增加在行数上，因此被试的搜索定位过程几乎没有受到影响，对中心目标的编码数量仅仅增加了一个，因此失误率出现突降；在按钮数量由 12 逐渐增加到 36 个时，失误率逐渐上升，在按钮数量为 42 个时，失误率降低。

　　由实验结果可知，随着按钮密度的增加，信息处理时间增加，在按钮数量达到 42 个时信息处理时间超过 1 s，不利于飞行员应急情况下的操纵；失误率随着控制面板上按钮密度的增加并不递增，而是与各种密度情况下按钮在面板上的排布相关，按钮布置矩阵符合人的认知规律，失误率会降低。

2.3　计算模型与实验结果对比分析

　　利用飞行员信息处理模型对目标的处理进程进行计算。

　　(1) 注意力转移时间为 85 ms。

　　(2) 注意力转移到目标之后，飞行员开始对目标进行编码，同时对目标进行搜索定位。由于飞行员控制面板上的按钮数量为大小写字母 52 个，数字 10 个，因此需要飞行员的输入数据共计 62 个，每个数据在飞行员陈述性记忆中出现的频率均一致，因此取 $f_i = 1/62$；实验中要求被试距离显示屏 40 cm（此高度与飞行员眼位点到控制面板的距离基本相等），因此 $d_i = 1.43°$，综上目标的编码时间 $T_2 = 248.8$ ms。

　　(3) 目标的搜索定位时间与面板内目标数量有关，目标数量 $N \in [1, 25]$ 时，目标编码落后于目标定位搜索过程，因此，取目标的编码时间；当 $N > 25$ 时，目标编码过程先于目标定位搜索过程，因此，取定位搜索时间。

　　(4) 执行过程服从 Fitts 定律，如式（4）所示，在反应的目标位于控制面板的中心，$A = 15$ cm，$W = 1$ cm，因此可知目标的执行时间为 458.7 ms。

　　综上所述，可得

$$T = \begin{cases} 85 + 458.7 + 248.8, & N \in [1, 25] \\ 85 + 458.7 + 63.2 \times \left(\left[\dfrac{m+1}{2} \right] - 1 + \left[\dfrac{n+1}{2} \right] - 1 \right), & N > 25 \end{cases}$$

采用信息处理模型对信息处理过程的仿真结果与实验结果的对比如图 5 所示。

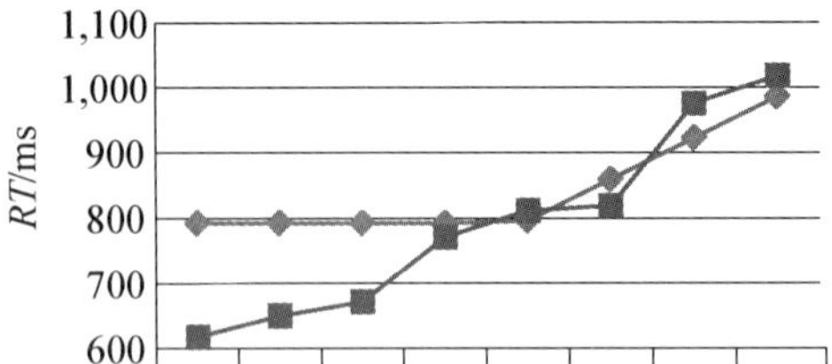

	1	3	9	12	25	30	36	42
◆ CAL	793	793	793	793	797	860	923	986
■ AVE	618	650	671	772	812	819	976	1,018

图 5　计算结果与实验结果的对比
CAL—计算结果，AVE—实验结果

由对比数据可以看出，在面板按钮数量为 1，3，9 时数据差异较大。对目标的反应过程即为点击目标的过程，服从 Fitts 定律；因此差异出现在认知过程的假设上，若假定在按钮密度较小时，飞行员仅将目标与扫视元素的特征进行对比，并没有完成目标的编码过程，将模型修正后与实验数据进行对比如图 6 所示。

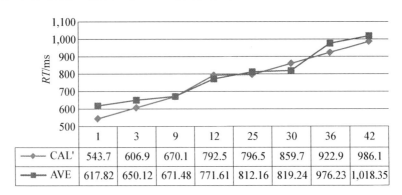

	1	3	9	12	25	30	36	42
◆ CAL'	543.7	606.9	670.1	792.5	796.5	859.7	922.9	986.1
■ AVE	617.82	650.12	671.48	771.61	812.16	819.24	976.23	1,018.35

图 6　修正后的模型与实验结果对比

修正后数据的吻合度明显好于修正之前，这表明在所要处理的信息量较少时，飞行员倾向于以特征的形式对信息进行对比搜索，若两个目标的特征完全一致，则认为是同一目标，也就完成了目标的定位过程，并不需要对目标进行详细的编码，完成语义上的认知。因此，可以认为目标的输入过程即为飞行员对目标特征的对比确认过程。

3　结论

飞行员信息处理模型以信息加工阶段理论为基础，结合飞行员处理信息的认知和操纵特点，能够准确地预测飞行员信息处理过程。结果分析表明，在驾驶舱界面设计时中央控制面板的密度不能超过 42 个/dm²，最好的操纵方向为飞行员右前方；

对操作程序的设计不仅要满足飞机的操作要求,而且也应该尽量满足飞行员的认知和操纵规律;飞行员诱发的飞行事故容易出现需要飞行员的认知和操作模式发生大的转变的地方,因此要加强这方面的培训。

飞行员是确保飞行安全的"最后一关",对飞行员与飞行界面之间的交互过程进行研究不仅对飞机驾驶舱交互界面设计如按钮的大小、方位、密度、控制方向等的设计具有指导意义,而且,通过对交互处理过程中飞行员特性的研究,对飞行员操作程序的合理性以及飞行安全评估也有指导意义。文中所建立的飞行员信息处理模型经过了一定的简化,后续研究中要对模型进一步的补充完善,研究飞行员在恶劣环境和应急条件下信息处理能力,并且实现飞行员认知和操纵模型的计算机仿真。

致谢

本项目获得国家重点基础研究发展计划 973(No. 2010CB734101)的资助,特表示感谢。

参考文献

[1] Slavucci D D. An Integrated Model of Eye Movements and Visual Encoding [J]. Cognitive Systems Research, 2001,1(4):201 - 220.

[2] Slavucci, D. D. A model of eye movements and visual attention [C]. Proceedings of the International conference on Cognitive Modeling. 2000.

[3] Anthony J. Hornof Cognitive modeling, visual search, and eye tracking. ONR Attention [C], Perception and Data Visualization Workshop George Mason University, 22 - 24 May, 2002.

[4] Byrne, M. D., & Anderson, J. R. Perception and action [M]. In J. R. Anderson & C. Lebiere (Eds.), The Atomic Components of Thought (pp. 167 - 200). Hillsdale, NJ: Lawrence Erlbaum Associates. 1998.

[5] Anderson, J. R. Daniel Bothell, Michael D. Byrne,. An Integrated theory of the mind [J]. Psychological Review 2004,3:1036 - 1060.

[6] Christian Schmiedl. An overview and evaluation of modern human interface devices January 27, 2010. http://academic. research. microsoft. com/Publication/13367837/.

[7] Jacob O. Wobbrock, Brad A. Myers, Htet Aung. The performance of hand posture in front- and back-of-device interaction for mobile computing [J]. International Journal of Human-computer Studies, 2008(66):857 - 875.

[8] Wobbrock, J. O., Cutrell, E., Harada, S., MacKenzie, I. S. An Error Model for Pointing Based on Fitts' Law [C]. Proc. CHI 2008, Florence, Italy, 2008, 1613 - 1622.

[9] 刘兴堂,李润玲. 人控制模型研究综述[J]. 系统仿真学报,1999,11,4:228 - 230.

[10] 于黎明,王占林,裘丽华. 人机控制与驾驶员模型研究[J]. 电光与控制,2001,1:2 - 8.

[11] Forester J, Bley D, Cooper S. E. Expert Approach for Performing ATHEANA Quantification [J]. Reliability Engineering and System Safty, 2004,83(2):207 - 220.

[12] David C. Foyle & Becky L. Hooey. Human performance modeling in aviation [M]. CRC

Press，2008.

[13] Foyle，D. C. ，Hooey，B. L. ，Byrne，M. D. ，et al. Human performance models of pilot behavior ［C］，Proceedings of the Human Factors and Ergonomics Society 49th Annual Meeting，Santa Monica：HFES. 2005.

[14] Leiden，K. ，Keller，J. W. ，&.French，J. W. Context of Human Error in Commercial Aviation ［R］. （Technical Report）. Boulder，CO：Micro Analysis and Design，Inc. 2001.

[15] 朱祖祥. 工程心理学教程［M］. 北京：人民教育出版社，2003.

驾驶舱飞行员认知行为一体化仿真建模

薛红军，庞俊锋，栾义春

西北工业大学航空学院，西安市 710072

摘要 运用ACT‐R认知理论对民机驾驶舱飞行员驾驶技能的获得、提取和运用的内在机制进行解释、建模，通过实验设计以及ACT‐R认知建模工具，对驾驶舱中飞行员告警信息感知、处理与决策等认知行为进行一体化仿真建模，并分别对实验操作、模拟仿真的过程和结果进行对比，充分说明ACT‐R理论适用于飞行员认知行为仿真建模。通过对飞行员高层认知过程的解释和仿真建模，推进驾驶舱人机工效一体化中飞行员仿真理论与建模研究，为驾驶舱的优化设计与评估提供深层次的支撑，为飞行员的培训及安全驾驶能力的提高提供指导。

关键词 飞行员，飞机驾驶，认知行为建模，ACT‐R

The Cockpit Pilot Cognitive Behavioral Integration Simulation Modeling

Xue Hongjue，Pang Juefeng，Lue Yichun

School of Aeronautics，Northwestern Polytechnical University，Shanxi Xi'an 710072

Abstract This paper provides some explanation and models for the inherent mechanism of pilot's driving skills acquisition，extraction and exertion on the base of the ACT‐R cognitive system. Through designing an experiment and application of ACT‐R cognitive modeling tool，the cognitive model of alarm information's perception，processing and decision-making is set up. Respectively，the process and results of the experiment，simulation were compared. It shows that the ACT‐R architecture applies the modeling of pilot. The pilot high-level cognition process explanation and the modeling provide help for the pilot simulation theory in the cockpit ergonomics；provide support for the cockpit optimized design；provide guidance for pilot training and driving ability enhancement.

Keywords pilot，aircraft driving，cognitive modeling，ACT‐R

1 引言

大量的调查实例和分析表明，在航空系统中，人的不安全行为导致的飞行事故占有很大的比例。在 20 世纪 90 年代，民用航空领域大约 90％的事故是人的因素所致。造成这种状况的原因：一方面是越来越多的高新技术运用到座舱设计中，导致人机匹配等问题[1]；另一方面飞行员本身的认知行为也成为影响飞行安全的重要因素，对后者的深入研究正日益引起航空工效学人员的关注。

1.1 认知理论

建立认知模型的技术通常称为认知建模，目的是为了从人的能力方面探索和研究人的思维机制，特别是人的信息处理机制。由于认知的复杂性，认知建模发展较为缓慢，目前有影响力、应用较多的认知体系架构主要有 ACT－R（Adaptive Control of Thought-Rational），SOAR（States，Operators And Reasoning），EPIC（Executive-Process Interactive Control）等，EPIC 可用于多任务执行，但没有学习强化和知识退化机制。SOAR，ACT－R 都可用于解决问题以及学习，但 SOAR 没有知识退化机制且其使用范围有限[2]。

ACT－R 是探讨人类认知过程工作机制的理论模型，揭示人类组织知识、产生智能行为的思维运动规律。ACT－R 属于符号主义处理架构的统一认知模型，可以近似地模拟人的全部认知行为，对人进行一体化仿真；ACT－R 已有大量实验信息可以直接被研究工作使用；该理论已用于很多领域，成功地解释了众多实践结果。基于以上的对比以及飞行员认知建模的需要，本文选择 ACT－R 认知体系作为飞行员认知研究的理论基础。

1.2 研究现状

在国内航空领域，飞行员认知研究主要集中在局部认知行为，如操作特性[3]、情景意识[4]、注意力分配[5]等方面，当前还没有飞行员认知行为一体化仿真建模的相关研究。

在国外航空领域，飞行员认知研究较为先进，与本文课题密切相关的是 NASA 的 HPM 项目。2003～2008 年，NASA 组织开展了人的能力建模项目（Human Performance Modeling Project，HPM），主要对复杂飞行任务中技术高超、训练有素的飞行员的认知和操纵能力进行仿真建模。研究人员对两个与飞行任务相关联的典型问题：机场地面滑行的操纵问题和综合视景系统进行研究，基于 ACT－R 认知体系构建了飞行员飞行进近的认知模型[6]。

2 ACT－R 认知架构

ACT－R 认知体系由 John. R. Anderson 于 1976 年提出，是系统阐述人脑如何进行信息加工活动的理论以及信息处理的整合模型，着重研究高级思维的控制过程，并不断吸收最新的认知心理学理论、研究成果、研究方法等对 ACT－R 体系和模型进行完善,整合和借鉴了其他较为先进认知行为的局部模型如 SOAR 的整个系统构架理念、EPIC 的感知输入与动作输出、HAM（Human Associative Memory）[7]的记忆模型、产生式系统表征的技能知识等。

　　ACT-R 主体系统的一般框架,由三个记忆部分组成:工作记忆、陈述性记忆和程序性记忆[8],如图 1 所示。

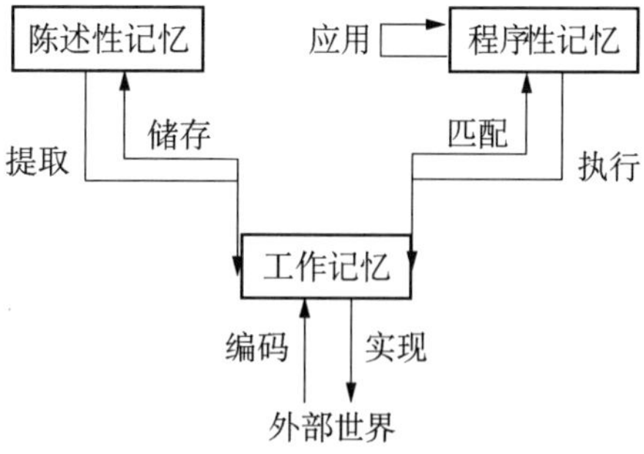

图 1　ACT-R 认知架构

　　ACT-R 理论作为飞行员认知行为研究的理论基础主要有以下三个特点:

　　1）针对性

　　ACT-R 属于符号主义处理架构的统一认知模型,可以近似地模拟飞行员的全部认知行为,对飞行员进行一体化仿真。

　　2）动态性

　　飞行员的飞行技能是一个成长的动态过程,学习假设（learning assumptions）理论基础使得 ACT-R 具有学习成长退化机制。

　　3）时效性

　　准确、快速是飞行技能的特点。ACT-R 程序性知识自动化表现在两个方面:一是速度;二是精确性。

3　ACT-R 认知架构下飞行员认知行为建模

　　一个飞行学员经过学习、训练、飞机操纵实践最后成为飞行专家,飞行员驾驶技能的成长是一个动态过程。飞行员驾驶技能是指顺利完成飞行任务的身心品质的合理组织方式,是心智技能和动作技能的有机结合。飞行员驾驶技能的成长分为驾驶技能获得、驾驶技能提取和运用两个阶段。

3.1　基于 ACT-R 的飞行员驾驶技能获得、提取和运用模型

3.1.1　飞行员驾驶技能获得模型

　　ACT-R 本质上是一种感觉理论,因为知识结构源于环境编码。飞行员运用一般性策略,如目标分解、逆向推理等从环境中编码知识组块,然后对与问题解决样例有关并需要进行变换的规则进行推导,重组大量的知识,并储存为长时记忆的认知资源[9]。

　　飞行员驾驶技能获得的三个阶段:

　　（1）陈述性阶段。在这一阶段,飞行员学习必要的理论知识或者基本概念、观摩教练的操作等获得有关飞机驾驶的陈述性知识,并以声明（statement）的形式储存在大脑的语义网络中,然后运用一般可行的程序来处理或理解这些知识。

　　飞行员在学习理论知识或者基本概念、观摩教练的操作等过程中,随着时间的

推移,对知识的理解越来越深,在 ACT‐R 认知体系中组块基本激活水平 B_i(base-level activation)的增长能够很好地模拟这一成长过程。基本激活水平随着学习次数的增长以及新知识与旧知识结合程度的提高而增长。因此,飞行员基本激活水平可用下式计算:

$$B_i = \ln\Big(\sum_{j=1}^{n} t_j^{-d} \Big) \tag{1}$$

式中:n 为组块的呈现次数;t_j 为从第 j 次呈现到当前的时间;d 为衰退参数。

(2)知识编译阶段。在此阶段,飞行员将声明形式表征的陈述性知识转化为以产生式规则表征的程序性知识,即操作技能。飞行员将分散的知识进行组合形成基本的驾驶技能,进而将基本的驾驶技能进行整合形成比较系统的驾驶技能,建立各种条件下问题情境与解决方法之间的联系,形成对应关系。

飞行员经过不断练习实践,对驾驶技能的运用会更加娴熟。在 ACT‐R 认知体系中,每一个产生式都与一个效用值相联系,产生式效用值(utilities)的增加能够很好地模拟驾驶技能的成长。每个效用值能够通过模型的反馈而成长,因此,飞行员的产生式效用值可用下式计算:

$$U_i(n) = U_i(n-1) + \alpha[R_i(n) - U_i(n-1)] \tag{2}$$

式中:$U_i(n-1)$、$U_i(n)$ 分别表示产生式经过第 $n-1$ 次、第 n 次运用之后的值;$R_i(n)$ 是它第 n 次运用获得的反馈;α 表示成长率。

(3)程序性阶段。当飞行员获得的各种驾驶知识与驾驶技能转化为解决问题的自动化过程后,提高认知技能激活速度的内在信息加工过程仍在进行。各种程序化、自动化过程不断优化,强化有效的问题解决策略和规则,进一步提高飞行员问题解决的认知加工速度。

飞行员在驾驶技能学习过程中会出现迁移现象,对于迁移,Anderson 等人提出了"共同要素理论"。基于"共同要素理论",我们认为飞行员两种技能之间发生迁移的条件是,它们之间必须共用相同的产生式规则,并且两种技能之间的迁移量,可以通过计算它们共用的产生式规则的数量来作出估计:①如果两种技能共用较多的产生式规则,它们之间将产生显著的迁移;②如果两种技能共用较少的产生式规则,即使它们共用相同的组块,它们之间也将产生很少的迁移或者没有迁移[10]。

3.1.2 飞行员驾驶技能的提取和运用模型

飞行员驾驶技能的提取和运用是飞行员身处的外部环境和飞行员内部认知加工的共同作用。飞行员根据特定目标要求或者特定情境,从大脑中提取知识进行一系列的操作来完成飞行任务,比如进行飞机降落时,他会从大脑中提取出有关飞机降落的知识进行操作。

当处在某种驾驶情境中,飞行员就要对环境信息进行筛选编码,把当前的情境与学习过的情境进行比较,从中选择出与当前情境最符的驾驶技能。基于 ACT‐R

认知体系结构,提出飞行员组块知识提取激活水平计算如下:

$$A_i = B_i + \sum_j W_j S_{ji} + \sum_l PM_{li} + \varepsilon \tag{3}$$

式中:B_i 是组块 i 的基本激活水平,它反映的是组块使用的崭新度和频率;W_j 是情境组块 j 的注意权重;S_{ji} 是从情境组块 j 与记忆组块 i 的关联强度;$\sum_l PM_{li}$ 表示局部匹配;ε 表示扰动。

飞行员所处的情境与过去某个学习过的情境相似时,有可能采取学习过的情境所对应的操作,也就是说由于相似性,飞行员很可能产生误判或者误操作,在 ACT-R 中激活水平计算中的局部匹配部分能够充分地体现这一点。

当然,飞行员驾驶技能的获得、提取和运用,是一个相互交织的过程,飞行员是伴随着"学中做、做中学"而不断成长。

3.2 基于 ACT-R 的飞行员认知行为一体化仿真建模

3.2.1 飞行员驾驶飞机的行动过程

飞行员驾驶飞机的行动过程是飞行员通过各种途径收集信息,并对信息进行加工处理做出判断决策,然后使用机上操纵装置,对飞机实施操纵控制以实现预定飞行目的的过程。飞机驾驶是人有目的的行动,按照认知心理学理论,飞机驾驶行动过程可分为感知发现、判断决策和实施操纵等三个阶段,如图 2 所示[11]。按照信息理论,感知发现是信息获取、储存的输入阶段,判断决策是信息的加工、处理阶段,实施操纵是输出阶段。

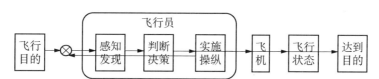

图 2 飞行员驾驶飞机行动过程阶段划分

在感知发现阶段,飞行员主要通过视觉、听觉、触觉和前庭功能等感觉器官,来感知发现飞行状态、收集信息。

在判断决策阶段,飞行员在感知发现的基础上,结合驾驶经验,运用一定的思维方式,经过分析做出判断,确定飞行目的和有利于达到飞行目的的操纵活动。

在实施操纵阶段,飞行员依据做出的操纵决策,通过手、脚等运动器官对飞机操纵装置实施具体的操作。

3.2.2 驾驶舱告警处理实验

飞行员驾驶飞机是感知发现、判断决策和实施操纵三个阶段或者三个子任务不间断地串联组合、连锁反应的过程。为了使实验及模型模拟具有代表性,任务材料的选取遵从以下三个原则:①任务材料必须包含飞行员驾驶飞机的三个阶段或者三个子任务;②任务材料必须便于认知建模;③任务材料必须简化抽象,使模型具有典型性,易于扩展到大部分具体的驾驶任务。本文以驾驶舱中出现频率最高的典型飞行任务"飞行员告警信息感知、处理与决策"为对象进行实验设计与建立仿真模型。

实验材料:驾驶舱主警告发出连续的告警声音,被试飞行员听到告警声音之后,查看显示组件主注意灯的颜色,如果被试飞行员发现主注意灯的颜色是红色,那么紧急作动进行按键操作消除告警;如果被试飞行员发现主注意灯的颜色是琥珀色,那么只需引起注意即可。

实验设备:驾驶舱仿真平台、运动捕捉系统、数据记录系统。

被试:飞行器设计专业研究生 10 名。

实验步骤:首先对被试的行为能力进行合理约束,然后让被试在驾驶舱仿真平台上进行实验操作,其行为序列通过仿真平台的运动跟踪系统进行摄像记录,其行为的详细数据通过数据记录系统进行记录。

3.2.3　模型建立

模型建立的关键是飞行员告警信息诊断和操作处理的数据库。基于 ACT - R 建模环境运用 lisp 语言进行程序编写来建立任务模型。图 3 所示为飞行员告警处理认知行为模型框架。图 4 所示为运用 ACT - R 建模工具建立模型。

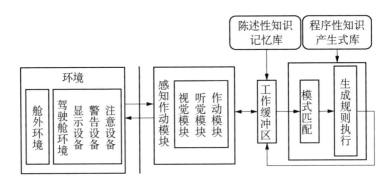

图 3　飞行员认知行为模型框架

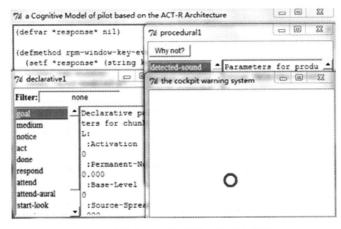

图 4　运用 ACT - R 建模工具建立模型

　　ACT-R建模工具中所包含参数的默认值为国外一般人群的实验数据,为了建立我国飞行员的仿真模型,建模中必须对视觉模块、听觉模块、作动模块等模块的参数进行设定。本文依据国家重点基础研究计划驾驶舱人机工效一体化中飞行员仿真理论与模型研究课题组所进行基础实验的信息数据来调整模型参数,从而使模型模拟顺利合理进行。课题组进行的基础实验包括:视觉实验、听力辨析实验、视听告警双任务干涉系列实验、信息激活强度关联实验、个体差异系列实验等。模型模拟结果如图5所示,调整模型参数 trace-detail 为 high 可得到更为详细的模拟结果。

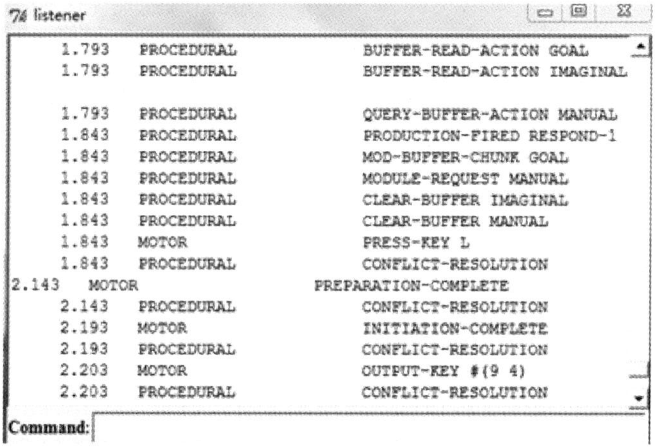

```
74 listener                                              ▭ ▢ ▨

        1.793    PROCEDURAL        BUFFER-READ-ACTION GOAL       ▲
        1.793    PROCEDURAL        BUFFER-READ-ACTION IMAGINAL

        1.793    PROCEDURAL        QUERY-BUFFER-ACTION MANUAL
        1.843    PROCEDURAL        PRODUCTION-FIRED RESPOND-1
        1.843    PROCEDURAL        MOD-BUFFER-CHUNK GOAL
        1.843    PROCEDURAL        MODULE-REQUEST MANUAL
        1.843    PROCEDURAL        CLEAR-BUFFER IMAGINAL
        1.843    PROCEDURAL        CLEAR-BUFFER MANUAL
        1.843    MOTOR             PRESS-KEY L
        1.843    PROCEDURAL        CONFLICT-RESOLUTION
2.143   MOTOR                      PREPARATION-COMPLETE
        2.143    PROCEDURAL        CONFLICT-RESOLUTION
        2.193    MOTOR             INITIATION-COMPLETE
        2.193    PROCEDURAL        CONFLICT-RESOLUTION
        2.203    MOTOR             OUTPUT-KEY #(9 4)
        2.203    PROCEDURAL        CONFLICT-RESOLUTION          ▼

Command:
```

图5　模型运行结果

3.2.4　对比分析

　　模型模拟可以确定飞行员单步认知行为的时刻,包括听到主警告的时刻、查看主注意灯的时刻、判断决策的时刻、按键操作的时刻。调整模型参数后可以得到更为详细的认知行为时间序列,如按键操作花费时间为 450 ms,其中 preparation-complete 为 300 ms, initiation-complete 为 50 ms, output-key 为 10 ms, finish-movement 为 150 ms。飞行员进行实验可以确定完成任务的总时间。飞行员操作和模型模拟单独进行,使用模拟结果对飞行员驾驶行为进行预测,以及通过模拟结果和飞行员的实际行为进行比较,来验证模型的有效性。

　　对比分析结果:

　　(1) 模型能够成功完成任务目标。由模型运行结果可知:出现琥珀色告警,模型能够引起注意;出现红色告警,模型能够进行按键操作以消除告警。

　　(2) 模型的单步操作和飞行员的单步行为相一致。飞行员的单步行为主要有听到主警告、查看主注意灯、判断决策、按键消除告警,从模型运行结果中抽取主要操作序列进行对比,可知两者一致。

　　(3) 被试飞行员平均花费时间为 2.022 s,模型模拟时间为 2.293 s。出现差异的主要原因是:①课题组进行基础实验有限,关键参数调整不完全;②实验简单,被试操作易达到自动化水平,时间花费减少。表1为被试操作与模型模拟对比。

表1　被试操作与模型模拟对比

类型	时间		动作
	被试	模型	
程序性知识	50 ms	50 ms	出现告警声音
听觉	—	385 ms	听到告警
程序性知识	—	485 ms	查看主注意灯
视觉	—	870 ms	红色
程序性知识	—	920 ms	判断
作动	2,022 ms	2,293 ms	按键操作

4　结论

　　基于 ACT-R 的飞行员认知行为建模,表明 ACT-R 认知体系适合飞行员认知行为仿真建模,有助于推动驾驶舱人机工效一体化中飞行员仿真理论与模型研究的发展,指导驾驶舱的人机工效设计与评估,预测飞行员的操纵失误率、评估飞行安全,验证飞行操作程序的合理性和正确性。

　　基于 ACT-R 的飞行员认知行为模型能够对飞行员高层认知和决策过程的工作机制进行解释,从而为探索飞行员组织知识、产生智能行为的思维运动规律提供了一种实现方法,有助于提高飞行员培训效率,降低飞行员培训成本,提高飞行员安全驾驶能力,降低随机失误的概率。

　　驾驶舱飞行员认知行为建模是一个庞大的系统工程,实验、理论、模型和工程应用综合在一起。若要对复杂飞行任务中的飞行员进行认知建模还需要做三方面的工作:①进行系统完整的基础实验,完善飞行员特有的信息数据以及模型准确性、有效性的验证实验;②扩展认知建模工具的操作模块;③建立飞机外部环境和驾驶舱控制的环境模型,环境模型可以为飞行员提供输入与输出反馈,从而形成回路,实现闭环仿真。

参考文献

[1]　郭小朝.飞机座舱显示——控制工效学研究近况[J].人类工效学,2001,7(4):34-37.

[2]　刘雁飞,吴朝晖.驾驶行为建模研究[D].浙江大学,2007.

[3]　王丽荣,庄达民,等.飞机拨动开关布置的人机工效分析[J].中国民航飞行学院学报,2004,15(2):3-6.

[4]　焦健,孙瑞山.飞行中情景意识探讨[J].中国民航飞行学院学报,2006,24:69-70.

[5]　柳忠起,袁修干,等.飞行员注意力分配的定量测量方法[J].北京航空航天大学学报,2006,32(5):518-520.

[6]　David C F, Becky L. Hooey. Human Performance Modeling in Aviation [M]. CRC Press/

Taylor & Francis Group，2008.

[7] Anderson J R，Bower G H. Human associative memory [M]. Washington：Winston and Sons，1973.

[8] Anderson J R. The architecture of cognition [M]. Cambridge，MA：Harvard University Press，1983.

[9] Anderson J R，Fincham J M，Douglass S. Practice and Retention：A Unifying Analysis [J]. Journal of Experimental Psychology. Learning，Memory，and Cognition，1999，(5)：1120 - 1136.

[10] Anderson J R，Schunn C D. Implications of the ACT - R learning theory：No Magic Bullets. In R. Glaser（Ed.），Advances in instructional psychology（Vol. 5）[M]. Mahwah，NJ：Erlbaum，2000.

[11] 丁邦昕. 飞机驾驶学[M]. 北京：蓝天出版社，2004.

LED 在民用飞机仪表板泛光照明中的应用

黄　瑜，林燕丹，姚　其，孙耀杰

复旦大学电光源研究所，先进照明技术教育部工程研究中心，上海 200433

摘要　民用飞机仪表板泛光照明作为一种应用场景，其照明系统设计必须满足特定的要求。LED 是世界上最先进的照明光源之一，当应用到仪表板泛光照明时，它在某些方面具有传统光源无法比拟的优势，同时也有一些仍待探讨的问题。本文讨论了 LED 作为仪表板泛光照明光源的应用，并尝试设计了一种大功率 LED 光学系统，实现了较好的照明效果。

关键词　民用飞机驾驶舱，仪表板泛光照明，LED 灯具

The Application of LEDs in Commercial Aircraft Instrument Panel Flood Lighting

Huang Yu，Lin Yandan，Yao Qi，Sun Yaojie

Institute for Electric Light Sources，Fudan University，Engineering Research Center of Advanced Lighting Technology，Ministry of Education，Shanghai 200433

Abstract　For the application of commercial aircraft instrument panel flood lighting，there are some specific requirements about the design of lighting system. The LED is one of the most advanced lighting sources in the world. When it is used in instrument panel flood lighting，the LED has some advantages over traditional light sources as well as problems that remain to be solved. In this paper，we have discussed the application and designed a high-power LED optical system that provides good illumination for the instrument panel.

Keywords　commercial aircraft cockpit，instrument panel flood lighting，LED fixture

1　引言

仪表板泛光照明是飞机驾驶舱照明的重要组成部分。飞机驾驶舱照明系统,主要是指通过对顶部板、仪表板、中央操纵台等区域的泛光照明设计以及飞行机组所需的局部照明设计,为飞行人员创造一个良好的视觉环境。仪表板泛光照明系统的合理设计涉及较多的实际工程问题,例如从人体工效学的角度要保证飞行员不易疲劳,而且保持较高的视觉绩效,从电气和机械的角度要做到符合驾驶舱应用要求等。其中照明光源的选用是一个比较突出的问题。在过去,传统光源主导了民用飞机仪表板泛光照明的应用。而 LED 作为世界上最先进的照明光源之一,其发展已经受到了各国的重视。一些厂商已经尝试把 LED 引入到飞机驾驶舱照明中。然而,关于 LED 在民用飞机仪表板泛光照明中的应用,还有很多值得研究的课题,包括其可行性如何、具体参数应该怎样选取等。图 1 所示为仪表板泛光照明布局。

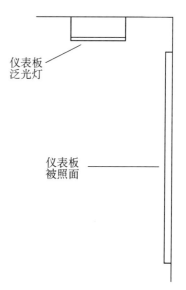

图 1　仪表板泛光照明布局

2　民用飞机仪表板泛光照明的指标和相关要求

总的来说,飞机驾驶舱照明系统的合理设计与布局要注重功能性与舒适性,旨在为飞行员创造一个良好舒适的视觉环境,对于飞行安全具有重要作用。各部分的灯具选择和布局设计必须满足特定的要求。针对仪表板泛光照明,主要看它的暗适应性、视觉效果、判读准确性[1]。相应地,必须根据实际情况对照明光色、亮度(由背光和泛光照明共同提供)、照度等指标提出具体的要求。涉及这些要求的标准可分为国外和国内标准两类。其中,国外的常用标准有美国机动车工程师学会标准(SAE)和美国军用标准(MIL),国内则为航空工业标准(HB)和国军标(GJB)。

纵观飞机发展的历史,人们选作仪表板泛光照明的光色主要有红光、白光、蓝白光和夜视绿[2]。究竟选用什么光色,主要取决于飞机的用途,同时还要考虑飞行速度、高度、驾驶员人机工效等方面的因素。目前主流的民用飞机普遍采用蓝白光(如空客公司产品)和白光体制(如波音公司产品)。而涉及亮度和照度的标准较多(如 SAE 的 ARP4103,AS7788 等),对具体指标的推荐也不尽相同[3—9]。对这些标准整理如表 1 所示。

表 1　各标准对仪表板泛光照明的要求

	国内标准		国外标准	
光色	无		ARP 1048	红、白
			ARP582	蓝白
照度/lx	HB6491-91	20~100	ARP 4103	最亮状态下为 538~1,614
亮度/(cd/m²)	无		ARP1048	17.1
			AS7788	3.5~10.3

3　LED 应用于飞机仪表板泛光照明的优势和问题

3.1　LED 应用于飞机仪表板泛光照明的优势

就驾驶舱照明应用而言,LED 较传统光源具有不少的优点。比如 LED 的功率较低,一般在 0.03~1 W,有利于飞机节能;LED 可以做到无红外线辐射,不会干扰某些机载设备的工作;LED 是固态光源,不像白炽灯那样有脆弱的灯丝,因此抗震性较强,适于较为恶劣的飞行条件。同时,由于工作稳定性高带来的飞机照明系统故障率的大幅降低,都使得飞机的维护成本减少。LED 在这些方面的卓越性能引起了相关机构的密切关注,比如美国联邦航空委员会(FAA)已经就 LED 在航空领域的应用问题开展了深入的讨论和研究。

3.2　LED 应用于飞机仪表板泛光照明的问题

应该说把 LED 应用于飞机驾驶舱还是一个新的话题。无论是新型的民航客机如 A380,B787 等使用 LED 作为驾驶舱照明光源,还是对原来使用传统光源的机型进行改造,在一定程度上也是一种尝试。把 LED 应用于仪表板泛光照明还需要进一步的研究。比如,如何根据驾驶员的实际需要来选择光源的亮度和色温;如何减少反光、防止各种眩光和视觉疲劳;目前驾驶舱照明主要表现为 LED、卤素灯和荧光灯共存[10],其中 LED 是使用直流电源(电流)控制,而卤素灯和荧光灯是电压控制(交流),所以在使用 LED 照明的时候也对照明电力控制系统提出了较高的要求。同时,LED 作为新型的照明光源,其本身的特性也有待提高。例如随着功率的增加,LED 灯具散热也会更困难,导致其寿命降低。

4　LED 仪表板泛光灯的应用实例

某机型采用了一款 LED 灯具为仪表板提供泛光照明。这款灯具的外观如图 2 所示。整体形状近似于长方体,长度、宽度和厚度分别约为 7,4 和 3 cm。它的光源为 48 颗小功率的白光 LED,灯具总体质量不超过 140 g。为配合飞机供电系统使用,这款灯具的工作电压为 28 V 直流电,工作电流约为 120 mA。为保证驾驶舱照明整体协调、舒适,同时能适用于不同的场景,设计者还在灯具上添加了调光功能。据悉该款灯具的标称寿命达到 10,000 小时以上。光学特性方面,灯具出光不对称,主要分布在法线单侧45°内(见图3),且该范围内最小光强为35 cd。此外,由于该款灯具形状普通、安装方便,也可以作为顶部板、断路器板泛光灯等使用。

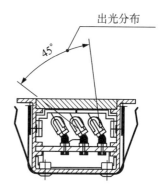

出光分布

45°

图2　某仪表板泛光灯外观　　　　　　图3　某仪表板泛光灯出光分布

5　LED仪表板泛光灯的设计

从图1中仪表板泛光照明的布局情况可以看到,仪表板泛光灯一般安装在仪表的侧上方,因此要求灯具朝向单侧出光。同时,由于灯具和仪表板被照面的侧向距离相对于仪表板的高度来说比较小,在光学设计上属于近场问题。这些因素都导致结构和光学设计成为仪表板泛光灯设计的难点。至于热学和电学设计,仪表板泛光灯与其他类型的灯具相比并没有太大的区别。

根据实际情况,取灯具和仪表板被照面的侧向距离为5 cm,仪表板的高度为15 cm,结合各标准对仪表板泛光照明的要求,可以列出仪表板泛光灯的光学设计目标如表2所示。其中光输出要求可实现从最暗到最亮状态的调节。

表2　仪表板泛光灯光学设计目标性能

项　目	规　格
光输出	最亮状态下照射在特定仪表板上的光通量为50 lm
灯具出光效率	70%以上
灯具利用系数	0.4以上
相关色温	3,000 K
显色指数	80以上

根据目标性能要求,本文选用某款功率为1 W的大功率白光LED两颗(其光效为65 lm@350 mA,配光为朗伯型),并提出一个可能的设计方案,其光强分布和照明效果分别如图4和图5所示。可见在最亮状态下,照射在仪表板上的光通量为53.6 lm,均匀度良好,各项指标均达到要求。

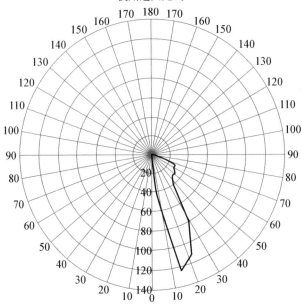

出光效率：0.80703，26291 条光线

最小值：0 cd，最大值：125.23 cd，总光通量：104.91 lm

图 4　光学设计方案光强分布

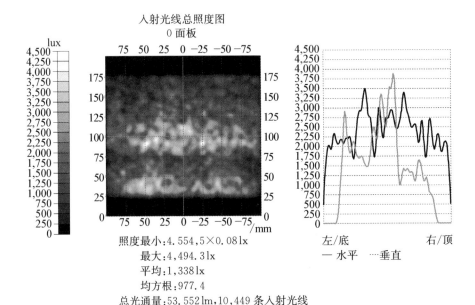

照度最小：4.554,5×0.08 lx

最大：4,494.3 lx

平均：1,338 lx

均方根：977.4

总光通量：53.552 lm，10,449 条入射光线

图 5　光学设计方案照明效果

6 结语

作为一种新兴光源,LED 正在逐步渗透到照明的每一个角落。对于 LED 在民用飞机仪表板泛光照明中的应用,还有很多实际问题值得研究和探索。本文从民用飞机仪表板泛光照明的基本要求出发,结合 LED 光源本身的特性,剖析了 LED 应用于民用飞机仪表板泛光照明的优势和问题。在实际应用中,光学设计是仪表板泛光照明的难点。本文设计了一种可能的方案,其光学性能较好地达到了仪表板泛光照明的要求。

参考文献

［1］ 陈方虎,岳红萍.飞机座舱导光板照明[J].电子机械工程.2004(4):12-14.

［2］ 蒋枫,汪慧丽.飞机驾驶舱照明的工程心理分析与工程设计实践[J].航空标准化.1980(5):1-3.

［3］ SAE ARP4103. Flight Deck Lighting for Commercial Transport Aircraft [S]. 2003.

［4］ SAE AS7788. Panels, Information, Integrally Illuminated [S].

［5］ SAE ARP1048 REV. A. Instrument and Cockpit Illumination for General Aviation Aircraft [S]. 2000.

［6］ SAE ARP582. Lighting, Integral, For Aircraft Instruments: Criteria for Design of Red Incandescent Lighted Instruments [S]. 2001.

［7］ GJB 455—88.飞机座舱照明基本技术要求及测试方法[S].

［8］ GJB 1394—92.与夜视成像系统兼容的飞机内部照明[S].

［9］ HB6491—91.飞机内部照明设备通用要求[S].

［10］ 程继金.自主创新,实现半导体光源技术在民航客机上的应用[N].中国民航报,2007.11.27.

大型客机驾驶舱气流热仿真及舒适性评价

王黎静，王昭鑫，何雪丽

北京航空航天大学航空科学与工程学院，北京 100191

摘要 驾驶舱的空气速度场和温度场直接影响驾驶员的热舒适和工效，是驾驶舱结构和环境控制系统设计中考虑的重要内容。本文对 B737－800 驾驶舱进行全尺寸三维建模，以高空夜航工作情况为例，采用商用 CFD（Computational Fluid Dynamics）软件 Fluent 计算该驾驶舱的流场和温度场，再用平均温度评价指标和 PMV 指标对驾驶舱进行温度环境客观评价，并采用问卷调查法对飞行员进行驾驶舱内部空气流速和温度舒适度的主观评价。数值仿真结果与主观评价结果都得出 B737－800 驾驶舱在高空夜航时基本处于舒适的环境，除躯干和脚部由于偏凉而略感不舒适。数值仿真结果与客观评价结果的一致性，验证了大型客机驾驶舱气流热仿真的准确性。

关键词 数值仿真，速度场，温度场，评价

Airflow thermal simulation and comfort evaluation of commercial airliner

Wang Lijing，Wang Zhaoxin，He Xueli

School of Aeronautic Science and Engineering，Beijing University of
Aeronautics and Astronautics，Beijing 100191

Abstract Air velocity field and temperature field of cockpit will affect the pilots' thermal comfort and performance directly，which are important parts needed to be considered in the design of cockpit structure and environmental control system. After full-size three-dimensional model of B737－800 cockpit being established，a commercial CFD（Computational Fluid Dynamics）software Fluent was used to calculate the flow field and temperature field of cockpit in high-altitude night work. Then the average temperature evaluation and PMV were used to evaluate cockpit temperature environment. A questionnaire was given to the pilot to obtain cockpit air flow and temperature comfort subjective evaluation. The results of simulation and subjective evaluation both indicate the B737 - 800 cockpit is a comfortable environment at night at high altitude, besides, the trunk and legs feel a little uncomfortable as a result of slight coolness. The agreement of simulation results and evaluation results validates the accuracy of the cockpit airflow thermal simulation.

Keywords Numerical simulation, velocity field, temperature field, evaluation

客机驾驶舱可视为一有内热源的有限封闭空间,舱内空调供气在其中作强迫对流换热,将舱内热源(如驾驶员、电子设备等)发出的热量带出舱外,在驾驶员周围形成适宜的微环境。飞机驾驶舱内空气流动和传热研究,涉及飞机舱室空调通风、人体热舒适性、能源合理利用、改善空气品质及降低气流噪音等,因此是飞机设计和使用部门非常关注的问题之一,也是人机环境系统工程领域一个重要的研究问题。

波音和空客公司在客机的设计研发前期,均对驾驶舱空气流动和传热进行了数值模拟[1—3],初步验证设计方案和指导后续工作。但在客机驾驶舱数值模拟的相关文献中,数值模拟的适用范围、计算条件和计算方法等都未给出,所以有必要对此进行研究。

本文对 B737 - 800 客机驾驶舱内空气的流动和传热进行数值模拟,进而得出以数值模拟结果等为计算依据的驾驶舱空气环境客观评价,并通过飞行员主观评价对数值模拟结果及客观评价进行验证。数值模拟计算按照 B737 - 800 客机的实际飞行情况,选择高空航、第三观察者不在驾驶舱内的工作情况进行研究,总结出数值模拟、客观评价与主观评价的关系,以供开展此类飞机驾驶舱初期的数值模拟设计、环境评价研究。

1 物理和数学模型

1.1 物理模型

图 1 给出了 B737 - 800 驾驶舱进气口的分布简图。驾驶舱空气出口位于脚蹬上部、中央仪表板后面。图 2 给出了 B737 - 800 驾驶舱及两位飞行员的结构简图。气流经位于驾驶舱内壁面的各进气口进入驾驶舱,在驾驶舱内做强迫对流换热后经飞行员脚的上部、仪表板的后部之后被抽出去。

为保证仿真计算的准确性,建模时驾驶舱和飞行员的几何模型尽可能与实际情况保持一致,特别是进气口的位置尺寸和几何尺寸。忽略舱壁连接处缝隙,即不考虑驾驶舱空气的泄漏。本文研究 B737 - 800 的飞行状态是夜间高空巡航,飞行高度 10,000m、飞行速度 Ma 数 0.785,此时飞机对驾驶舱提供 300kg/h 的供气分配。

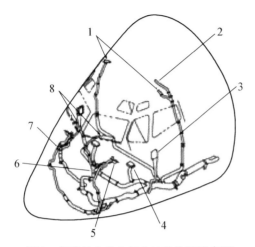

图 1　驾驶舱内各空气出口的位置示意图

1—顶部空气进口；2—随机人员单独空气进口；3—
侧壁空气进口；4—座椅下部空气进口；5—机长单独
空气进口；6—机长脚部空气进口；7—副驾驶脚部空
气进口；8—风挡空气进口

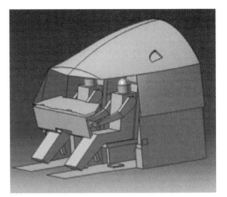

图 2　驾驶舱简化结构示意图

1.2　数学模型

本文假设驾驶舱内空气为不可压缩、牛顿黏性流体,流动为稳态湍流。考虑人
体代谢热、设备发热,但不考虑驾驶舱内辐射换热的影响。驾驶舱内空气流动和传
热的控制方程可描述如下:

连续性方程

$$\frac{\partial \rho}{\partial \tau} + \nabla \cdot (\rho U) = 0 \tag{1}$$

动量方程

$$\frac{\partial(\rho U)}{\partial \tau} + \nabla \cdot (\rho U \times U) = \nabla \cdot p + \nabla \cdot \left[\mu_{\text{eff}} U + (\mu_{\text{eff}} \nabla \cdot U)^{\mathrm{T}}\right] \tag{2}$$

式中:μ_{eff} 为有效黏性系数;p 为修正压力。

能量方程

$$\frac{\partial(\rho T)}{\partial \tau} + \nabla \cdot (\rho U T) = \nabla \cdot (\lambda \nabla T) + S_{\mathrm{T}} \tag{3}$$

式中:λ 为导热系数;S_{T} 为源项。

湍流动能 k 方程

$$\frac{\partial(\rho k)}{\partial \tau} + \nabla \cdot (\rho U k) - \nabla \cdot \left(\frac{\mu_{\text{eff}}}{\sigma_k} \nabla k\right) = G - \rho \varepsilon \tag{4}$$

湍流动能耗散率 ε 方程

$$\frac{\partial(\rho\varepsilon)}{\partial\tau} + \nabla \cdot (\rho U\varepsilon) - \nabla \cdot \left(\frac{\mu_{\text{eff}}}{\sigma_\varepsilon} \nabla\varepsilon\right) = \frac{\varepsilon}{k}(C_{\varepsilon1}G - C_{\varepsilon2}\rho\varepsilon) \tag{5}$$

式中:k 为湍流动能;ε 为湍流动能耗散率;$C_{\varepsilon1}$,$C_{\varepsilon2}$,σ_k,σ_ε 为常数;G 为湍流动能生成项。

上述方程为空气流动和传热的基本方程,由于假设的驾驶舱内的空气为稳态湍流,故方程中的时间项没有意义,且流体的密度 ρ 为定值。

1.3 边界条件

驾驶舱内流动和传热边界主要包括空气进口、空气出口、飞行员服装附近表面和各个壁面。各种飞行条件下旅客机舱内空气的温度、湿度、气流速度都应保持在允许范围或规定值[4]。故根据飞行和环境状况驾驶舱进气温度取为 288 K,气流流量取为 300 kg/h,各进口气流量分别取为:顶部空气进口流量为(58.5 kg/h)×2;侧壁空气进口为(45 kg/h)×2;座椅下部空气进口为(21 kg/h)×2;脚部空气进口为(6 kg/h)×2;风挡空气进口为(19.5 kg/h)×2。驾驶舱壁面的流动边界条件采用固体壁面无滑移边界条件,应用标准壁面函数决定靠近壁面附近流体的流动行为。采用试验测试的驾驶舱壁面温度数据作为驾驶舱壁面边界条件(这里根据驾驶舱分区给定壁面温度)。

1.4 网络生成和求解策略

本文采用非结构化四面体网格[5]划分计算区域。非结构化网格对复杂外形具有很强的适应能力和很好的灵活性,便于网格的自动生成、自适应处理及并行计算实施。节点数约 45 万个,体单元数约 158 万个。为了更多获得驾驶舱各壁面边界层信息,对各壁面进行了附面层控制和网格控制,对各进气口和飞行员附近需要关注的区域进行网格局部加密。

采用标准 k-ε 模型封闭湍流控制方程,控制微分方程离散时采用有限容积法,计算中能量方程和动量方程采用二阶迎风格式离散,连续方程、k 和 ε 方程采用一阶迎风格式离散。采用 RMS(Root Mean Square)残差值对速度、压力的监测值判断是否收敛,计算迭代共 1,600 步,至 800 步时 RMS 到达 1×10^{-4},速度和压力监测值趋于稳定,获得较理想的结果。

2 仿真结果及分析

2.1 驾驶舱分析截面的选择

数值模拟中,主、副驾驶员周边空气流动和传热的情况相同,所以只对主驾驶员周围空气进行分析。选取如图 3 所示 5 个截面对数值模拟结果进行分析,它们的位置情况如下:

截面 1——主驾驶员正面,距主驾驶员前胸垂直距离约 5~8 cm;

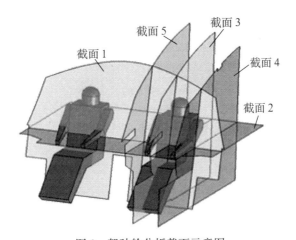

图 3　驾驶舱分析截面示意图

1—截面 1；2—截面 2；3—截面 3；4—截面 4；5—截面 5

截面 2——与驾驶舱地板平行，距地板约 80 cm；

截面 3——过主驾驶员正中心、与飞机左右对称面平行；

截面 4——主驾驶员左侧与截面 3 平行，距主驾驶员左侧表面的垂直距离约 4 cm；

截面 5——主驾驶员右侧与截面 3 平行，距主驾驶员右侧表面的垂直距离约 4 cm。

2.2　速度场分析

300 kg/h 供气状态下，图 4 给出对应 5 个截面的速度等值线图。在主驾驶员身体附近的区域，截面 1 上最大速度为 0.43 m/s；截面 2 上最大速度为 0.45 m/s；截面 3 上最大速度为 0.35 m/s；截面 4 上最大速度为 0.66 m/s；截面 5 上最大速度为 0.36 m/s。除截面 4 上人体附近最大风度超过 0.5 m/s，其余人体附近区域的截面

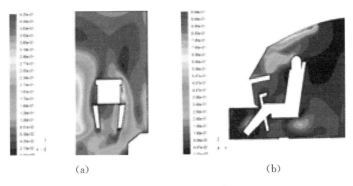

(a)　　　　　　　　　　　(b)

图 4　部分截面速度等值线图

(a) 截面 2；(b) 截面 3

上风速都符合飞行员舒适标准。截面 4 上人体附近风速过大原因是 B737 - 800 驾驶舱侧壁空气进口处没有安装导流板,进气直接吹向驾驶员腰部,导致腰部空气速度超过指标。

2.3 温度场分析

图 5 为部分截面温度等值线图。在主驾驶员身体附近的区域,截面 1 平均温度为 18.8℃,最高温度为 19.9℃,最低温度为 17.5℃;截面 2 平均温度为 18.9℃,最高温度为 20.1℃,最低温度为 17.9℃;截面 3 平均温度为 19.1℃,最高温度为 20.1℃,最低温度为 16.9℃;截面 4 平均温度为 19.3℃,最高温度为 19.8℃,最低温度为 16.6℃;截面 5 平均温度为 18.9℃,最高温度为 19.8℃,最低温度为 17.4℃。数值仿真计算出来飞行员身体附近区域的 5 个截面的平均温度值,均在大型客机驾驶舱空气温度设计要求最佳值 18~22℃ 之间;且截面 1 至截面 5 的温差分别为 2.4℃,2.2℃,3.2℃,3.4℃,2.4℃,基本符合驾驶舱温度不均匀度小于 3℃ 的要求,截面 3,4 温差稍大的原因和侧壁空气进口没有加导流板的设计有关。

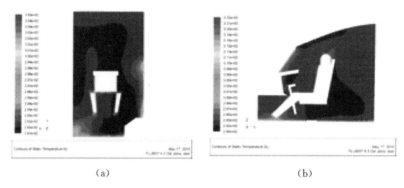

(a) (b)

图 5 部分截面温度等值线图

(a) 截面 2;(b) 截面 3

3 客观评价

根据供气量为 300kg/h 状态的速度场和温度场的计算结果,采用预测平均热感觉指标(Predicted Mean Vote,PMV)[5]对该驾驶舱的环境进行评价。

$$PMV = (0.303e^{-0.036M} + 0.028)\{M - W - 3.05 \times 10^{-5} \times$$
$$[5,733 - 6.99(M - W) - P_a] - 0.42(M - W - 58.15) -$$
$$1.7 \times 10^{-5}M(5,867 - P_a) - 0.001,4M(34 - t_a) -$$
$$3.96 \times 10^{-8}f_{cl}[(t_{cl} + 273)^4 - (t_r + 273)^4] - f_{cl}h_c(t_{cl} - t_a)\} \quad (6)$$

PMV 代表了大多数人对同一环境的冷热感觉,反映了某一环境的热舒适度。PMV 热舒适评价指标综合考虑了环境与人的 6 个因素,包括:人体活动情况(新陈代谢率)、衣着情况(服装热阻)、空气温度、空气相对湿度、空气流速、平均辐射温度。PMV 指标与上述 6 个因素之间的计算关系式见文献[6]。PMV 值与热感觉的对应关系如表 1 所示。

表 1　PMV 值与热感觉的对应关系

热感觉	热	暖	稍暖	舒适	稍凉	凉	冷
PMV 值	+3	+2	+1	0	−1	−2	−3

由驾驶舱空气流动和传热数值仿真得到人体各节段邻近服装表面的环境温度(19.90℃)和空气流速(0.28 m/s),然后再结合飞行员活动情况、衣着情况。结果代入计算式,可得相应的 PMV 值,如表 2 所示。

表 2　PMV 评价指标计算结果

工作状态	按数值仿真模拟结果计算所得结果	稍凉	舒适	稍暖
300 kg/h 供气	−0.415	−1	0	1

由表 2 得结论:模拟结果得出的 PMV 值介于−1 和 0 之间,即在稍凉和舒适感觉之间。这与仿真结果的分析相吻合,即飞行员对驾驶舱整体感觉舒适,但飞行员身体局部位置偏凉。

4　飞行员主观评价

通过直接测定飞行员对驾驶舱热环境的主观感受,来与数值模拟和客观评价做对照补充,验证数值模拟和客观评价的有效性。

对 B737‐800 飞行员做该机型夜间高空巡航时热舒适性的主观评价。邀请国航某飞行大队中的 10 位 B737‐800 飞行员(B737‐800 人均飞行时数:7,000 小时;性别:男;年龄:29～37 岁)进行主观评价问卷调查,调查飞行员在 B737‐800 飞机高空夜航时的热环境评价[6]。结果汇总如表 3～7 所示(所有表中的数字均代表人数)。

表 3　热感觉主观评价数据统计表

	凉	稍凉	不冷不热	暖	稍暖
头部		2	5	1	2
躯干	3	4	3		
手部	1	1	7		1
脚部	3	1	5		1

表4　情感主观评价数据统计表

	舒适	稍不舒适	不舒适	很不舒适
头部	7	3		
躯干	2	4	2	2
手部	8	2		
脚部	4	3	1	2

表5　热偏好主观评价数据统计表

	凉一点	稍凉一点	无变化	稍暖一点	暖一点
头部	1	3	4	2	
躯干			4	6	
手部		1	8	1	
脚部		2	3	4	1

表6　个体可接受性主观评价数据统计表

是	不是
6	4

表7　个体耐受度主观评价数据统计表

完全可耐受	稍难耐受	相当难耐受	很难耐受	无法耐受
6	4			

　　根据数据的数字特征及其分布的统计学特性进行评价结果数据分析[7]如表8所示。

表8　评价结果数据分析表

	飞行员感觉	飞行员感情评价	飞行员热偏好
头部	不冷不热	舒适	无变化
躯干	稍凉	稍不舒适	稍暖一点
手部	不冷不热	舒适	无变化
脚部	不冷不热	舒适	稍暖一点

　　60％的飞行员将热环境判定为在个体水平上是可接受的。60％的飞行员认为驾驶舱环境完全可耐受,40％认为环境是稍难耐受。可以得出结论,B737－800飞行员在高空夜航时总体上处于舒适的环境,躯干和脚部由于偏凉而略感不舒适。该结论与数值模拟结果和客观评价结论相一致。

5 结论

（1）本文完成 B737 – 800 驾驶舱空气流动和传热的数值模拟，得出的数值模拟结果符合大型客机驾驶舱空气设计要求，且与飞行员主观评价结果相一致。

（2）根据数值模拟结果求得驾驶舱 PMV 指标，该客观评价得出 B737 – 800 驾驶舱处于舒适区的范围内，与飞行员主观评价结果相一致。

（3）根据以上两个结论，得出本文驾驶舱的模型设计、简化与驾驶舱边界条件、计算条件的处理均为合理的，本文的驾驶舱数值模拟方法和客观评价方法可运用于此类机型驾驶舱的初期设计研究中。

参考文献

[1] Fred A，Matthew J. Warfield. Numerical analysis of airflow in aircraft cabins [J]. SAE，1991，911441：1294 – 1304.

[2] Alan W D，Drew B L. Commercial airplane air distribution system development through the use of computational fluid dynamics [R]. AIAA – 1992 – 0987：1 – 9.

[3] Singh A，Hosni M H，Hortsman R H. Numerical simulation of airflow in an aircraft cabin section [J]. ASHRAE Transactions，2002，108(1)：1005 – 1013.

[4] 寿荣中,何慧珊. 飞行器环境控制[M]. 北京：北京航空航天大学出版社,2006.

[5] 陶文铨. 数值传热学[M]. 西安：西安交通大学出版社,2004.

[6] 谢东,王汉青. 不同气流组织下夏季空调室内热舒适环境模拟[J]. 建筑热能通风空调,27(3)：66 – 69.

[7] 国家标准 GB/T 18977—2003. 热环境人类工效学、适用主观判定量表评价热环境的影响[S].

基于信息处理和事故链原理的结构化飞行员差错模型

杜红兵[1]，刘　明[1]，靳慧斌[2]

1. 中国民航大学 安全科学与工程学院，天津 300300
2. 中国民航大学 民航安全科学研究所，天津 300300

摘要　分析民航业中常见人为差错研究方法，指出其局限性。在此基础上对民航飞行人为差错进行系统分析，基于信息处理过程和事故链原理，从微观的认知过程和宏观的 HFACS 框架两方面构建了一种新的飞行员人为差错分析模型，并对模型中的情景分析、差错模式与分类、认知机理、差错恢复和差错成因等几大模块进行阐述，重点分析了认知机理在差错形成中产生的关键影响，为民航飞行员人为差错的产生机理、差错分类和差错成因的分析提供了一种有效的方法，最终使得对飞行员人为差错的分析结果更加全面、深入和准确。

关键词　人为差错，事故链原理，差错模式，认知模型，差错成因

Structured Pilot Error Model Based on Information Processing and Accident Chain Principle

Du Hongbing[1] , Liu Ming[1] , Jin Huibin[2]

1. College of Safety Science and Engineering, Civil Aviation University of China, Tianjin 300300
2. Research Institute of Civil Aviation Safety, Civil Aviation University of China, Tianjin 300300

Abstract　Summarize and analyze the major existing research methods about human errors in civil aviation industry, Point out their shortcomings. Systematically analyze the human errors of civil aviation pilot, based on Information processing and Accident chain principle, build a new human error model of pilot from the micro-cognitive processes and macro-HFACS framework, and introduce some module such as Scenario analysis, error models and classification, cognitive mechanism, error recovery and error causes. Analyze the key effect of the cognitive mechanism of the errors generated. The purpose of this study is to provide an effective method for the analysis of cognitive mechanism, error classification and error causes of pilot error, to make the result of human error analysis more comprehensive, thorough and accurate.

Keyword　human error, accident chain principle, error model, cognitive model, error cause

随着航空设备可靠性和自动化水平的提高,飞机座舱中的动态显示信息越来越多,飞行员的主要工作已由操作控制向监视转移,由飞行技术引发的飞行事故大大减少,而人为差错造成的不安全事件的比例却呈上升势头。国际航空运输协会的统计资料显示,所有飞行事故的 80% 都与人的不安全行为有关。所以,人为因素是当今航空飞行安全的最大隐患,同时也是提高飞行安全水平最有效的手段。

美国于 1997 年开发了主要用于事故分析的 HFACS 系统,欧盟针对空管人为因素开发了 HERA - JANUS 系统,欧美各国已经开始采用各种客观、量化的方法来研究人为差错问题。我国学者罗帆对民航机组行为差错进行了心理分析,认为机组行为差错多是属于心理性质的,是人、机、环境因素相互作用的结果,探讨了人为因素与机组行为差错的原因;孙瑞山等运用 HFACS 方法对 CFIT 事故树中得出的基本事件进行分类分析,得出了不安全行为后面还隐藏着更多组织管理上的隐性差错[1, 2]。

虽然我国的航空人为因素研究已经取得了一定进展,但是总的来看还处于起步阶段,尤其是对于紧系民航安全的飞行员人为差错研究还不够深入,仅停留在飞行事故的调查分析层面,缺乏对民航飞行员人为差错认知机理方面的研究。本文在分析人为差错常见分析模型方法的基础上,构建出基于信息处理过程和事故链原理的结构化人为差错分析模型,并对模型的主要功能结构模块进行了分析,希望能对飞行员差错产生机理分析起到一定的指导作用。

1　人为差错研究方法

1.1　人为差错概念

人为差错(Human Error)也称为人因差错或人因失误,近年来,人们对人为差错的问题研究逐步深入,对人为差错的概念也有了更深的认识,并从不同的角度给予了多种阐述。大多数研究者对人为差错的界定是:偏离某种规范或标准和操作者的无意行为,即人为差错是操作人员在各种因素的影响下无意中出现的某种偏离。

1.2　民航常用人为差错分析方法

国内外人为差错理论以及相关模型在民航飞行安全中的应用成果十分有限,其中在民航中得到广泛应用的主要有 E. Edwards 于 1972 年提出的 SHEL 模型,Scott A. Shappell 和 Douglas A. Wiegmann 在 Reason 模型基础上开发的应用于飞行事故调查的 HFACS[3],波音公司推出的 MEDA[4] 和欧盟 EUROCONTROL 开发的 HERA - JANUS[5]。

　　由于这些方法都有自己的特点和适用范围,表 1 就以上四种方法进行了简单的比较分析。

表 1　民航常用人为差错分析方法特点和使用范围

模型/方法	优点	缺点	主要应用领域或适用范围
SHEL	很好地分析人与系统中的硬件、软件、环境等界面之间的关系及存在的缺陷	对人在感知、知觉、决策和行为等认知机理原因分析有很大的局限性	常用于分析人因错误的来源,也是事故调查和安全管理的工具
HFACS	人为差错事件调查与分析效果好,能够发掘出组织层面的差错诱因	对人的认知过程的分类不够充分和重视,分类缺乏全面性和一致性	事故调查和分析的工具以及飞行机组原因的分类分析
MEDA	能够全面调查分析错误形式和导致差错的所有因素,操作简单	主观性太强、可信度低,且仅仅关注差错外在表现形式	在维修差错调查方面取得了显著的效果和认可
HERA - JANUS	能够综合各方面因素找到人为差错的心理学机制,对差错识别很具体详细	对调查员要求太高,不利于模型的推广,人因术语太多	在空管领域人为差错分类分析取得广泛的应用

　　上述模型都有其局限性,并不适于对飞行员人为差错进行全面和准确的分析。因此,本文提出一种结构化人为差错分析模型,从微观层面分析飞行员人为差错产生的内部认知机制,更加准确地对飞行员差错进行分类,更加全面地对飞行员差错成因进行归类。

2　基于信息处理和事故链原理的结构化差错模型

　　过去的研究中,对于人为差错的研究方法和理论模型在一些行业取得了一定的成果,但对于特殊的民航业来说,照搬一些现有的差错分析模型显然不能很好地分析出人为差错产生机理和成因。下面主要从微观层面的信息处理过程和宏观层面的事故链原理两个角度对于飞行差错进行分析。

2.1　飞行员信息处理过程

　　传统的信息处理模型分为四个阶段:感知—知觉(注意力、记忆)—计划/决策—执行/操作。从飞行员的个人心理和生理状态,结合有关的任务和情境资料,分析飞行员的情景意识,确定认知差错形成的内在机理和差错表现形式。人为差错内在机理主要描述差错形成的心理机制,本模型阐述了在差错成因影响下差错产生的内在作用机制[6, 7],如图 1 所示。

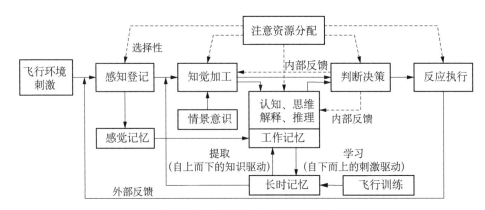

图1　飞行员信息加工处理过程图

飞行员的认知过程可分为三个阶段：

（1）对环境信息的感知。感觉器官将接收到的刺激转化为视觉、听觉、触觉、嗅觉等各种感觉信息。对于飞行员而言，为实现情境认知，首先应对飞行环境中各相关成分的状况、属性及动态特性进行感知。需要精确地觉察飞机本身的信息（如空速、位置、高度、航线、航向等）、气候变化、飞行动态监控、紧急通告及其他的有关信息。飞行员在感知信息的过程中，一定要注意两点：一是感知的信息不正确、不完整或失真；一是潜意识当中把正确的信息曲解，如常见的飞行错觉。

（2）中央加工或对信息的转换。大脑对感觉到的信息进行变换整合加工，提取刺激特征加以组合输出有用的信息。长时记忆系统将输入的刺激信息与系统中的信息进行比较与模式匹配，一部分信息被激活，这部分被激活（称为主动记忆）的信息送入工作记忆系统。本阶段认知活动最为复杂，涉及很多心理活动，如记忆的提取与存取；注意的分散与集中；对特定的飞行情景进行分析推理；保持较高的情景意识，这个反应通常是非常短暂的一个过程。飞行员需要良好的记忆力，特别是在某段时间内对空中交通动态的短期记忆能力，关键是对一些冲突信息的识别记忆。集中注意力、强迫记忆是提高短期记忆力的较好方法，另外，利用各种设备的提醒、标注功能（包括飞行进程单的摆放）和机组内的相互监控、提醒是防止疏漏和遗忘的有效"防火墙"。

（3）对信息做出反应选择和判断决策。一般来说，当飞行员意识到某件事情已经发生或一件预料中的事情没有发生时，就需要进行飞行决策。飞行决策的定义包括两部分：一是分析并建立所有飞行处境信息之间的联系，分析每一种可供选择的方案，并确定每种方案可望达到目标的程度；二是根据分析选择一种方案并在飞机当前处境允许的时间内果断地执行这个方案[8]。

飞行决策是飞行员的一种能力，这种能力的发挥，既取决于其辨别力、知识储备、临场发挥和综合分析等内在素质，也受外界情况的影响。对于航空飞行这样的环境，飞行员可能没有这么多时间，肯定不允许飞行员仔细地考虑动作的多种结果。理性选择决策需要大量的数据,而飞行员往往要考虑其不确定性,如天气条件、故障内部情况、复飞可能性等。在时间紧迫、信息模糊不清、目标定位不良以及条件不断变化的情况下,当备选方案实质上差别不大时,飞行员会利用自己的飞行经验估计形势,快速做出反应,寻求第一个可行的备选方案做出决策,而并不是对不同的备选方案进行比较之后去追求最佳的处理方案。在不正常情况时,飞行员总是选择那些熟悉的操作程序、步骤,这样的选择决断快,遇到的"抵抗"少,处置的速度就快。

在这些过程中产生的差错主要是人机界面设计中存在的问题与飞行员潜意识和习惯共同作用的结果。第一阶段是飞行员对各种动态信息监测和接收,此过程产生的差错主要是飞行员的注意力不集中及驾驶舱设计布局不当所引起。对信息进行分析是结合已有的知识进行判断推理的过程,此过程产生的差错主要是飞行员所拥有的知识在内容、结构和应用上的欠缺。由此可见在不同阶段产生的人的差错的内在机理也是不同的。

2.2　事故链原理

事故链原理认为,一个事故是因若干个环节在连续时间内出现缺陷,由众多连续的缺陷构成了整个安全体系失效。基于 HFACS 框架的四个层次失效分析差错形成,组织影响、不安全监督、不安全行为前提条件三方面作为触发器,激发了不安全行为的发生,在突破防御机制的屏障后形成不安全事件的后果。模型对事故人为因素进行分析时,采用逆推方式,即从结果事件入手,推出导致结果事件的不安全事件。采用 HFACS 框架作为总的触发器,可把不安全事件的多因素简化,只要控制其中任何一个因素就能有效地预防事故发生[10, 11]。

针对飞行员差错没有非常有效的模型加以深层次地分析研究,对于飞行安全至关重要的飞行员来说,有必要对飞行员的差错用有效可靠的模型进行分析研究,所以针对驾驶舱的特性,以及结合现有人为差错分析模型,本文构建了基于信息处理过程和事故链原理的结构化差错模型来对飞行员的人为差错进行定性研究[9, 10],详见图 2。

模型在结构上结合微观的信息处理过程和宏观的事故链原理对飞行差错进行分析,模型的上半部分主要基于个人的认知差错发展演变过程进行分析,按照事故因果发展方向分析了差错发生的过程,模型下半部分主要基于事故链序列,对人为差错导致的不安全事件进行回溯性分析,从飞行不安全事件结果出发,往前分析事件的触发因子,即差错成因[11, 12]。

3　基于信息处理和事故链原理结构化差错模型模块分析

3.1　微观层面分析

模型从微观层面对个人认知差错进行分析,主要包括飞行情景分析、差错模式与分类研究、差错恢复、差错成因和反馈机制等结构化模块。

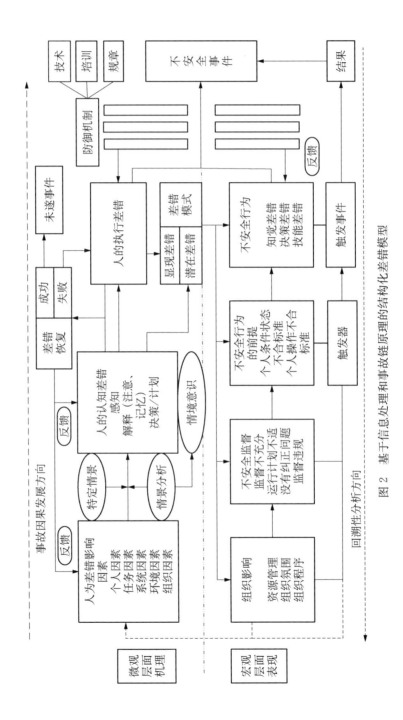

图 2　基于信息处理和事故链原理的结构化差错模型

1）飞行情境分析

分析特定情景下要实施的任务，难度、复杂度等，以及各个任务的子任务，需要遵守哪些程序，执行哪些操作。首先对差错事件发生的行为条件进行资料的收集与整理，不能漏掉一些潜在的重要信息，然后分析差错事件发生的具体行为条件与行为之间的关系，并做出评价，这是分析的基础。

飞行中的人为差错基本上都在驾驶舱这个特定的环境中发生的，对于突发事件和工作任务，飞行员必须在驾驶舱的基础上完成一系列的决策与动作。各个飞行阶段的工作负荷也不一样，这就要求在分析飞行员的人为差错时必须充分考虑飞行员所处的情景，所接受的任务，飞行员的情境意识能力等因素。

2）差错模式与分类

差错模式通常指潜在差错和显现差错两类型。本文主要侧重在微观方面对个人的认知差错进行研究，认知功能的潜在差错按照飞行员差错产生的内在机理可分为感知差错、信息处理差错（注意力分配不当、记忆偏差）、决策差错、操作差错。

感知差错，主要是由于飞行员注意力分散而对信息感知出现了偏差、疏忽或者观察不全面的情况。信息加工是判断的基础，信息处理差错会在很大程度上导致判断差错。操作的动作选择来自于判断，在很大程度上，判断差错是操作差错的基础。不少情况下的操作差错，都是由于反应、判断差错所造成，当然，由于飞行员操作能力和技术水平有限导致的飞行差错也为数不少。我们常见的典型飞行差错有飞行错觉、情景意识不足、飞行决策不充分、没有及时复飞、动作不充分、不遵从飞行标准程序等。

3）差错恢复

在发生人为差错的时候，系统和程序会有恢复措施，判断此情况下是否有效，能否阻止不安全行为的发生。如果在飞行事故产生之前，能使用差错恢复策略发觉并纠正已发生或正在发生的差错，那么产生的也只是未遂事件或未遂事故，从而防止事故带来严重的伤害和损失。该模块在飞行驾驶舱中有着很显著的体现，为了对各种差错进行控制，我们会采取各种措施来阻断差错的进一步发展，比如设备仪器的报警，机组间的交叉检查，观察员位置的设立，任务命令的复诵等。

4）差错成因

基于特定情景从个人、任务、系统设计、环境、组织五方面分析影响人为差错形成的具体因素[13]。

人是任务完成的决定因素。飞行员自身的各种因素将是在执行任务过程中发生差错与否的成因。这些因素包括：人的生理、心理、个性、情绪、能力、技能和经验等。

任务的性质是引起人的差错发生的重要原因之一，对于飞行安全来说尤其如此，飞行不同阶段有不同的任务程序，不同的情境有不同的应急程序，任务的复杂程度、重复性、执行时间以及完成任务所需时间长短等任务因素均会成为差错的诱因。

　　系统设计主要包括人机关系、硬件和软件。系统将直接影响飞行的整个工作过程,驾驶舱系统的设计直接影响飞行员的认知、决策和操作水平。因此,系统设计是人为差错原因分析中不可忽略的一个重要因素。

　　飞行环境分为工作环境和社会环境。工作环境如湿度、温度、噪声、光线、振动、天气、放射性、工作场所空间、距离以及工作和休息时间等。社会环境包括上下级关系、同事关系、家庭关系,社会环境的消极方面也会对飞行员差错产生很大影响。

　　组织因素也是人因差错研究关注的焦点。组织因素主要包括管理因素、交流沟通、责任划分、监督和安全文化等,可以从深层次分析出差错的潜在根源,而一线飞行员的差错是由组织管理的不足或缺陷长期积累而导致的结果。

　　5)反馈机制

　　包括系统环境的外部反馈和认知过程中的内部反馈。飞行员会接受这些反馈信息并对反应行为和结果进行监控和核查。一旦与预期要达到的目标不符合,就会立即采取改正行为。因为人的认知和行为经常会产生差错,对反馈信息的监控和核查就显得非常重要。正因为有了这些反馈过程,飞行员才能不断地纠正从感知到操作整个过程中所产生的差错。

3.2　宏观层面分析

　　模型在宏观层面主要通过 HFACS 框架,利用事故链原理分析四个层次失效导致的不安全事件。

　　1) HFACS 框架

　　HFACS 是根据里森“瑞士奶酪”理论模型,定义模型中的洞的含义后提出的。HFACS 描述了四个层次的失效,每个层次都对应于里森模型的一个层面,包括不安全行为、不安全行为的前提条件、不安全的监督和组织影响。根据事故链原理,当在飞行运行环节中所有失效同时发生时,就会产生飞行不安全事件的后果。

　　2) 差错模式

　　在宏观层面的显性差错主要就是指不安全行为以及由此导致的不安全事件,而潜在差错主要指组织潜在差错,该模型将组织潜在差错分为公司组织决策失误、各级管理缺陷、潜在的运行前提失效和潜在防护机制失效四种模式。

　　3) 防御机制

　　民航系统的防护机制一般可归为技术、培训和规章。防护机制通常是控制潜在状况以及人的行为能力过失后果的最后一道安全网。大多数防范具有危险后果的安全风险的缓解策略均基于强化现有防护机制或者建立新的防护机制,是人为差错导致不安全事件的最后屏障。

4　结构化差错分析方法

　　上面分析了结构化差错分析的结构模型与结构模块,在此基础上下面介绍回溯性分析方法对飞行差错事件的分析。此分析方法由六个步骤构成,如图 3 所示。此

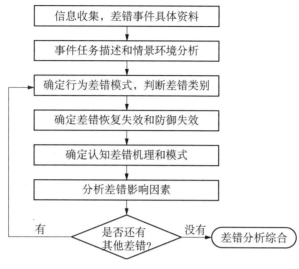

图 3　差错事件/事故分析过程

方法以结构化的步骤形式向差错分析人员展示了它的易操作性,且避免使用专业性很强的术语。

5　结论

(1) 本文在分析了现有民航人为差错分析模型优缺点的基础上,结合飞行员认知心理过程和 HFACS 框架,从微观的认知内在机理和宏观的人为差错外在表现两方面建立了基于信息处理过程和事故链原理的结构化差错分析模型。

(2) 从微观层面对模型的情景分析、差错模式与分类研究、差错恢复、差错成因和反馈机制等结构化模块进行了分析;从宏观层面对 HFACS 框架四个层次、差错模式、防御机制等进行分析。

(3) 根据结构化差错分析模型,提出了结构化的飞行差错事件/事故分析过程。

(4) 模型主要从飞行员的心理及行为环境出发,对人为差错进行定性分析,克服了现有飞行员人为差错模型研究方法存在的空白和不足,更好地研究飞行员差错的产生以及对人为差错的控制和预防起到指导作用,达到提高民航安全水平的目的。

参考文献

［1］　刘汉辉,孙瑞山,张秀山.基元事件分析法[J].中国民航大学学院报,1997,15(3):1-9.

［2］　丁晓华.飞行机组人为因素对飞行安全的影响研究[D].西安:西北工业大学,2005.

［3］　Shappell S A, Wiegmann D A. The Human Factors Analysis and Classification System-HFACS [R]. U. S. Department of Transportation,2000.

[4] 中国民用航空总局航空器维修人的因素课题组. 人的因素案例集——民用航空器维修差错[M].北京:中国民航出版社,2003.

[5] Barry K, Mark R, Dirk S. 人的因素在空中交通管理中的影响[M].闫少华,陈治怀,等译,北京:中国民航出版社,2008.

[6] 邓娟. 对空中交通管制员的信息加工错误的分析[J].西南民族大学学报,2004,1:407-410.

[7] 威肯斯 C D,李 J D. 人因工程学导论[M].第二版.刘乙力,张侃,等译.上海:华东师范大学出版社,2007.

[8] Robert L S, Otto H M, Maclin M K. Cognitive Psychology [M]. 8th. ed. Shao Zhifang Direct. Beijing: China Machine Press, 2010.

[9] 李鹏程. 一种结构化的人误原因分析技术及应用研究[D]:湖南:南华大学,2006.

[10] Reason J. Human Error [M]. New York: Cambridge University Press, 1990.

[11] Reason J. A framework for classifying errors. In: Rasmussen J, Duncan K D, Leplat J, New Technology and Human Error [M]. Wiley, Chichester, UK, 1987.

[12] Reason J. A System approach to organizational error [J]. Ergonomics, 1995,38(8):1708-1721.

[13] 程道来,杨琳,仪垂杰. 飞机飞行事故原因的人为因素分析[J]. 中国民航飞行学院学报,2006,17(6):3-7.

Research on Modeling Cognitive Behaviors of Pilot Processing Alarm in Cockpit

Luan Yichun[1] , Xue Hongjun[2]

1. College of Aeronautic Northwest Polytechnical University Xi'an 710072
2. College of Aeronautic Northwest Polytechnical University Xi'an 710072

Abstract A general process about modeling the perception, attention, decision-making behaviors of cockpit pilot processing alarm was proposed. Based on information process stage model and ACT – R theory, modeling was combined with description about modeling tasks of processing alarm, cognitive behavior of pilot and operational components. A serial of fundamental experiments of specific operational tasks were developed to support modeling. Validity and exactitude of the model was proved through matching the model perform those tasks with pilot practice. The modeling aim was to improve flying safety.

Keywords modeling cognitive behaviors, information process stage model, ACT – R, process alarm

1 Introduction

Warning system is one of the important factors influenced flying safety in cockpit. Modeling cognitive behaviors of cockpit pilot processing alarm was named cognitive modeling, aimed to explore theirs mind-action mechanisms especial perception, attention and decision-making courses in an emergency. Based on information process stage model and ACT – R theory, combined with description about modeling tasks of processing alarm, cognitive behavior of pilot and operational components, a model of cognitive behaviors of pilot was built. Experimental and simulating methods were used to study pilot perception course involved information content, alarm display mode, process logic and display devices layout of alarm, pilot decision-making course involved information fused, analyzed, diagnosed and action. Base on above research, cockpit warning system was optimized to improve flying safety. We can't explore the brain's actual operation like detect other body parts, so we could check whether the model was validity and exactness from how it keep cognitive theory, the reality and accurate of its prediction about what happen to tasks.

2　The information process stage model

Cognitive modeling mostly use information point of view and information process theory to study how human attend, select, comprehend and memorize, and make decision to act according to information [1]. Information was emphasized to study system functions by information process method and system was thought through information required, transfer, transform and process to act. Information process of system was made qualitative and quantitative analysis to build the model which represent operational of system. By this, cognitive behaviors was divided many stages or modules including percept, consciousness, attention, memory, decision-making, thought and action. Information was inputted from visual receptor, aural receptor, feeling receptor, proprioceptor or other receptors, and was processed by stages such as consciousness, attention, memory, decision-making, thought, at last outputted action. Fig. 1 was the information process stage model [2].

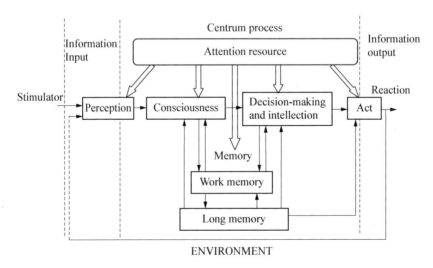

Fig. 1　Human information process stage model

Based on information process theory, the information process stage model drew some good ideas from control theory, information theory and system theory.

It made researchers analyzing problem through information flow from the
environment to mind then to the environment, and made mind study based on a
great deal of experiments and become more rigorous and theoretical consistent, but
mechanical defect was outstanding. In addition, single modules' function, work
mechanism, information configuration and representation had more studied, but
the connection and coupling between modules had few studied.

There was a gap between the information process stage model in cognitive
psychology and modeling cognitive behaviors on computer. Cognitive psychology
and the information process stage model was semantic description, but modeling
cognitive behaviors must transform those semantic relations into mathematic
formulas and parameters which implemented on computers. So a great deal of
study had to do to actualize that transformation.

3   Modeling cognitive behaviors of pilot processing alarm in cockpit

3. 1   Orientation of the Model

Modeling cognitive behaviors of cockpit pilot processing alarm was focus on
brain action mechanism in information process, and had a distinct difference with
two kinds of intelligence system which is expert support system and other is brain
model for product's creative design. Expert support system such as fault diagnosis
expert system of the civil aviation aircraft mainly built a database from a great deal
of fault information and got some reference information by index words to support
assist for human. Brain model for product's creative design was focus on
creativity. The outstanding point of cognitive behaviors of cockpit pilot processing
alarm was:

- Strongly goal-directed processing;
- Extreme time pressure;
- A rich body of declarative knowledge;
- A great deal of practice;
- Grow into master.

Pilots could become masters from greener by a great deal of training and
practice. Their knowledge acquired and skilled enhance went through declarative
stage, knowledge compile stage and procedural stage. So cognitive behaviors of
pilot processing alarm may show many views, but dynamic, growing, and auto-
learning features were outstanding. There were distinct differences in response
time, accuracy, etc. between novice and expert.

3.2　A General Process of Modeling Cognitive Behaviors of Pilot Processing Alarm in Cockpit

Modeling cognitive behaviors of pilots was a huge systems engineering which integrated with experiments, models and theory, and engineering applications. Experiments included fundamental ergonomical experiments and verification experiments to prove model whether was exact and validity. Model and theory was based on cognitive psychology, cognitive science, and ACT - R theoretical system. Engineering applications were cockpit pilot processing alarm of civil aircraft. The associated architecture of those parts was shown in Fig. 2.

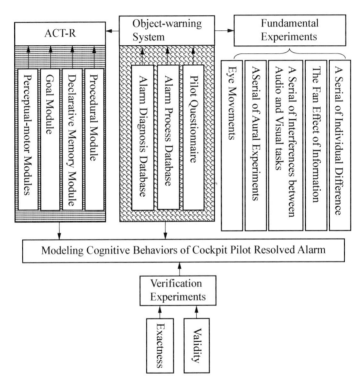

Fig. 2　A general process modeling cognitive behaviors
of pilot processing alarm in cockpit

1) Object

Modeling cognitive behaviors mostly aimed at exploring human brain action when implementing some specific tasks or solving problems, not a general or single cognitive behavior. So it must have an object that was cockpit pilot processing alarm information. The course of pilot processing alarm includes fault diagnosis, intelligence estimate and operational process. The cognitive behaviors of pilot on commercial airplane could assist to establish the alarm diagnostic database and

operational process database. Fault diagnosis or intelligence estimate was the
pilot's mental process and couldn't be obviously observed. We should combine
with the alarm information content, alarm display mode and refer the pilot
questionnaire to get those. Pilot questionnaire on the significance of that was this.

2) Theory and model

A unified model structure based on information processing stage model was
established. Like the information processing stage model, the model will comprise
several modules. Working mechanism, information representation, information
input and output of each module would be established. Information conversion and
coupling between two modules would be importance. There were two types of
cognitive models, the local architecture and unified cognitive model. Local
architecture used computers to simulate one particular aspect of brain model;
however, unified cognitive model could explain how all the components of the mind
worked to produce coherent cognition. Wegman put forward the concept of Unified
Architectures in cognitive science [3]. So far there were three well-known unified
architectures, two of which were symbolic processing framework, SOAR (state,
operator and results) named by Newel [4] and ACT – R (adaptive control of
thought-rational) put forward by Anderson [5]; Another connectionist
architecture was, a parallel distributed processing (PDP) by Rumelhart and
McClelland. Bogdan summarized the framework of classical cognitive science as
"ICM Method", that "information, capacity and mechanisms to approach", in
order to understand information processing mechanism for how to run, people
must first understand the operation of the program, the ability to define; in order
to understand a process, one must identify in advance the information of the
program execution tasks, as well as in the implementation and problems
encountered. Therefore, This approach to proceed from top to bottom: from the
information and task to the ability of implementing tasks and then to run the
program function mechanism [6].

ACT – R was cognitive system which was to develop continuously for
exploring human cognitive behavior and widely used in many fields [7 – 8]. It was
based on the study of the cognitive psychology and cognitive psychology; a large
number of experiments were designed to support it. Its' resource management
tasks were successfully implemented by linguistic, conceptual manners. Finally
some theories of cognitive psychology which reflected one particular aspect of brain
model as local architectures were proved in unified brain; the exactness and validity
of the model was verified through matching human implement practical tasks and

model simulation [9]. ACT – R should be a powerful tool for simulation human cognitive behaviors and prediction of performance.

Model consisted of two kinds of manipulative modules, symbolic module and subsymbolic module. Symbolic module was the statement of facts, based on extensive statistics derived from ergonomic experiments. It was important concepts and theories of cognitive psychology and information processing stage model representing. Subsymobolic module included learning and performance mechanisms, declarative mechanisms and procedural mechanisms which manipulated through many time-related formulas parallel processing. Symbolic work was controlled by subsymbolic module. Most subsymbolic module was used to simulate human learning process [10]. Learning was achieved by knowledge cells collection and transformation. These knowledge cells could be combined to produce a complex cognitive process.

Fig. 3 was the architecture of ACT – R model and it liked the information process stage model. Tab. 1 was function and modules match between two models. The environment played an importance role in those models and formed the main. In ACT – R model, the environment established the structures of problems and played an important role in the learning process. These structures could assist learning of chunks, and promote the formation of production rules, reemphasize

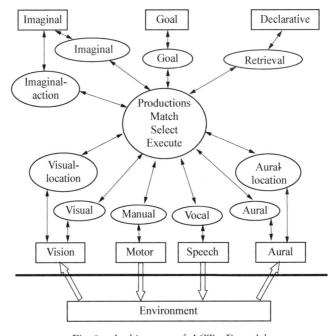

Fig. 3　Architecture of ACT – R model

the key step in understanding of human cognition-the importance of analyzing
essential characteristics of the environment.

Tab. 1   Match modules between information process stage model and ACT - R model

Information process stage model	ACT - R modular structure
Perception (vision, aural, esthesia, taste, proprioceptor)	Vision, aural
Instantaneous memory	Visual-location buffer, visual buffer, aural-location buffer, aural buffer
Consciousness (context relevant information to current task)	Imaginal module
Attention (intent, motivation)	Goal module
Work memory	Imaginal buffer, imaginal-action buffer, goal buffer, retrieval
Long memory	Declarative modules
Decision-making and intellection	Production
Motor (speech, manual, other body action)	Speech module, motor module

Modeling cognitive behaviors of pilot processing alarm would use ACT - R
theory to simulate pilot psychological process and exterior actions. Psychological
process would include goal, imaginal, memory and intellection; Exterior actions
would include visual, aural, speech, manual motor. The scenes of alarm
processing tasks were built by programming language, then we used built-in
cognitive theory in ACT - R and coupled with necessary assumption and knowledge
description of alarm processing tasks to establish ACT - R cognitive architecture
model. The comparison of the model operational results and experimental results
will verify the validity of the model; then the model compatible with human
cognitive was used to instruct study, and achieved the prediction, guidance and
control purposes of tasks.

Modeling cognitive behaviors of pilot processing alarm was based on the
descriptions of tasks of warning information processing, model of cognitive
behaviors of pilots and cognitive description of control components. That included
four sections: procedural knowledge representation and content, declarative
knowledge representation and content, operational process and goals about alarm
processing. Alarm information processing tasks consisted of a great deal of
elements of cognitive tasks or operating rules which were no longer decomposed.
These smallest units were closely consecutive and formed sequences. The database
of typical cognitive behaviors of processing alarm was based on rules and sequence
steps of information diagnosed and manipulated which were extracted from courses

of pilot processing alarm and pilot's interview, those were divided into procedural knowledge and declarative knowledge about alarm processing. The perception control theory [11 - 12], means-end method were used to create the goal hiberarchy of pilot processing alarm, then the goal hiberarchy instructed to build operational procedural rules jointing means and control procedural rules process. Response time of some specific rules could get by basic experiments, most of the operating rules used interior parameters of ACT - R model which were obtained by experimental means early. Finally response times of steps were added up and compared with reaction times of pilots practice on alarm tasks.

When pilot began training, they followed every step in the instructions. After some periods of time, some strategies were generated in their mind and could integrate some separated declarative and procedural knowledge into chunks, simplify by eliminating redundant steps, shorten response time and heighten veracity. The model could do so. Those production rules which often were employed could be strengthened, united and integrated; those production rules which few were employed could be weaken and abated. The ultimate results could be showed reaction time and accuracy. The change of reaction time can be compared with pilots' to verify whether the model was exact.

3) Experiments

Modeling cognitive behavior of pilot processing alarm required a lot of basic experimental support to obtain operational time of some specific minimum tasks. Those experiments include:

● A serial of experiments of eye movement. Several factors influencing the divert time of pilot's point of view would be obtained by statistical analyzing experimental data; the influencing law of formula for reaction time would be explored and fit for instructing optimizing layout of alarm displays in cockpit.

● A serial of experiments of aural distinguish. Experiments on the pilot' memory and retrieval capacity of sounds were carried out. Reaction time of pilot disposing specific alarm sounds and alarm speech were tested and compared.

● A series of experiments of audio-visual alarm dual-task interference.

● Information associated with activation strength test.

● A series of experiments of individual differences.

In addition, the complex alarm experiments would be designed to verify the accuracy of the model. Pilots would process those typical and importance alarms information in test platform. Processing course and the important parameters, such as reaction time, were recorded to compare with the course and parameters of model processing alarm information to verify the accuracy of the model. The difference between the novice and experts which had trained and long-term actual flight was clarified to explore internal psychological mechanism of expert converting knowledge into skill and provide experimental guidance for pilot training to quickly and accurately grasp the alarm information processing skills.

4   Conclusion

Modeling cognitive behavior of pilot process alarm was a huge systems engineering which included experiments, models and theory, and engineering applications. In this paper, the general process of modeling cognitive behavior of pilot had put forward with various contents working logistically and supporting each other. Basic experiments supported theoretical models; experiments which pilot tested in the experimental platform could be verified the accuracy and exactness of the model; the database of pilot process alarm information could support for establish the structure and task scenes of model; at last the model would provide guidance for optimal design of the cockpit warning system and pilot training program.

Acknowledgment

This research was supported by National Basic Research Program of China (973 Program), 2010 CB734101. I would like to thank Su Run'e for discussions relevant to preparation of the last section of this article and Zhang Yugang for his comments on the manuscript.

References

[1] Shi Zhongzhi. Intelligence Science [M], Beijing: Tsinghua University Press, 2005.
史忠植. 智能科学[M]. 北京:清华大学出版社,2005.

[2] Zhu Zuxiang, Engineering Psychology Tutorial [M], Beijing: People's Education Press, 2003.
朱祖祥. 工程心理学教程[M]. 北京:人民教育出版社,2003.

[3] Wagman. M. Cognitive psychology and artificial intelligence: theory and research in cognitive science [M]. Connecticut: Praeger Publishers, 1993.

[4] Newell, A. Unified Theories of Cognition [M]. Cambridge, Mass: Harvard University Press. 1990.

[5] Anderson, J. R. The architecture of cognition [M]. Cambridge, Mass: Harvard University Press. 1983.

[6] Radu J. Bogdan. Grounds for cognition: How goal-guided behavior shapes the mind [M],

Erlbaum, Hillsdale, N. J. , 1994.

[7]　Liu Yanfei. Researches on driver behavior modeling [D]. Zhejiang University, 2007.
　　　刘雁飞. 驾驶行为建模研究[D]. 杭州:浙江大学,2007.

[8]　Liu Yanfei, Wu Zhao-hui, Driver behavior modeling in ACT－R cognitive architecture [J].
　　　Journal of Zhejiang University, 2006. 40(10):1657－1662.
　　　刘雁飞,吴朝晖,驾驶 ACT－R 认识行为建模[J]. 浙江大学学报,2006. 40(10):1657－1662.

[9]　Anderson, J. R. , et al. , An Integrated theory of the mind [J]. Psychological Review,
　　　2004. 111(4):1036－1060.

[10]　Anderson, J. R. , et al. , Information-Processing Modules and their Relative Modality
　　　Specificity [J]. Cognitive Psychology, 2007. 54:185－217.

[11]　Powers, W. T. Behavior: the Control of Perception [M]. Chicago: Aldine. 1973.

[12]　Hendy, K. C. , Beevis, D. , Lichacz, F. , and Edwards, J. L. , Analyzing the cognitive
　　　system from a perceptual control theory point of view [R]. In, Cognitive systems
　　　engineering in military aviation environments: Avoiding cogminutia fragmentosa: A report
　　　produced under the auspices of the Technical Cooperation Programme Thechnical Panel
　　　HUM TP－7 Human Factors in Aircraft Environments. Human Systems IAC, Wright-
　　　Patterson AFB, OH. 2002.

[13]　Wagman. Cognitive psychology and artificial intelligence: theory and research in cognitive
　　　science [M]. Connecticut: Praeger Publishers. 1993. pp. 20－20.

[14]　Newell A. Unified Theories of Cognition [M]. Cambridge, Mass: Harvard University
　　　Press. 1990.

[15]　Anderson J R. The architecture of cognition [M]. Cambridge, Mass: Harvard University
　　　Press. 1983.

[16]　Radu J. Bogdan. Grounds for cognition: How goal-guided behavior shapes the mind,
　　　Erlbaum, Hillsdale, N. J. , 1994.

[17]　Liu Yanfei. Researches on driver behavior modeling. Zhejiang University, 2007.

[18]　Liu Yanfei, Wu Zhao-hui, Driver behavior modeling in ACT－R cognitive architecture [J].
　　　Journal of Zhejiang University, 2006. 40(10):1657－1662.

[19]　Anderson, J. R. , et al. , An Integrated theory of the mind [J]. Psychological Review,
　　　2004. 111(4):1036－1060.

[20]　Anderson, J. R. , et al. , Information-Processing Modules and their Relative Modality
　　　Specificity [J]. 2007:185－217.

[21]　Powers, W. T. Behavior: the Control of Perception [M]. Chicago: Aldine. 1973.

[22]　Hendy, K. C. , Beevis, D. , Lichacz, F. , and Edwards, J. L. , Analyzing the cognitive
　　　system from a perceptual control theory point of view. In, Cognitive systems engineering in
　　　military aviation environments: Avoiding cogminutia fragmentosa: A report produced unde
　　　the auspices of the Technical Cooperation Programme Thechnical Panel HUM TP－7 Human
　　　Factors in Aircraft Environments [R]. Human Systems IAC, Wright-Patterson AFB, OH.
　　　2002.

An Intelligent Lighting Control System Based on Ergonomic Research

Yandan Lin, Wenting Cheng, Chunze Wu, Yaojie Sun

Institute for Electric Light Sources, Engineering Research Center of Advanced Lighting Technology of Ministry of Education, Fudan University, Shanghai 200433

Abstract   In this paper an intelligent lighting control system is introduced with adjustable illuminance, uniformity, correlated color temperature (CCT) and duration and transition of different modes, whose control strategy is based on the ergonomic research. It is verified by ergonomic experiments to be a suitable adjustable system to meet different situation of human needs.

Keywords   intelligent lighting control, human factors, visual performance, health, psychology

1   Introduction

Comfortable lighting and intelligent lighting are the important trends of lighting technology [1]. It is important to ensure that the lighting environment not only meet human beings' visual need [2], but also meet their psychological and healthy need [3 – 5]. However the essence of most current intelligent lighting control system is the application of a field-bus technology, communication technology, automatic control technology and power adjustment technology in lighting field. The aim of these systems is mainly to meet the requirement of illumination adjustable or/and colour of the lighting system transformable. It does not consider that the nature of light is to create a lighting environment that can fit human being, and that the lighting control system should be designed to fulfill the needs of human being's visual requirement, the comfort of subjective feeling, and in accordance with the responses of the body's circadian rhythms. Therefore, it is really necessary to provide LED intelligent lighting a correct ergonomic strategy based on whole analysis of the human beings' need.

After the research of people's behaviors and requirements [6 - 11], we designed the static mode lighting strategy and dynamic lighting strategy, then optimized configuration of the parameters such as the illumination, uniformity and color temperature, etc. Results of ergonomics experiments showed that the optimization can increase the users' acceptance to the environment.

2 Design Principle

The evaluation of lighting quality evaluation will gradually transit from single visual effect evaluation to dual evaluation of visual and non visual effect [12]. Visual effect emphasizes on functionality while non visual effect correlates closely with human health. In order to combine all these effects into the control strategy, we design the principle map of ergonomic control strategy, shown in Fig. 1. The main idea is to analyze the human being's visual, psychological and physiological need, and transform these need into the control strategy of adjusting illuminance, luminance, CCT, CRI (Color Rendering Index), glare and so on.

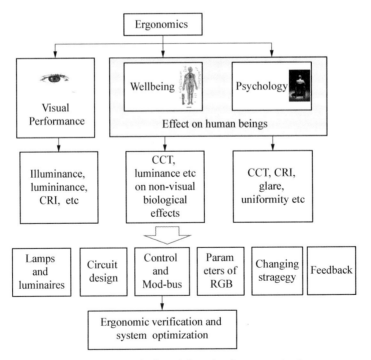

Fig. 1　Principal of applying visual ergonomics in designing the lighting control system

Based on our premier ergonomic research [13], illuminance, illuminance uniformity, CCT are selected as the most important parameters, which affect people's visual comfort, visual function and task performance. Values of these parameters can be set at a certain value in some range in order to meet different needs of different visual tasks. Each combination of different certain values can be stored as different pre-set static control modes. Transition between different 2 modes can be adjusted by controlling the speed of changing to meet different requirements of application.

3 System design

3.1 System configuration

Our designed system consists of an Industrial Personal Computer (IPC) and different data collection boards. Different light sources including LEDs, halogen lamps and fluorescent lamps can be controlled by combining the relative electrical drivers into the data collection boards. The procedure of control is given in Fig. 2.

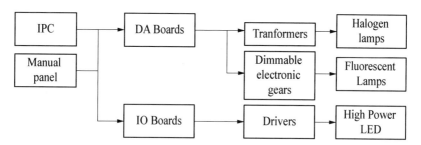

Fig. 2　Control procedure of the designed system

The communication between IPC and data collecting boards is designed with PCI bus control. Control strategy is built in the CPU of IPC and sent through the PCI control bus to the electrical drivers of different kinds of lamps. The reason of choosing the PCI control bus is to ensure the high communication speed, anti EMI performance and for further extension possibility of the system as in the future the intelligent lighting system will be with more and more functions.

The DA boards of PCI - 1724U with 14 - bit 32 multiple isolated analog distributors are used in this system. Output of the DA distributors can be DC signals of $\pm 10\,V$ or $0\sim 20\,mA/4\sim 20\,mA$, respectively sending the control signals to the dimmable electronic control gears of fluorescent lamps to realize an dimmable output between $1\%\sim 100\%$, and to the transformer of a halogen lamps to realize the dimming output between $0\%\sim 100\%$.

The IO boards of PCI‑1730 with 16 digital channels are used in this system. Each channel adopts the photon coupled isolators to ensure the safety of the system and has an output of a voltage of $5\sim40\,\mathrm{V}$ and the maximum current of $200\,\mathrm{mA}$. After receiving the control command from the IPC, the IO board channels will send out independent high speed PWM dimming signals, later amplified by the driver boards to control the independent output of different colorful LED lamps, thus realizing the adjustable illuminance and color or CCT of light.

In order to realize the visual function of the lighting system, we adopt not only RGB LED but also dimmable fluorescent lamps (with CCT of 3,000, 4,000 and $6,400\,\mathrm{K}$) and halogen lamps to ensure enough light levels and better mixed effects of the colors and high CRI. Optical design is carried out to ensure the biggest uniformity of lighting and to reduce glare.

3.2　Setting of controlled parameters' values

This system is targeting the application where not only visual tasks are needed, but also people's comfort feeling and wellbeing are considered especially when the duration of visual tasks are relative long, e. g. classroom lighting and office lighting. Therefore the need of human visual performance is very important, which requires an illuminance level of about $300\sim500\,\mathrm{lx}$ with uniformity of higher than 0.7 and CRI of 80, according to the Chinese standard GB50034‑2004. For adjusting a psychological and physiological effect, CCT $(2,800\sim6,500\,\mathrm{K})$ or even some special colorful effects out of the CCT range should be realized with light levels up to $1,000\,\mathrm{lx}$. Therefore the values should reach the maximum value of the visual task requirement, and should be adjustable at any certain value required by the users between the minimum and maximum data.

Depending on the visual adaptation theory a maximum speed of $100\,\mathrm{lx/s}$ to change the illuminance between 2 different modes is selected, to release the uncomfortable results of the lighting changing on human beings. Values of the speed can be set at any values between 0 and the maximum value.

4　Ergonomic verification

An ergonomic within-subject design is used to verify the above designed system and their effects on observers' rating, respectively on different modes with certain parameter values, and on the changing modes between different scenes. For each rating, a classic 7-score table is used as shown in Tab. 1 to evaluate the final feeling on the scene while different values of the parameters of the dynamic control strategy are setting. For describing the static scenes those words in Tab. 1 are accordingly changed to evaluate the degree of like, comfort, brightness, atmosphere, wide, color satisfaction and acceptance.

Tab. 1　A 7-Score rating table for the ergonomic research

	−3	−2	−1	0	1	2	3	
		←←		same		→→		
dislike								like
uncomfortable								comfortable
nervous								relax
narrow feeling								wide feeling
unfriendly								friendly
depressive								exciting
unremarkable								impressive
too quick changing								unnoticeable change
tired								alertness

4. 1　Effect of scene parameter values on the rating

Four scenes with different values of illuminance and CCT are selected for rating. The values of key parameters in these 4 scenes are shown in Tab. 2.

Tab. 2　Key parameter values of the 4 scenes

Scene	1	2	3	4
Illuminance/lx	500	250	100	800
Uniformity	0. 5	0. 33	0. 7	0. 8
CCT/K	4,000	3,000	2,700	6,400

These scenes were selected from people's daily life. For example, Scene 1 simulated a commonly used indoor general lighting with enough illuminance level and media color temperature; Scene 2 simulated lighting for restaurant, with warmer color and lower illuminance; Scene 3 could be used as bedroom lighting; Scene 4, with a high CCT and high illuminance, could be used in offices to increase people's alertness.

Analysis on the variance of the average rating scores shows that there is a statistical significant difference at the level 0. 05 ($F = 11. 91 > 3. 01$), as shown in Tab. 3.

Tab. 3　Anova of the results

Source	Sum of Squares	df	Mean Square	F	Sig.
Within-subjects	3	452. 46	150. 82	11. 91	<0. 05
Between subjects	24	304. 01	12. 67		
Total	27	756. 47			

An F-test based on the evaluation for the 7 factors of the scenes are made, which results shown that all the evaluation reach a significant difference at the level 0. 05 ($F > 0. 31$), as shown in Tab. 4.

Tab. 4　F values of the results for each factor

	like	comfort	brightness	atmosphere	wide	Color satisfaction	acceptance
F-values	8. 63	25. 77	9. 92	31. 2	22. 75	13. 52	11. 9

Fig. 3 shows the subjective ratings are strongly affected by the illuminance values under the 4 pre-set scenes. With the increase of the illuminance values, a feeling of wider and higher acceptance is reported by the observers.

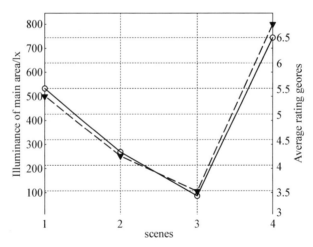

Fig. 3　Different subjective rating under different scenes
with different illuminance values

4. 2　Effect of dynamic strategy

Duration of transition between 2 scenes is set as 4 values of 1 min, 30, 10 and 5 s for comparison. Based on the experiments, the changing mode between different scenes is selected as Tab. 5 suggests. As the different values of parameters will have different effects on the human evaluation, these setting is according to the characteristics of human beings' need.

Tab. 5   Suggestion for changing between scenes

Premier illuminance/lx	Duration of changing time	Changing rate/(lx/s)	CCT changing	Illuminance target/lx
500	1 min	5	Neutral white-warm white	200
200	30 s	3	Warm white-yellowish	100
500	1 min	5	Neutral white-warm white	200
100	5 s	140	yellowish-coolwhite	800

Subjects were asked to do rating on 9 items as Tab. 1 shows. The rating results showed that under different changing strategy, people had similar rating on several items such as "like" and "atmosphere", but had different perception of the items such as "impression" and "alertness". That means different changing rate and duration time does matter. In some applications such as changing to Scene 2 and 3, it's better that lighting changes gradually and makes people unnoticeable, but in others lighting changes suddenly to wake up people or make them alert, for example in Scene 4. Different dynamic lighting control strategy should be adopted under these considerations.

5   Summary

Based on the system design and the results of verification experiments, we can conclude that, lighting control strategy has an important effect on people's perception. Static control strategy for the 4 pre-set scenes can ensure both visual and emotional requirements of people under different application. Besides, the strategy for dynamic changing also satisfies the users.

Finally two kinds of lighting control strategies are mixed up and used in the system. One is static mood lighting scene, and the other is dynamic transition between lighting scenes. With adjustable illuminance, uniformity, CCT and duration time, four scenes are used for different applications in daily life, whose control strategy is based on ergonomic research. Between each two scenes, suitable control strategy is used to achieve a dynamic transformation which makes the user feel natural and comfortable. This should be the trend for future intelligent lighting control system, combining the human needs with lighting control technology.

Acknowledgment

This work is supported by the National Basic Research Program of China (973 Program No. 2010CB734102); Ministry of Science and Technology of China (Project No. 2009GJC00008); Science and Technology Commission of Shanghai

Municipality (Project No. 09DZ1141004); SHMEC project (2006 – 24).

References

[1] Aries M. Human Lighting Demands [M]. Technische Universiteit Eindhoven, 2005.

[2] Pang Y. Vision and Lighting [M]. Peking, 1996.

[3] R. Boyce P. Human factor in Lighting [M]. Taylor &. Francis, 2003.

[4] Rea M S. The IESNA Lighting Handbook [M]. IESNA, 2000.

[5] Simeonova M. Healthy Lighting for the Visual, Circadian and Perceptual Systems [R]. Hospital Engineering &. Facilities management, 2004.

[6] Su L. Research on Color Psychology Based Color Selection Algorithms [M]. Harbin: Harbin Engineering University. 2007.

[7] Tsao J Y. Solid-State Lighting: An Integrated Human Factors, Technology, and Economic Perspective: proceedings of the Proceedings of the IEEE, 2010[C].

[8] Wang S. Study of the Evaluation System for Official Light Environment [R]. China Academy of Building Research, 2007.

[9] Yang G. Vision and Visual Environment [M]. Shanghai: Tongji University Press, 2002.

[10] Zhu H, Zhang Y. An exploration to color psycology based on experimental psycology [J]. China Packaging Industry, 2008,7:48 – 51.

[11] Ju J, Lin Y. The Non-visual Biological Effect of Lighting and its Practical Meaning [J]. China Illuminating Engineering Journal, 2009, 20(1):25 – 28.

[12] Yao Q, Ju J Q, Lin Y D. Discussion on the visual and non-visual biological effect of different light sources [J], China Illuminating Engineering Journal, 2008,19(2):14 – 19.

Flight-Safety-Oriented Virtual Pilot Modeling with the Embodiment of Distributed Human Factors

Yin Tangwen, Fu Shan

School of Aeronautics and Astronautics, Shanghai Jiao Tong University,
Shanghai 200240

Abstract   The highest priority in aviation is safety. Air accidents are attributed to human factors to some degree. The paper presents an innovative flight-safety-oriented virtual pilot with the embodiment of distributed human factors to facilitate human-centered aircraft design and evaluation. The idea is inspired by the essence of flight safety and the virtual pilot takes a distributed form to cover human factors influencing flight safety. The virtual pilot is supposed to be not only convenient for control — adaption to tasks, environment, capabilities, and constraints, but also convenient for analysis — evaluation of capability, measurement of workload, prediction of aircraft performance, judgment of handling quality, and disclosure or estimation of flight safety.

Keywords   Human Modeling, Distributed Control, Virtual pilot

1   Introduction

Virtual pilot modeling was actually initiated by aerodynamicists and control engineers for the analyses of the complete aircraft control loop. Duane McRuer explored the model of the human controller which was a transfer function of time delay model of human (TDMH) [1 - 3, 10]. Pilot modeling has shifted or advanced to the stage of computational modeling of human behavior — human performance modeling (HPM), including task network modeling, computational cognitive modeling, vision modeling, and human error modeling [3 - 5]. HPMs are no longer human controllers described by classical control theory in manual control systems. HPMs are usually been applied to support error prediction and safety analysis as well as to control objects in various degrees of automation [4 - 8].

The growth of automation in cockpit or flight deck brings increasing complexity that leads to the lack of global model of automation upon which the pilot feedback can be designed [9]. The lack of appropriate representation of automation makes it both impossible to model the pilot who works as a supervisor most of the time and to integrate the model with highly automatic systems. While almost all available HPMs are derived from rule of thumb data or experimental data. Such a conclusion can be drawn from [3] by inspecting the data needed to create the HPMs therein.

HPMs are capability descriptive models while TDMHs are compensatory control models. There are huge gaps among the two kinds of human models. Based on the above understanding and the extensive review on flight safety [10], aircraft performance [11], handling quality [10 - 15], and human factors [4, 6, 8], the paper tries to fill those gaps, and tries to handle control and analysis together.

2   Role of Virtual Pilot

2.1   Human-Centered Aircraft Design and Evaluation

The foremost steps and arrangements in aircraft design are concept designs where overall parameters are to be determined, followed by the preliminary design where local parameters are to be defined, and then the detailed design where specific parameters are to be fixed. Since there is no physical aircraft in reality, any considerations on human factors have to be based on former experience and any possible defects on design that related to human factors or conditions which leads to human errors are hard to be identified until flight test or even more sadly late in commercial operation.

Such traditionally designed aircraft would bring about high risk, even more severe consequence. So it is necessary to take human factors in to account at all aircraft design and evaluation stages, especially at the very early stages, which means that the total system including human, aircraft and environment should be thoroughly considered at every stage to guarantee safety and superiority. It is necessary to adopt a human-centered aircraft design and evaluation method which is a whole new idea compared with traditional ones.

2.2   Theory of Pilot-in-The-Loop Complex Flight System

The confronted object in human-centered aircraft design and evaluation is no longer the aircraft by itself. It's a complex flight system with pilot in the loop in which various environment factors can be considered. The aircraft, the pilot and the environment in such a system are all absolutely theoretical with no physical existence.

The pilot-in-the loop complex flight system consists of aircraft parameters, pilot parameters, environment parameters and interactions among pilot, aircraft and environment. An integrated model of the pilot-in-the loop complex flight system in virtual existence should be developed to predict and analyze its overall behavior resulted by such parameters and interactions under various conditions, and ultimately to facilitate human-centered aircraft design and evaluation.

2.3 Modeling of Pilot-in-The-Loop Complex Flight System

The ever possible way to predict and analyze the pilot-in-the loop complex flight system's overall behavior is to build a simulation model of such a system. The model built should be able to predict the system's overall behavior as accurately as possible and be able to analyze the system's overall behavior as thoroughly as possible. To be more specific, the model could be able to assure the reoccurrence of the system's overall behavior, the reducibility of the causes of the system's overall behavior and the decomposability of the controls needed for certain system behavior.

Aircraft modeling, virtual pilot modeling and environment modeling are at the core of pilot-in-the-loop complex flight system modeling. Aircraft modeling includes but not limited to aircraft dynamics modeling, aircraft flight deck modeling and aircraft control modeling. Environment modeling includes but not limited to atmosphere modeling, weather modeling, and airfield modeling.

2.4 Indispensable Virtual Pilot in Pilot-in-The-Loop Model

Virtual pilot modeling is far more complicated than aircraft modeling and environment modeling. Virtual pilot modeling concerns but not limited to error prediction capability, scheduling and multitasking capability, attention and situation awareness capability, memory and learning capability, workload measurement capability, capability to predict aircraft performance, capability to judge handling quality, and capability to evaluate flight safety.

Virtual pilot is the proactive part of the pilot-in-the-loop complex flight system, while aircraft is the reactive part and environment is merely the constrictive part.

Considerations on human factors are essential for the need and the enforcement to study the pilot-aircraft-environment complex system as an integrity. Apparently, virtual pilot modeling is very important and necessary for the modeling of pilot-in-the-loop complex flight system. Virtual pilot with embodiment of human factors is indispensible in any human-centered aircraft design and evaluation.

3   Essence of Flight Safety and Human Factors influencing Flight Safety

3. 1   Essence of Flight Safety

The pilot schedules the aircraft's performance according to the aircraft's handling quality, the aircraft's performance margin and the pilot's capability, to make sure whenever a flight task is accomplished the aircraft is within safety margin — the space needed to manipulate the aircraft is within valid space and the aircraft's fuel is sufficient [10].

3. 2   Distributed Human Factors Influencing Flight Safety

Researches on human factors in aviation have been increasingly carried out, and some major aspects of human factors with significant impacts to flight safety are indentified. Human factors influencing flight safety include but not limited to:

(1) Pilot's control time and control precision.

(2) Visual scene of the cockpit and out-of-the-window area.

(3) Pilot's workload and workload increment.

(4) Pilot's fatigue and fatigue related error.

(5) Pilot's prediction on parameter variation.

(6) Pilot's situation awareness.

(7) Pilot's fast recognition on unacceptable conditions.

(8) Pilot's detection and isolation of fault transient.

Both the pilot's supervision over and the pilot's control to the aircraft are distributed, and the influence to the aircraft caused by the pilot's supervision and control are also distributed. For the pilot's constrained capability, the pilot cannot handle all of the supervision and control at the same time, especially in an persistent manner. That means human factors influencing flight safety are inherently distributive.

4   Virtual Pilot Modeling

4. 1   Safety Orientation and Functional Requirement

Air accidents are caused by unsafe pilot behavior or unsafe aircraft status. Such behavior or status often arise from pilot's shortcomings which are induced by bad environment conditions combined with various human factors. Therefore, it's necessary to reveal objective factors related to human factors and the relevant conversion mechanism from the perspective of human factors. The virtual pilot with the embodiment of distributed human factors to be modeled here holds a unique orientation of flight safety.

As for functional requirements, the virtual pilot to be modeled here should not only be convenient for control — adaption to tasks, environment, capabilities and constraints, but also be convenient for analysis — evaluation of capability, measurement of workload, prediction of aircraft performance, judgment of handling quality, and disclosure or estimation of flight safety as well as human factors analysis and classification, investigation of instability or loss of control, estimation of critical control inputs and dangerous maneuver boundaries, and rapid recognition of faulty situations [16].

In order to simulate reasonable control behavior, the virtual pilot should have two parts, one for the responsibility of stability augmentation and performance prediction of the pilot-aircraft-environment complex flight system, and the other for the judgment of acceptance of handling quality.

In order to disclose potential flight safety hazards, the virtual pilot should be able to establish the connection between pilot parameters and handling quality parameters, and to establish the connection between aircraft performance parameters and flight safety parameters, in case the aircraft is out of safety margin or the pilot is beyond its ability.

Furthermore, the virtual pilot should be able to be effectively integrated with aircraft dynamics which could be high order differential nonlinear equations, and be capable to handle various extreme flight conditions or rare flight conditions which are valuable for research on flight safety.

4.2　Framework and Rationale behind

1) Framework

The pilot-in-the loop complex flight system representing the overall situation as illustrated in Fig. 1 is an integrated architecture which can be used as a fast all-digital simulation framework. The framework can be divided into nine parts: virtual pilot, flight deck, aircraft dynamics, human factors, environment, flight plan, aircraft performance, handling quality as well as flight safety. Of the nine parts, the first three parts are executive entities, and the remaining parts are content of investigation or context of research.

The framework is a computer model of the pilot-in-the loop complex flight system and a tool for comprehensive analysis of human factors, aircraft performance, handling quality as well as flight safety. Both the computer model and the analysis tool are essential in human-centered aircraft design and evaluation.

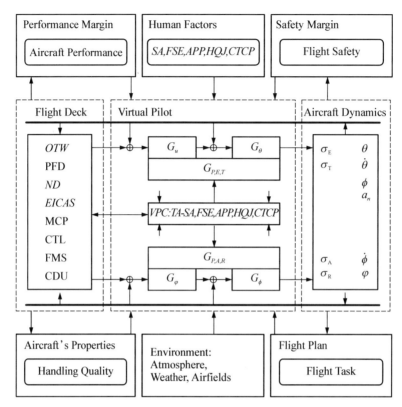

Fig. 1 The framework of the pilot-in-the loop complex flight system and the interrelation among virtual pilot, flight safety, aircraft performance, handling quality, aircraft dynamics, flight deck, environment, flight task and human factors. The interrelation among these parts manifests the links between the virtual pilot's capability and the distributed parameters which stand for control strategies and control effects respectively. The safety oriented virtual pilot with the embodiment of distributed human factors is at the core of the pilot-aircraft-environment complex flight system. Note: symbols and acronyms used are listed in Tab. 1 and Tab. 2.

Tab. 1 Symbols and Meanings

Symbol	Meaning
θ	pitch angle
$\dot{\theta}$	rate of pitch angle
ϕ	roll angle
$\dot{\phi}$	rate of roll angle
a_n	normal acceleration
φ	heading
δ_E	deflection of elevator control surface
δ_T	throttle offset

(continued)

Symbol	Meaning
δ_A	deflection of aileron control surface
δ_R	deflection of rudder control surface
$G_{P, E, T}$	compensatory controller of δ_E and δ_T
$G_{P, A, R}$	compensatory controller of δ_A and δ_R
G_u	compensatory controller of forward displacement
G_θ	compensatory controller of pitch angle
G_ϕ	compensatory controller of roll angle
G_φ	compensatory controller of heading angle

Note: only the most essential control variables and compensatory controllers are listed here.

Tab. 2   Acronyms and meanings

Acronym	Meaning
VPC	Virtual Pilot Capability
TA – SA	Task Allocation and Situation Awareness
FSE	Flight Safety Estimation
APP	Aircraft Performance Prediction
HQJ	Handling Quality Judgment
CTCP	Control Time and Control Precision
OTW	Out of the Window
PFD	Primary Flight Display
ND	Navigation Display
EICAS	susceptibility
MCP	Mode Control Panel
CTL	Controls
FMS	Flight Management System
CDU	Control Display Unit
TDMH	Time Delay Model of Human
AvSSP	Aviation Safety and Security Program
HPM	Human Performance Modeling
DA	Decision Maker

Note: the VPC module consists of the TA – SA module, the FSE module, the APP module, the HQJ module, and the CTCP module. The VPC module, the $G_{P, E, T}$, compensatory controller and the $G_{P, E, T}$ compensatory controller consist of the distributed virtual pilot.

The framework is a hybrid human supervisory control system that combines continuous aircraft dynamics with discrete pilot events through multiple feedback and feed forward loops.

Continuous aircraft dynamics can be simulated via tensor modeling and matrix coding as introduced in [17]. The most difficult task lies in modeling the virtual pilot, including the simulation of reasonable pilot behavior which can be expressed by a series of discrete events and the simulation of feedback or feed forward loops which can be described by the control laws needed and the constraints imposed on.

Due to its inherently interdisciplinary nature, the coupling between continuous dynamics and discrete events is very difficult to be fully understood and simulated [18]. The coupling in real world and the gap in virtual world are the both sides of the same thing. Discrete event is usually not the goal in itself, hence the considering continuous systems with discrete switching events are kind of hybrid switching systems which are referred to as supervisory control [19 - 24]. Intelligent and cooperative capabilities among logic-based decision makers (DA) and distributed controllers to construct and select a continuously parameterized family of candidate controllers can be distributed among decision makers at different levels.

2) Rationale behind

The virtual pilot schematized in Fig. 1 is inspired by the essence of flight safety and the inherence of distributive human factors as specified in section III of the paper. Research and insight on flight safety and human factors helped in revealing the interactive mechanism among the nine parts illustrated in Fig. 1, and more importantly, in revealing the inner drive and explicit capability of the virtual pilot.

A distinctive hybrid virtual pilot with a distributed structure was devised. The virtual pilot capability (VPC) module is a discrete event dynamic system that covers the dominant capabilities such as task allocation and situation awareness (TA - SA), flight safety estimation (FSE), aircraft performance prediction (APP), handling quality judgment (HQJ), control time and control precision (CTCP). All of the modules are decision makers at different levels. The VPC module behaves like a sensory and decision-making part at the top level. The $G_{P, E, T}$ module and the $G_{P, A, R}$ module are continuous controllers that produce adaptive operations such as adjustment of airspeed, vertical speed, altitude, heading, or autopilot.

It is the virtual pilot capability module that drives the hybrid system. Great endeavor was made with the ultimate expectation to create digital models as capable as humans. As a project of the NASA Aviation Safety and Security Program (AvSSP), the Human Performance Modeling (HPM) Project was a stride forward [3]. The HPMs proposed lay emphasis on human capability or performance such as decomposition of tasks, computation of cognitive and generation of behaviors.

Such HPMs will come down to empirical models which require lots of experimental data, because there are no feed foreword control links and no feedback loops between decision-making parts which generate control strategies and knowledge parts which reflect problem solving status. The VCP module is no longer the case.

The $G_{P, E, T}$ module and the $G_{P, A, R}$ module are so special that they can either be regarded as optimal effectors of the virtual pilot or as time delay compensatory controllers of the aircraft dynamic system. In either case, the two modules are time delay models of humans with adaptive ability endowed by the VPC module and compensatory effect to the aircraft dynamic system. Such dual characters help to fill the gap between modern human performance modeling and classical time delay modeling of human. While in classical time delay modeling of human, any modeled human is only capable of handling simple tasks at lower intelligent level with specific optimization criteria and control objects.

4.3 Emergence Mechanism

Define the set of human factors as

$$H = \bigcup_{i=1}^{I} h_i \tag{1}$$

The h_i above are distributed human factors as listed in section 3. Define the set of manipulations as

$$M = \bigcup_{j=1}^{J} m_j \tag{2}$$

The m_j above are manipulations needed to fulfill certain flight tasks. Define the set of states in safe flight as

$$S = \bigcup_{k=1}^{K} s_k \tag{3}$$

The s_k above are states that the space needed to manipulate the aircraft is within valid space and the aircraft's fuel is sufficient. The set of states in unsafe flight can be expressed as $\overline{S}$ which represents the violation of flight safety or excess of safety margin. States in unsafe flight are often shifted from various syndromes of flight accidents that can be defined as

$$\widetilde{S} = \widetilde{S}_Q \cup \widetilde{S}_P \cup \widetilde{S}_T \tag{4}$$

The $\widetilde{S}_Q$ above represents the syndrome of worse handling quality that the aircraft is unable to be controlled or unable to achieve the required control precision. The $\widetilde{S}_P$ above represents the syndrome of poor aircraft performance that the aircraft is unable to execute the commands. The $\widetilde{S}_T$ above represents the syndrome of weak of pilot capability that the pilot is unable to carry out the work as planned.

Three kinds of situations may emerge: the normal situation, the emergent situation, and the occurrence of accident. The above situations can be expressed as

$$T_N: H \times M \times S \to S \qquad (5)$$

$$T_E: H \times M \times S \to \tilde{S} \qquad (6)$$

$$T_A: H \times M \times \tilde{S} \to \overline{S} \qquad (7)$$

Equation (5) represents the normal implementation of the flight plan. Equation (6) represents the occurrence of syndromes of flight accidents, while (7) represents the actual occurrence of flight accidents. Equation (5), (6) and (7) are the emergence mechanisms that cover all the activities in the pilot-aircraft-environment complex flight system. Equation (6) and equation (7) are the major concerns of the paper, especially, equ. (6) is the foremost important emergence mechanism that will be utilized in the following steps of virtual pilot modeling. The outcomes of equ. (6) are well recognized as macro shifts or bifurcations and the outcomes of equ. (7) are well recognized as catastrophes in mutation theory.

4.4 Key Steps in Virtual Pilot Modeling

Prior to present the detailed steps in flight-safety-oriented virtual pilot modeling, here briefly sketches the key steps:

(1) Design the distributed framework of the virtual pilot;

(2) Specify the capability of the virtual pilot;

(3) Specify the distribution of capability;

(4) Specify the constraints on capability;

(5) Specify the characteristic parameter of capability;

(6) Specify the parameter range via inverse simulation;

(7) Verify the virtual pilot via simulation.

4.5 Detailed Steps in Virtual Pilot Modeling

The key steps sketched above actually describe the fulfillment of the gap between modern human performance modeling and classical time delay modeling of human, and the virtual pilot modeled is the synthesis of human performance model and time delay model of human. More comprehensive steps in virtual pilot modeling are as follows.

1) Design the distributed framework of the virtual pilot

(1) Using flight task as a point of departure, specify the situation awareness needed and the capability supposed to be available in the TA – SA module;

(2) Specify the human factors that influence flight safety according to the situation awareness needed;

(3) Specify the control time and control precision needed and the capability supposed to be available in the CTCP module according the flight task and the human factors;

(4) Specify the controls needed and the capability supposed to be available in the $G_{P, E, T}$ module and the $G_{P, A, R}$ module according to the human factors and the control time and control precision;

(5) Specify the feed forward links and the feedback loops among the FSE, APP, and HQJ modules according to the human factors and the controls.

2) Specify the capability of the virtual pilot

(1) Specify the highest adaptive capability that supposed by be available by the virtual pilot based on the worst handling quality;

(2) Specify the extreme values of parameters that reflect the virtual pilot's highest adaptive capability;

(3) Specify the lowest adaptive capability that supposed by be available by the virtual pilot based on the best aircraft performance;

(4) Specify the extreme values of parameters that reflect the virtual pilot's lowest adaptive capability.

3) Specify the distribution of capability

(1) Specify the capability distributions within the distributed framework;

(2) Specify the types and characteristic parameters of the distributed capability of the virtual pilot's.

4) Specify the constraints on capability

(1) Specify the constraints on the distributed virtual pilot's capability of information processing;

(2) Specify the constraints on the distributed virtual pilot's capability of timely responses;

(3) Specify the constraints on the distributed virtual pilot's capability of handling extreme conditions.

5) Specify the characteristic parameters of capability

(1) Specify the compensatory parameters based on control time and control precision;

(2) Specify the response time parameters based on the SA on OTW and the flight deck;

(3) Specify the lead time compensatory parameters base on workload and increment of workload;

（4）Specify the gain compensatory parameters base on fatigue status and fatigue related errors;

（5）Specify the adaptive identification algorithm based on predictions of the change of parameters;

（6）specify the predictive algorithm based on the segregations of fault transitions through estimation and decision.

6）Specify the parameter ranges via inverse simulation

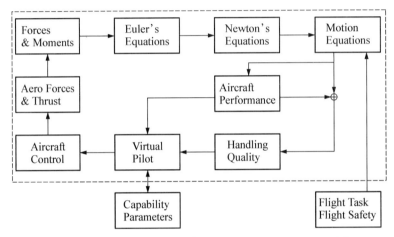

Fig. 2　Inverse Simulation [25] for the specification of capability parameters, using flight task and flight safety as input

（1）Specify the upper limits of the capability parameters according to unacceptable conditions and optimization criteria based on handling quality judgment;

（2）Specify the lower limits of the capability parameters according to unacceptable conditions and optimization criteria based on aircraft performance prediction.

7）Verify the virtual pilot via simulation

（1）Select a group of arbitrary values of the characteristic parameters of capability within the valid range specified previously in step 6;

（2）Use the selected values as the distributed capability;

（3）If there are capability macro shifts compared with the capability specified in step 2, go to step 2;

（4）Run the simulation and estimate whether the aircraft is within flight safety margin through the surveillance of the occurrence of situations represented by （6）and（7）;

（5）If the aircraft is within flight safety margin, modeling is over, otherwise go to step 5.

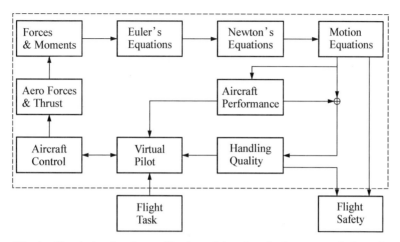

Fig. 3   Simulation for the verification of the virtual pilot, using flight task
as input, and using flight safety as evaluating indicator

5   The Integration of the Virtual Pilot with Simulated Aircraft and Environment

A demonstrative simulation platform which combines the virtual pilot with
simulated aircraft and environment was built as shown in Fig. 4. A typical descent
and landing scenario was simulated with the resultant vertical velocity profile and
altitude profile shown in Fig. 5.

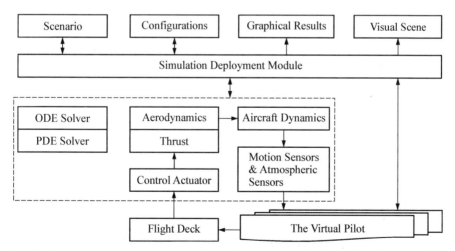

Fig. 4   The integration of the virtual pilot with simulated aircraft and environment. The
simulation platform could be configured to do scientific research on human factors
and flight safety in aviation

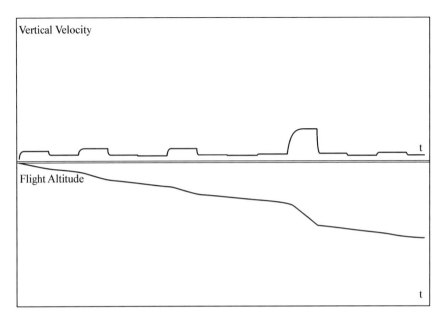

Fig. 5   The resultant vertical velocity profile and altitude profile of a primitive descent
         and landing scenario simulated by the simulation platform with the integration
         of the virtual pilot proposed in the paper. Note: the vertical velocity profile
         was drawn in the upper section and the altitude profile was drawn in the lower
         section

The variation and changing patterns of such profiles' along with the profiles by themselves are very informative and very significant for the analysis of flight safety, aircraft performance, handling quality, human factors, and pilot modeling. For instance, the simulation of descent and landing scenario might provide some insights in system safety assessment to identify system failure conditions [26 - 27].

6   Conclusion

Virtual pilot modeling is very valuable in human-centered aircraft design and evaluation. Close attention was paid on flight safety and human factors in trying to make the virtual pilot modeled been both convenient for control and convenient for analysis. As this is a working paper, further development of the virtual pilot is still in progress.

References

[1]   McRuer D T, Krendel E S. The human operator as a servo system element [J], Journal of
         the Franklin Institute, 1959,267(5):381 - 403.

[2] McRuer D T. Human dynamics in man-machine systems [J], Automatica, 1980,16(3):
237 - 253.

[3] Foyle D C, Hooey B L. Human performance modeling in aviation, 6000 Broken Sound
Parkway NW: CRC Press, 2008.

[4] Leiden K, Laughery K R, Keller J, et al. A Review of Human Performance Models for the
Prediction of Human Error [M]. Moffett Field, CA: NASA, 2001.

[5] Robert G, Hockey J, Compensatory control in the regulation of human performance under
stress and high workload: A cognitive-energetical framework [J], Biological Psychology,
1997,45:73 - 93.

[6] Leiden K, Laughery K R, Keller J, et al. Context of Human Error in Commercial Aviation,
Moffett Field [M], CA: NASA, 2001.

[7] Quintana R, Camet M, Deliwala B. Application of a predictive safety model in a combustion
testing environment [J], Safety Science, 2001,38:183 - 209.

[8] Kontogiannis T, Malakis S. A proactive approach to human error detection and identification
[J], Safety Science, 2009,47:693 - 706.

[9] Vakil S S, Hansman Jr R J. Approaches to mitigating complexity-driven issues in
commercial autoflight systems [J], Reliability Engineering and System Safety, 2002,75:
133 - 145.

[10] Hodgkinson J. Aicraft Handling Qualiyies, Reston, VA: American Institute of Aeronautics
and Astronautics, Inc. 1999.

[11] Eshelby M E. Aircraft Performance: Theory and Practice, Reston, VA: American Institute
of Aeronautics and Astronautics, Inc. 2000.

[12] MOOIJ H A. Criteria for Low-Speed Longitudinal Handing Qualities of Transport Aircraft
with Closed-Loop Flight Control Systems [M], Martinus Nijhoff Publishers, AD Dordrecht,
The Netherlands, 1985.

[13] Lombaerts T J J, Chu Q P, Mulder J A, et al. Modular flight control reconfiguration design
and simulation [J], Control Engineering Practice, 2011,19:540 - 554.

[14] Sachs G. Path-attitude decoupling and flying qualities implications in hypersonic flight [J],
Aerospace Science and Technology, 1998,1:49 - 59.

[15] Teofilatto P. Preliminary aircraft design-lateral handling qualities [J], Aircraft Design,
2001,4:63 - 73.

[16] Goman M G, Khramtsovsky A V. Computational framework for investigation of aircraft
nonlinear dynamics [J], Advances in Engineering Software, 2008,39:167 - 177.

[17] Zipfel P H. Modeling and Simulation of Aerospace Vehicle Dynamics, Reston, VA:
American Institute of Aeronautics and Astronautics, Inc. 2000.

[18] Liberzon D. Switching in Systems and Control, New York, NY: Birkhäuser Boston, c/o
Springer-Verlag New York, Inc. 2003.

[19] Witsenhausen H S. A class of hybrid-state continuous-time dynamic systems [J]. IEEE
Trans. Automatic Control, 1966,11(6):665 - 683.

[20] Ramadge R J, Wonham W M. Supervisory control of a class of discrete event processes [J],
SIAM J. Control and Optimization, 1987,25(1):206 - 230.

[21] Lin F, Wonham W M. Decentralized supervisory control of discrete-event systems [J],
Information Sciences, 1988,44,3:199 - 224.

[22] Moor T, Raisch J. Supervisory control of hybrid systems within a behavioural framework [J], Systems & Control Letters, 1999,38,3,26:157 - 166.

[23] Persis C D, Santis R D, Morse A S. On the supervisory control of multi-agent product systems: Controllability properties [J], Systems & Control Letters, 2007,56,2:113 - 121.

[24] Zaatri A, Oussalah M. Integration and design of multi-modal interfaces for supervisory control systems [J], Information Fusion, 2003,4,2:135 - 150.

[25] Thomson D, Bradley R. Inverse simulation as a tool for flight dynamics research — Principles and applications [C], Progress in Aerospace Sciences, 2006,42:174 - 210.

[26] Lombaerts T J J, Chu Q P, Mulder J A, et al. Early warning and prediction of flight parameter abnormalities for improved system safety assessment [J], Control Engineering Practice, 2011,19:540 - 554.

[27] "Certification Specifications for Large Aeroplanes, CS25, Amendment 8" [S], European Aviation Safety Agency, 2009,pp. 2 - f - 136 - 2 - f - 139.

The Research of Crew Workload Evaluation Based on Digital Human Model

Yiyuan Zheng, Shan Fu

School of Aeronautics and Astronautics, Shanghai Jiao Tong University,
Shanghai 200240

Abstract　The present paper presented the requirements of a digital human model used in the research of human factors in flight deck design and the methods to assess the workload, and established one model to evaluate the crew workload based on certain task.

Keywords　Digital Human Model, Human factors, Fuzzy Model Identification

1　Introduction

Aircraft accidents have always been the major safety crisis to the aviation industry. According to NASA's statistics, over 70% of these accidents could be attributed to the performance of man. Therefore the research of human factor is important to the safety of the aircraft. To correct the errors h happened in preliminary design phrase would cost much more time and efforts in detail design phrase, thus the human factor should be considered seriously at the beginning of the design, especially in the cockpit design, for over 90% flight operations are accomplished there.

Currently, the majority of the researches of human factor in aviation were carried out with the concept of aviation psychology. Although these measures are fairly reliable, it results from the accidents and catastrophes. Therefore, it is reasonable to study the human factor from the perspective of engineering design with digital human as the role of pilots.

2　Construction of Digital Human

Currently, digital human models have been wildly used in different fields like training, risk assessment and UI design. There are lots of benefits of using them in the engineering design and human factors engineering, such as shorter design time, lower established cost and increasing quality. It narrows the gap between engineering and ergonomics, and realizes human-centered concept in the design process.

2. 1　Requirements of Digital Human

Digital human model in the aviation industry, especially researching human factors in the flight deck, is an emerging technique. Based on the specifications and activities of the pilots, the development of such digital human model should consider the following features related to the physical and behavioral attributes or requirements of real pilot:

Coverage: The range of pilot selection is 5%–95% of normal people [1].

Visibility: The pilot's angle of view of outside sitting in normal position is at least 15°.

Comfort: Whether the seat is comfortable enough directly influence the behaviors of the pilot.

Accessibility: Make sure that the digital human models could reach and operate each switch easily.

Ingress and egress: Make sure the target could get in and out the cockpit easily.

Cooperation: Whether the performances of the pilot or the co-pilot could affect the other one that would influence the performances of aircraft?

Decision making: Such digital human model should have basic mental characteristic of human.

2. 2　Construction of Digital Human

Beside the above requirements, the established digital human model should subject to the basic physical parameters of pilots. As a special career, pilot has its own specifications, as Tab. 1 giving out the main dimensions of male pilots in China.

Tab. 1　Main dimensions of male pilots in China

Monitoring Items		Monitoring Items		Monitoring Items	
Height/mm	1,720	Eye height/mm	1,606	Shoulder breath/mm	388
Weight/kg	71	Shoulder height/mm	1,402	Sitting hip breath/mm	349
Upper Arm/mm	320	Sitting height/mm	932	Total head height/mm	233
Forearm/mm	238	Sitting eye height/mm	819	Hand length/mm	186
Thigh/mm	500	Sitting shoulder height/mm	614	Foot length/mm	254

According to the parameters in table and concerned requirements, we built up a digital human with certain parameters in Fig. 1.

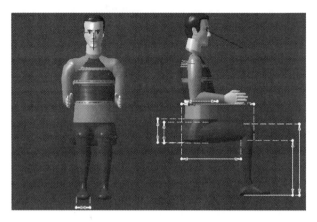

Fig. 1   the Standard Chinese male pilot digital human model

The visibility and accessibility analyses were carried out in a model of prototype of MA60, and the results were satisfied.

3　Crew Workload Research

The definition of workload can be described as the portion of human resources the operator required to perform a task under certain environmental and operational conditions. The resources reflect as physical and mental requirement of the operator. Appropriate workload might improve the efficiency of the operator, however, excessive workload definitely results in severe performance decrements.

Workload evaluation techniques are typically classified into three categories: subjective assessment, physiological methods and task performance measures [2].

Task performance measures basically separate into two types: primary task measures and secondary task measures. Primary task is frequently used as the research of generalization of the study, while the secondary task itself has no practical means and only serves to measure the work load of the operator.

In order to make the primary task measures reliable, it is necessary to well-plan the parameters that should be measured. Frequently, primary task measures are worked out with combination of experimental and operational assessment, like speed, response time, accuracy and error rate are often served to evaluate the primary task performance.

We used primary task measures to evaluate the workload. The primary task was continuous portion of the operations of taking off which included 2 operations to mode control panel, 2 operations to overhead panel and 1 operation to CDU, and the parameters we selected to indicate the workload were average time to fulfill such task and the angle of rotation of shoulder joint of digital human, where Average Time is calculated based on the MOD method [3], and Angle of Rotation relies on Human Posture Analysis in CATIA. After such simulation, we compared the results with data from experienced pilot who achieved the same task under the condition of semi-physical simulation to assess that whether the established digital human model was suitable for workload research.

3.1 Measurement of features of Digital Human and determination of evaluation types

Simulated the digital human models to obtain the average time and angle of rotation of shoulder joint by forcing the models to accomplish the above task, the data as Tab. 2:

Tab. 2 Measurement Results of Digital Human Model

Digital Human Model	Average Time/s	Angle of Rotation of Shoulder Joint/(°)
Standard Human Pilot Model	7.5	366

To giving out a comparative reference, we gathered five experienced pilots to discuss the type of the average time and the angle of rotation of shoulder joint according to their flight experiences. The consensus was showed in Tab. 3.

Tab. 3 the type of behavior

Type	Average Time/s	Angle of Rotation of Shoulder Joint/(°)
Bad	>7.8	>390
Common	7.3 - 8.2	355 - 395
Good	6.8 - 7.6	340 - 370
Excellent	<7.2	<350

3.2 Fuzzy Model Identification

Then, to analyze the workload, we used the Principle of Maximum Membership Degree. Assuming that the parameters of Average Time and Angle of Rotation of Shoulder Joint are following the normal fuzzy set, therefore, the relevant function of membership degrees is [4]:

$$
A_j = \begin{cases} 0, & |x_j - \bar{x}_j| > 2s_j \\ 1 - \left(\dfrac{x_j - \bar{x}_j}{2s_j}\right)^2, & |x_j - \bar{x}_j| \leqslant 2s_j \end{cases}
$$

Where x_j is the parameter of Average Time and Angle of Rotation of Shoulder Joint, $\bar{x}_j$ is the average value of consensus, and $2s_j$ is the standard deviation, then compare each $A_i(x)$ by the following function to determine the membership:

$$A_i(x) \approx \frac{1}{2}\sum_{j=1}^{2}A_j(x_j)$$

3.3　Results

According to the calculation, the membership degree of DH to 'Common' is 0.744,4, which is higher than the degree to 'good' that is 0.449,9. Therefore, we classified the performance of Digital Human to 'Common'.

The results of the model are in accordance with the expectations, and we thought the model could be used to evaluate the crew workload. In order to make the model more human, the mental part would be added to it.

4　Conclusion

In this paper, the author introduced the digital human models used to evaluation the crew workload of pilot, and established one to accomplish a given task. The results are in line with the expectations.

Acknowledge

This research work was supported by National Basic Research Program of China −(973 Program No. 2010CB734103).

References

[1]　Peter A. Van der Meulen, Perry Diclemente Ergonomic Evaluation of an Aircraft Cockpit with RAMSISI 3D Human Modeling Software. Tecmath of North America (2001).

[2]　Farmer, E. and Brownson, A. Revier of workload measurement, analysis and interpretation methods. European Organisation for the Safety of Air Navigation (2003).

[3]　Huang Sheng. Modular Arrangement of Predetermined Time Standard. Repute (2009).

[4]　The Methods and Applications of Fuzzy Math. HUST (2003).

A Method of Analysis Integrating HCR and ETA Modeling for Determining Risks Associated with Inadequate Flight Separation Events

Ruishan Sun, Yunfei Chen, Xinyi Liu, Tianchi Peng, Lu Liu

Civil Aviation University of China, Tianjin

Abstract　This paper proposes an event tree to analyze the risk of inadequate flight separation based on the HCR model. We explore how the consequences of such an event depend on factors such as abilities of pilots and air traffic controllers (ATCs), their nervousness, and the efficiency of human-machine interaction. We also discuss possible measures to control risks, which would be practical to improve civil aviation safety.

　　Keywords　inadequate flight separation, HCR model, Event Tree Analysis (ETA)

1　Introduction

　　With the rapid development of aviation transportation industry, the capacity of air traffic is increasing constantly and the airspace also will become more congested. An effective measure for this jam is reducing flight separation such as the implementation of Reduced Vertical Separation Minimum (RVSM). Many countries have carried this solution into practice step by step, but we should see that it also brings the challenge to pilot and controllers' operation. Because the event of inadequate flight separation will happen if they make some operational errors.

　　In fact the incidents of traffic conflict happened frequently and the risk of inadequate flight separation always exists, if the potential hazard of collision is not detected and effective actions are not taken timely, it could lead to a mid-air collision. It is a disaster when happens, the collision between Boeing 757 and TU –

154 over south Germany on July 1, 2002 was such a sad tragedy (Brooker, 2008;
De Carvalho, Gomes, Huber, & Vidal, 2009).

Therefore, there is a great need in analyzing how the abilities of pilots and
ATCs and the efficiency of human-machine interaction affect the risk analysis
results of inadequate flight separation, and effectively assessing the safety risk.
Based on this, we can implement regulations and measures to decrease the odds of
near-misses and mid-air collisions as well as improve aviation safety.

Currently most of research on inadequate flight separation occurrences focused
on modeling the near miss and calculating the collision probability. There are few
works finished on analysis of human factors and human errors in these incidents
(Xiaohao, Dongbin, & Xiong, 2008; Xiaohao, Dongbin, Xiong, & Xiuhui, 2008;
Zhaoning & Jimin, 2010). Event Tree Analysis (ETA) is a system safety analysis
method which is highly effective in determining how various initiating events can
result in accidents of interest. Human Cognitive Reliability (HCR) is an effective
way to quantify the numerical relationship between human and machine interfaces.
In this study, the ETA and HCR method were integrated together to analyze and
calculate risks brought by inadequate flight separation. The event tree of
inadequate flight separation was given out and the calculation process of error rate
was also illustrated. The case study proved the applicability of this method, and
some constructive suggestions were concluded finally.

2　Methodology

2.1　Safety Risk Analysis

Safety risk analysis (SRA) consists of a variety of methods, e. g., risk
matrix, fuzzy comprehensive evaluation method, fault tree analysis (FTA), event
tree analysis (ETA), and any other methods (Global Aviation Information
Network, 2003; Johnson, 2003; Kumamoto & Henley, 1996). These methods
can all help the Air Traffic Control system to identify hazards, assess risks and
investigate accidents.

A risk matrix is the most popular analysis method in civil aviation industry. It
has two factors: risk severity and risk probability. Each factor has several levels,
e. g., high, medium, and low. This method is more useful when the risk of an
accident is small, and the results are mainly subjective. However, it is not
applicable to severe-consequence rare events (e. g., aviation accidents).

The fuzzy comprehensive evaluation method is another widely used method. It
applies the fuzzy set theory in mathematics to establish an expert system on a given

problem. Supported by Flight Safety Foundation, Hadjimichael and McCarthy (2009) developed the Flight Operations Risk Assessment System (FORAS). This assessment is performed using a mathematical model which synthesizes a variety of inputs, including information on crew, weather, aircraft, and so on. The system will identify those elements that contribute most significantly to the calculated risk. It can be used to reduce the risks caused by Controlled Flight into Terrain (CFIT), Loss of Control (LOC), Runway Incursion (RI), and etc. It is good for multi-level problems, but depends highly on subjectivity (Place, 2005).

FTA is a failure analysis in which an undesired state of a system is analyzed using backward logic to combine a series of lower-level events. This analysis method is mainly used in the field of safety engineering, especially for complex and large systems to quantitatively determine the probability of a safety hazard. However, the required data is hard to collect and the computation is very complicated.

ETA is an inductive procedure that shows all possible outcomes resulting from an accidental event, taking into account whether installed safety barriers are functioning or not, and additional events and factors. By studying all relevant accident events, the ETA can be used to identify all potential accident scenarios and sequences in a complex system. Design and procedural weakness can be identified, and probabilities of various outcomes from an accidental event can be determined. ETA is generally applicable for almost any type of risk assessment application, but used most effectively to model accidents when multiple safeguards are in place as protective features. ETA is highly effective in determining how various initiating events can result in accidents of interest.

2.2　Human Reliability Analysis (HRA)

HRA was first used for safety analysis of nuclear plants in 1960s. Swain and Guttmann (1983) developed the technique for human error rate prediction (THERP), based on an HRA event tree model, which decomposed all crew behaviors into different developing processes, identified the path of failure in the event tree and then carried out quantitative calculation. Later, Kirwan (1994) showed that the main purpose of HRA should be assessing risks caused by human faults and trying to control such human faults. Therefore, he divided HRA into three processes: fault identification, fault frequency determination and design of fault avoidance methods.

The first generation of HRA methods emphasize analyzing, predicting and reducing human faults, analyze and assess human reliability by qualitative and quantitative means on statistics. They can be used to design and improve safety of a system by declining the probability of important human faults to an acceptable threshold. HRA is related to the field of human failure theory and classification research, human reliability data collected (including field data and simulator data), as well as statistical analysis and forecasting techniques of human failure probability based on experts. The main HRA models are as follows: ASEP, CM, DNE, HCR, MAPPS, MFSM, OAT, PC, SLIM, SHARP, and THERP (Embrey, Humphreys, Rosa, Kirwan, & Rea, 1984; Hall, Fragola, & Wreathall, 1982; Hannaman, Spurgin, & Lukic, 1985; Hannaman, Spurgin, Joksimovich, Wreathall, & Orvis, 1984; Potash, Stewart, Dietz, Lewis, & Dougherty, 1981; Samanta, O'Brien, & Morrison, 1985; Seaver & Stillwell, 1984; Siegel, Bartter, Wolf, & Knee, 1984; Swain & Guttmann, 1983).

The second generation of HRA methods further explores human behavior processes. In particular, they studied the mechanisms and probabilities of human errors in different stages of cognitive activities including observation, diagnosis, decision and action. The second generation of HRA models is focus on the integrity of the interaction between people and machine, the impact of psychological process and environment. In addition, the effects of team spirit are often considered . So the second generation of HRA models are properly complies with the idea of human-machine-environment system engineering. Contemporary popular second-generation HRA models are as follows: CES, IDA, ATHEANA, and CREAM (Cooper, et al. , 1996; Hollnagel, 1998; Smidts, Shen & Mosleh, 1997; Woods, Roth, Pople, & Embrey, 1987).

HRA requires different analysis methods for different scenarios and industries. Up to date, these various reliability models have been rarely applied to air traffic control.

The human cognitive reliability (HCR) model belongs to the first generation of HRA methods. Hannaman et al. (1984) proposed a way to quantify the numerical relationship between non-response probability and response time based on human-machine interfaces, human cognition and mean response time (Humphreys, 1995). HCR takes cognitive psychology as a basis, establishes a model to study dynamic cognitive processes (including inspection, diagnosis and decision), and explores the mechanism of human errors. Compared with other models of HRA, HCR has its own advantages. Firstly, the approach explicitly models the time-dependent nature of HRA (Humphreys, 1995). Secondly, it is a fairly quick technique to carry out and has a relative ease of use (Humphreys, 1995). Finally, the three modes of decision-making, knowledge-based, skill-based and rule-based are all modeled (Humphreys, 1995). These characteristics are propitious to the safety analysis of civil aviation. In this paper, we will use HCR and ETA to study the faults of pilots and air traffic controllers (ATCs) in an event of inadequate flight separation.

3　Analysis on An Event of Inadequate Flight Separation

3.1　Build an Event Tree

An event tree describes the process of people's behavior and activities which are described in form of two-state event tree and time-sequence, based on task analysis. Generally, when we analyze human failure through event tree, each branch node only has two probabilities (success or failure). Using the follow scenario as an example, an event tree will be constructed based on this principle.

In 2008, a B737 – 800 received a TCAS alert upon reaching 30,000 feet. A voice warning of "TRAFFIC" and a visual warning of "TA" on the display panel suggested that there was an approaching aircraft at the same altitude (about 6 nautical miles ahead), and that these two aircraft would collide in 20 seconds if they maintained their current directions and speeds. After effective communication and some collaborative work between the air traffic controller and the pilots, this hazard was avoided (Aviation Safety Office, 2007). A scenario was designed based on this event, i. e., if the ATC discovers the conflict, he or she will correct the error, and if the air traffic controller doesn't discover the conflict, TCAS warns, the pilot will take action. Using this logic, we can construct the event tree in Fig. 1.

3.2　Analysis Model of Human Error Rate

In this scenario, the pilot and the ATC are working with a specific human-machine interface and under time pressure, which accords to the scope of HCR model. Therefore, we adopt HCR to study the possible errors of the pilot and the ATC. The formula of the HCR model (Tiemin, Xingkai, Gongzhi, 2005) is given by

$$E = e^{-\left[\frac{(t/T_{0.5})-B}{A}\right]^C} \tag{1}$$

Where t represents the available time to choose and execute appropriate actions, $T_{0.5}$ is the average necessary time to choose and execute appropriate actions, and A, B and C are coefficient representing the behavioral level of a person (see Tab. 1).

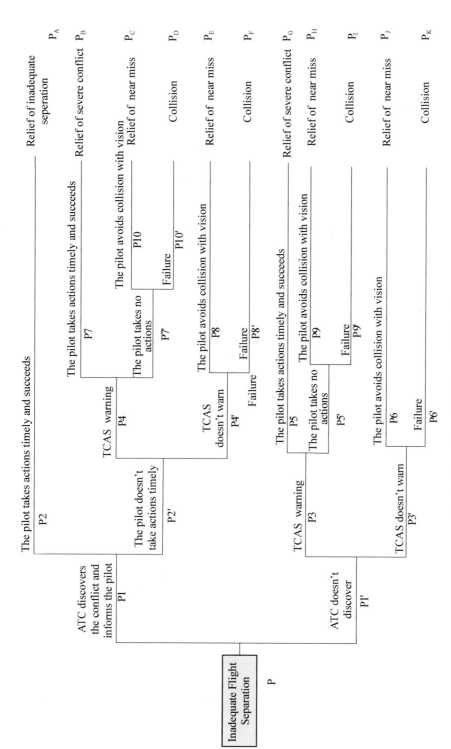

Fig. 1 Event Tree of Inadequate Flight Separation

Tab. 1　Coefficient A, B, C

Behavioral Level	A	B	C
Skill-based	0.407	0.7	1.2
Rule-based	0.601	0.6	0.9
Knowledge-based	0.791	0.5	0.8

The more abstract and novel the behavior, the higher the probability of error. Skill based behaviors are the most routine, consisting simply of stored patterns. Rule based behaviors are slightly more complex, using "if-then" logic. Knowledge based behaviors are the most abstract.

This model considers only cases where $t/T_{0.5} \geqslant B$; when $t/T_{0.5} < B$, it is impossible to finish the job effectively, and the probability of error is considered to be 1. Moreover, with regard to the values for the coefficient A, B and C, due to the differences in crew abilities, human-machine interfaces and regulations across different countries, parameters obtained from experiments of airplane flight simulators will differ. We use the experiment data provided by IAEA, since studies show they are generally applicable for the ordinary rule-based jobs (Xuhong, & Xiangrui, 2007).

The pilots' operations are mostly skill-based, while ATCs' are rule-based. According to Tab. 1, $A = 0.407$, $B = 0.7$, $C = 1.2$ are for pilots' operations, while $A = 0.601$, $B = 0.6$, $C = 0.9$ are for ATCs'. The available time to choose and execute appropriate actions t can be obtained by simulated experiments and analysis. The average necessary time to choose and execute appropriate actions $T_{0.5}$ can be calculated by the following equation(Tiemin et al., 2005):

$$T_{0.5} = \overline{T_{0.5}}(1+k_1)(1+k_2)(1+k_3) \tag{2}$$

Where $\overline{T_{0.5}}$ = Average necessary time to choose and execute appropriate actions under a standard state;

k_1 = coefficient for operators' abilities; k_2 = coefficient for operators' nervousness; k_3 = coefficient for efficiency of human-machine interaction See Tab. 2.

Tab. 2　Coefficient k_1, k_2, k_3

	Condition	Value [31]	Meanings
k_1	Proficient	−0.15	plentiful operating experience
	Average	0.00	ordinary operating experience
	Beginner	0.40	lack of operating experience

(continued)

	Condition	Value [31]	Meanings
k_2	Urgent	0.60	high-strung, personnel have been threatened
	Fairly nervous	0.28	stressful, accidents may occur
	Optimal	0.00	optimal nervousness, appropriate load
	Relaxed	0.20	without sign, low Alertness
k_3	Excellent	−0.22	own emergent support in an emergency
	Good	0.00	display of comprehensive information
	Average	0.51	display without comprehensive information
	Inferior	0.78	display but do not accord with the ergonomics
	Very poor	0.92	the operator can not see a direct display

3.3 Calculation of Human Error Rate

In the scenario given above (See Fig. 1), the cruising speed of an airplane is about 500 knots (kn). When two airplanes are flying in opposite directions, their relative speed is about 1,000 kn.

When the hazard of this scenario happens, the distance between the two airplanes is 6 n mile, leaving only 21.6 seconds(s) for the ATC and pilot to deal with it.

Once ATC discovers two airplanes within an inadequate flight separation, immediately action is taken to increase the separation: the air traffic controller discovers the conflict → the air traffic controller decides how to adjust the separation → the air traffic controller informs the pilot and sends instructions → the pilot answers and reads back the instructions → the pilot moves the flight controls → the airplane's flight path changes.

According to repeated measurements, it takes 2 s for ATC to identify conflict and make decision of adjustment; 1.5 s for ATC to inquire and send instruction; 1.5 s for pilot to answer and repeat; 0.4 s for pilot to control the lever; 2 s for the airplane to change flight path after pilot's operation (Qinggui, 2005). The average total time for the whole procedure of relieving this conflict is 7.4 s.

The total available time to choose and execute appropriate actions t is 21.6 s. We then analyze t, $\overline{T_{0.5}}$ and human error rate for the four basic events in this whole event tree.

If ATC discovers the conflict and informs the pilot, and then the pilot takes actions and succeeds. This includes Event 1 and Event 2.

Event 1. ATC discovers the conflict and informs the pilot:

It includes two basic actions: ATC discovers the conflict and decides to adjust the separation, ATC inquires and sends instructions of separation adjustment, $\overline{T_{0.5}} = 2 + 1.5 = 3.5\,\text{s}$, Corresponding $t = 21.6\,\text{s}$.

Event 2. The pilot takes actions timely and succeeds after ATC commanded:

It includes three basic actions: ① pilot answers and repeats the ATC's instructions; ② pilot controls the lever; and ③ the airplane changes flight path. $\overline{T_{0.5}} = 1.5 + 0.4 + 2 = 3.9\,\text{s}$; corresponding t should be the total available time (21.6 s) subtracted by the actual time of Event 1. However, the actual time of Event 1 is unknown, thus we assume: actual time for a former event/total available time = time for basic actions which the former event include/average necessary time to relief the conflict (in a standard state). Then for Event 2, $t = 21.6 - 21.6 \times 3.5/(3.5 + 3.9) = 11.384\,\text{s}$.

Event 3. The pilot takes actions timely and succeeds after TCAS warning:

If ATC doesn't discover or ATC discovers but the pilot fails to deal, the pilot takes actions and succeeds after TCAS warning, the progress is: the pilot hears and judges the TCAS warning → the pilot informs the air traffic controller (2 s) → the air traffic controller knows and makes decision of adjustment (2 s) → the air traffic controller inquires and sends instructions of separation adjustment (1.5 s) → the pilot answers and repeats (1.5 s) → the pilot controls the lever (0.4 s) → the airplane changes flight path (2 s). Therefore, the whole procedure takes 9.4 s on average. $\overline{T_{0.5}} = 9.4\,\text{s}$; $t = 21.6\,\text{s}$, equal to the total available time.

Event 4. A pilot avoids collision with vision:

If both ATC and TCAS failed, a pilot avoids collision with vision. In a good weather condition, a pilot can see another coming aircraft within an efficient distance of no more than 5 n mile. The total available time for a pilot to choose and execute appropriate actions to avoid collision is at most 18 s (Potash et al., 1981).

A great amount of statistical has shown that it takes 2.35 s (Potash et al., 1981) on average, it takes 2.35 s (Potash et al., 1981) for a pilot to shift his or her vision from outside the cockpit to the display panel inside the cockpit and back outside the cockpit again (see Tab. 3; values from Qinggui, 2005). Generally, it takes about 2 - 5 s for a pilot to decide the best course of action and begin to manipulate the flight controls. The actual time of completion depends on individual abilities, with some needing as few as 2 - 3 s while others need at least 4 - 5 s (Boling, 1999). The current study adopts the average value of 4 s. Therefore, the average necessary time for a pilot to finish Event 4 is $\overline{T_{0.5}} = 2.35 + 4 = 6.35\,\text{s}$. The total available time is $t = 18\,\text{s}$ (See Tab. 4; values from Qinggui, 2005).

If the available time for the pilot to choose and appropriate actions is less than the average necessary time for proper operations (6. 35 s), a mid-air collision will happen no matter what actions are taken.

Suppose that, at the beginning, the air traffic controller and pilot are both at the optimal nervousness, i. e. , in Event 1 and Event 2, $k_2 = 0. 00$. Considering the nervousness of the pilot will increase after he hears TCAS warning or finds the conflicting airplane by visual, we assume in Event 3 and Event 4 $k_2 = 0. 28$. Calculate the human error rate in different combinations of k_1 and k_3 (See Tab. 5 through 8) different hypothetical situations can be constructed. For example, if the operator is a beginner and the human-machine interaction is extremely poor, $t/T_{0.5} < 0. 7$ for Event 3, and the chance of collision is considered to be 1.

Tab. 3   Time for looking from outside of cockpit to instruments and then looking back

Action	Time/s [32]
Turn head and eyes turn to the flight instruments	0. 225
Align the sight and instruments	0. 07
Focus on and instruments	0. 50
Read figures	0. 80
Turn head and eyes to outside of the cockpit	0. 255
Look at the coming plane	0. 50
Total	2. 35

Tab. 4   Time parameters for single event

No	Event	t/s	$\overline{T_{0.5}}$[32]/s
1	ATC discovers the conflict and informs the pilot	21. 6	3. 5
2	The pilot takes actions timely and succeeds after ATC commanded	11. 384	3. 9
3	The pilot takes actions timely and succeeds after TCAS warning	21. 6	9. 4
4	The pilot avoids collision with vision	18	6. 35

Tab. 5   Rate of human errors for event 1

k_3 \ k_1	−0. 15	0. 00	0. 40
−0. 22	$1. 53 \times 10^{-5}$	$7. 67 \times 10^{-5}$	$1. 12 \times 10^{-3}$
0. 00	$1. 64 \times 10^{-4}$	$5. 99 \times 10^{-4}$	$5. 15 \times 10^{-3}$
0. 51	$4. 83 \times 10^{-2}$	$7. 70 \times 10^{-3}$	$3. 43 \times 10^{-2}$
0. 78	$7. 78 \times 10^{-3}$	$1. 69 \times 10^{-2}$	$6. 16 \times 10^{-2}$
0. 92	$1. 13 \times 10^{-2}$	$2. 34 \times 10^{-2}$	$7. 86 \times 10^{-2}$

Tab. 6   Rate of human errors for event 2

k_3 \ k_1	−0.15	0.00	0.40
−0.22	7.17×10^{-7}	1.40×10^{-5}	1.30×10^{-3}
0.00	5.37×10^{-5}	4.743×10^{-4}	1.29×10^{-2}
0.51	6.27×10^{-3}	2.28×10^{-2}	0.16
0.78	2.31×10^{-2}	6.52×10^{-2}	0.30
0.92	3.85×10^{-2}	9.84×10^{-2}	0.39

Tab. 7   Rate of human errors for event 3

k_3 \ k_1	−0.15	0.00	0.40
−0.22	1.13×10^{-3}	5.65×10^{-3}	6.43×10^{-2}
0.00	1.17×10^{-2}	3.76×10^{-2}	0.22
0.51	0.15	0.29	0.74
0.78	0.29	0.49	0.97
0.92	0.38	0.60	1

Tab. 8   Rate of human errors for event 4

k_3 \ k_1	−0.15	0.00	0.40
−0.22	8.04×10^{-5}	6.59×10^{-4}	1.60×10^{-2}
0.00	1.70×10^{-3}	7.91×10^{-3}	7.97×10^{-2}
0.51	4.83×10^{-2}	0.12	0.44
0.78	0.12	0.24	0.67
0.92	0.17	0.32	0.79

3.4   Equipment Reliability

The key equipment in this case is TCAS, so we only consider the reliability of TCAS in this paper. Assume that it has two independent TCAS systems. According to maintenance manual, the probability of TCAS failure in normal operation is 0.02 (Yunfei, 2011).

3.5   Risk Analysis

According to statistics of civil aviation of China, there were 61 events of inadequate flight separation from 1998 – 2007. The total flight time was 21.801×10^6 h (flight hours) (Yun, Yunxiao, & Xiaochun, 2004), the odds of inadequate flight separation are $61/21,801,000 = 2.80 \times 10^{-6}$ /h.

Based on the human error rates (that we calculate by means of the HCR

model), we can get the probabilities of each branch on the event tree. According to
Civil Aviation Administration of China (CAAC) accident and incident standard
(GB14648 – 93, MH/T 2001 – 2008) this paper classifies the consequences of
inadequate flight separation into four categories: conflict of inadequate flight
separation, serious flight conflict (incident), near-miss (serious incident), and
collision (accident). On the basis of different combinations of coefficient for
operators' abilities, coefficient for operators' nervousness, coefficient for efficiency
of human-machine interaction and 4 basic events, we can get more than 100
analysis results of the event tree. Limited to the length of the article, here we
choose the two figures to illustrate: probabilities of consequent events when we
① fix $k_1 = 0.00$, and change the value of coefficient for efficiency of human-
machine interaction; and ② fix $k_3 = 0.00$ and change the value of coefficient for
operators' abilities (See Fig. 2 and 3).

　　(1) When there is efficient human-machine interaction, the accident odds are
acceptable even the operators are beginners. As the skill of the operators grows,
the probability of an accident (or an incident) will decrease; when the operators
are highly skillful, every curve becomes flat. As the operators' abilities changes
from "beginner" to "average", the probability of serious incident is decreasing even
faster than that of incident (See Fig. 2).

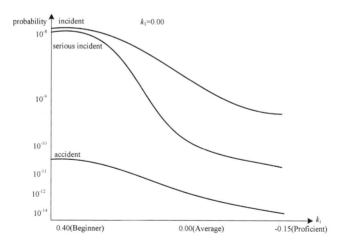

Fig. 2　The Influence of Operators' Ability

　　(2) The efficiency of human-machine interaction has an enormous impact on
the risk of inadequate separation events. Improving it can effectively decrease
incident probability, increase the probability of reliving a conflict. Whether design
of the human-machine interfaces follow principals of human factors; whether the

display equipment and warning system are efficient; whether both pilot and ATC have an emergency support, are very critical to cope with an inadequate separation event. For a generally trained operator, when there is comprehensive display and good human-machine interaction, accident probability can decrease from unacceptable to acceptable (See Fig. 3).

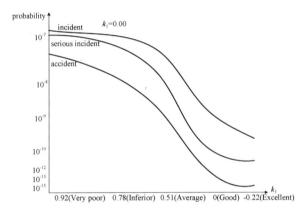

Fig. 3   The Influence of Human-machine Interaction

3.6　Risk control

(1) Design scenario trainings for ATCs and pilots together. They should be trained to get prepared for a real event of inadequate flight separation. When training, special attention should be paid to: the efficiency of communication between ATCs and pilots; ATCs should closely surveillance the airplane during the whole process and pilots should follow strictly standard operations as much; ATCs and pilots should all have a good understanding about the priority between ATCs' instructions and TCAS's advisory.

(2) Improve mentality of ATCs and pilots. ATCs and pilots should be kept in a good and stable metal status. Both ATCs and pilots need to avoid negative mood such as self-satisfaction, slackness, impetuosity, fear and hypertension. In particular, ATCs need to be quick and alert; pilots need to be concentrated, clear-minded, quick and in full control of the airplane all the time.

(3) Improve the efficiency of human-machine interactions. The design of ATC devices and airplane control system should be improved, according to principals of human factors. Such designs should be optimized to fit the operators' physical and mental comfort, display necessary and sufficient information timely, reduce the probability of illusion and misjudgment, and provide effective emergency support.

4 Conclusions

A safety risk analysis method which applies HCR to ETA has been developed in this paper and is applied to the risk analysis of inadequate flight separation.

HCR, basing SRK framework and taking into account the person's cognitive processes and the interaction of human-machine interface, can describe and predict human errors. In addition, HCR is simple-expressioned and clear-concepted. Thus, this risk analysis, combining HCR and ETA, can be applied to the risk analysis of complex human-machine-environment system.

Meanwhile, the case study shows that the application of this method is also very difficult in many cases. In order to develop it into an effective risk analysis method, some research needs to be done. Firstly, the construction rules of event tree which take into account human behavior should be researched. Secondly, $\overline{T_{0.5}}$ in HCR is a key parameter and it is difficult to be calculated sometimes, so the engineering method to get $\overline{T_{0.5}}$ should be researched and developed. Thirdly, alternatives which are more appropriate for human behavior analysis and reliability analysis should be researched when HCR parameters can't be obtained or the method is limited by some conditions.

Acknowledgments

This paper job is supported by the NFSC (Grant No. 60979009) and Major State Basic Research Development Program (973 Program, Grand No. 2010CB734105).

References

[1] Brooker P. The Uberlingen accident: Macro-level safety lessons [J]. Safety Science, 2008, 46:1483 – 1508.

[2] De C P, Gomes J, Huber G, Vidal M. Normal people working in normal organizations with normal equipment: System safety and cognition in mid-air collision [J]. Applied Ergonomics, 2009,40(3):325 – 340.

[3] Xu X H, Li D B, Li X. Research on Safety Assessment of Flight Separation [J], Acta Aeronautica Et Astronautica Sinica; 2008,06:1411 – 1418.

[4] XU X H, Li D B, Li X, et al. Research on Collision Risk of Vertical Separation Minima Based on EVENT Model [J]. Journal of Civil Aviation University of China; 2008,04:1 – 5, 17.

[5] Zhang Z N, Liu J M. Assessment of Collision Risk of Vertical Separation Based on CNS Performance [J]. Journal of Civil Aviation University of China. 2010,01:5 – 8,37.

[6] Global Aviation Information Network, Guide to Methods & Tools for Safety Analysis in Air

Traffic Management，USA，2003.

[7]　Kumamoto H，Henley E. Probabilistic Risk Assessment and Management for Engineers and Scientists，IEEE Press，USA，1996.

[8]　Johnson C W. Failure in Safety-Critical Systems：A Handbook of Accident and Incident Reporting，University of Glasgow Press，Glasgow，Scotland，2003.

[9]　Michael Hadjimichael，A fuzzy expert system for aviation risk assessment，Expert Systems with Applications 36(2009)6512 - 6519.

[10]　Simon Place. ，"Review of Risk Assessment Techniques for Flight Operations"，2005.
　　　a) URL：http：//www. airport-int. com/categories/techniques for flight operations/review of risk assessment techniques for flight operations. Asp.
　　　b) D. Swain，H. E. Guttmann，Handbook of Human Reliability Analysis with Emphasis on Nuclear Power Plant Applications. Final Report. NUREG/CR - 1278，Washington，DC (USA)，1983.

[11]　Kirwan，B. ，A guide to practical human reliability assessment. Taylor & Francis (Bristol，PA)，1994.

[12]　L. M. Potash，M. Stewart，P. E. Dietz，C. M. Lewis，E. M. Dougherty Jr. ，Experience in Integrating the Operator Contributions in the PSA in Actual Operating Plants. Proceedings of the ANS/ENS Topical Meeting on Probabilistic Risk Assessment，port Chester，NY，1981.

[13]　A. Seaver，G. Stillwell，Procedures for Using Expert Judgment to Estimate Human Error Probabilities in Nuclear Power Plant Operations. NUREG/CR - 2743，Washington，DC (USA)，1984.

[14]　W. Hannaman，A. J. Spurgin，Y. Lukic，Human cognitive Reliability Model for PSA Analysis. Draft report NUS - 4531，EPRI Project RP2170 - 3，San Diego (USA)，1985.
　　　A. L. Siegel，W. D. Bartter，J. J. Wolf，H. E. Knee，1984，Maintenance Personnel Performance Simulation (MAPPS) Model：Description of Model Content，Structure，and Sensitivity Testing. NUREG/CR - 3626.

[15]　P. K. Samanta，J. N. O'Brien，H. W. Morrison，1985. Multiple Sequential Failure Model：Evaluation of and Procedure for Human Error Dependency. NUREG/CR - 3837.

[16]　R. E. Hall，J. Fragola，J. Wreathall，Post Event Human Decision Errors：Operator Action Tree/Time Reliability Correlation. NUREG/CR - 3010，Washington，DC(USA)，1982.

[17]　E. Embrey，P. Humphreys，E. A. Rosa，B. Kirwan，K. Rea，SLIM _ MAUD：An Approach to Assessing Human Error Probabilities Using Structured Expert Judgment，Vol. I：Overview of SLIM - MAUD，Vol. II：Detailed Analyses of the Technical Issues. NUREG/CR - 3518，Washington，DC(USA)，1984.

[18]　W. Hannaman，A. J. Spurgin，V. Joksimovich，J. Wreathall，D. D. Orvis，Systematic Human Action Reliability Procedure (SHARP)，EPRI NP - 3583，Research Project 2170 - 3，June 1984.

[19]　D. Woods，E. M. Roth，Jr H. E. Pople，D. Embrey，1987 Cognitive environment simulation：an artificial intelligence system for human performance assessment NUREG/CR - 4862.
　　　a) Smidts，S. H. Shen，Mosleh A. ，1997. IDA cognitive model for the analysis of NPP operator response under accident condition.

[20]　S. E. Cooper，et al. ，1996，"A technique for human error analysis (ATHEANA)，

NUREG/CR – 6350.
[21] Hollnagel, Erik, 1998, Cognitive reliability and error analysis method CREAM.
[22] Xie Hongwei, Sun Zhiqiang etc. , An Overview of Typical Methods for Human Reliability
      Analysis. Journal of National University of Defense Technology, 29(2007), pp. 101 – 107.
[23] Humphreys, P. (1995). Human Reliability Assessor's Guide. Human Factors in Reliability
      Group.
[24] Aviation Safety Office, Aviation Safety Report for Civil Aviation of China, CAAC
      Press, 2007.
[25] Liu Tiemin, Zhang Xingkai, Liu Gongzhi, Application Guide for Safety Assessment
      Methods, Chemical Industry Press, 2005.
[26] Yuan Leping, Chen Yue, Human Error Models in ATM, Air Traffic Management, 2010,
      pp. 35 – 37.
[27] He Xuhong, Huang Xiangrui, HRA in Industrial Systems: Principles, Methods and
      Applications, Tsinghua University Press, 2007.
[28] Wang Hongde, Gao Wei , Study on Erroneous Operation due to Human Factor Based on
      Human Cognitive Reliability (HCR) Model [J]. China Safety Science Journal, 2006,16:51 –
      56.
[29] Liu Qinggui, Commander Vision — Theory and Practice of Flight Safety, CAAC Press,
      2005.
[30] Xu Boling, Lessons Learned — Flight Safety Review of Civil Aviation in New China. CAAC
      Press, 1999.
[31] Chen Yunfei, Research on Applying HRA Methods in Safety Risk Analysis of Air Traffic
      Management [D]. Civil Aviation University of China, 2011.
[32] Luo Yun, Fan Yunxiao, Ma Xiaochun . Risk Analysis and Safety Assessment. Chemical
      Industry Press, 2004.

第三篇　实验研究

材料熟悉程度对长时记忆
提取过程影响研究

张玉刚，薛红军，李云锋

西北工业大学 航空学院，西安 710072

摘要　本文针对长时记忆提取过程的自动性和控制性成分展开了研究，通过设置"分散注意"实验条件进行记忆提取实验，以飞机"警告信息-所属系统"范式为研究材料，获得不同材料熟悉程度与记忆提取的自动性和控制性关系。结果表明随着被试实验次数的增加以及对材料越来越熟悉，提取时间逐步减少，分散注意对记忆提取的影响逐步减弱直至消失，记忆提取表现为控制性向自动性逐步过渡。

关键词　自动性记忆提取，控制性记忆提取，实验范式，提取时间

Analysis of Material Familiarity and Long-term Memory Retrieval

Yugang Zhang，Hongjun Xue，Yunfeng Li

College of Aeronautics，Northwestern Polytechnical University，Xi'an 710072

Abstract　The article analyzed automatic and controlled memory retrieval from human long-term memory. The relation between material familiarity and memory retrieval performance were acquired using "divided attention" during retrieval and "warning information — relative system" material in the experiment. The result showed the memory retrieval time decreased with subjects familiar with materials gradually. It was concluded that influence of divided attention on memory retrieval weakened step by step and disappeared evenly，memory retrieval revealed a transition from controlled to automatic memory retrieval.

Keywords　automatic memory retrieval, controlled memory retrieval, experiment paradigm，retrieval time

1　引言

人的记忆按照信息加工观点包括编码（识记）、存储（保持或遗忘）、提取（回忆或

再认)三个基本过程。Koler 提出的记忆编码与提取重复观点认为,人的大脑是根据记忆材料编码时的操作或活动来回忆的,所以记忆编码和提取过程应是互相联系的[1]。因而 Craik 提出,编码过程是对材料的知觉和理解,提取过程则是试图重现这种初始过程,并随着编码过程在深度、广度和精细程度上的变化而变化。因此,编码和提取过程,除了完成它们各自不同的心理任务之外,在实质上是十分相似的[2]。针对上述观点不同学者从不同的角度通过实验试图证明编码和提取过程是有区别的[3—7],并提出记忆提取过程具有自动性和控制性。本文试图通过记忆提取实验从材料熟悉程度角度获得提取过程自动性和控制性成分间的关系。

2　记忆提取的自动性与控制性

Baddeley 等在他们的记忆实验中设置了"分散注意"实验条件,以分别影响编码和提取过程。结果发现,分散注意作用在编码阶段时,降低了后来的记忆成绩;而作用在提取阶段时,后来的记忆成绩基本不受影响。根据上述实验结果对编码和提取是两个相似过程的观点提出了质疑,并提出了提取过程在很大程度上是自动化的观点[3]。Engle 和 Conway 考查了短时记忆和长时记忆的提取之间的差异。实验结果表明,短时记忆的提取时间随记忆材料容量的增大而增加;而长时记忆的提取时间却不随记忆材料容量增大而变化。这种结果说明,短时记忆的提取是一个系列搜索的过程,所以它所需时间会随记忆材料容量增大而增加,而长时记忆的提取只是一个寻址过程,它只提取一个记忆材料的地址。因此,长时记忆提取不是一个控制性的资源有限过程,而是一个自动的过程[4]。Craik 等考查了分散注意对自由回忆、线索提示回忆和再认记忆中编码和提取的影响。结果表明,当分散注意作用于编码阶段,并强调记忆任务时,记忆作业受到较大影响;而当分散注意作用于提取阶段,记忆作业基本不受影响。这种结果说明,编码阶段受到策略性控制的影响,注意的减少,造成记忆作业的下降;而提取阶段不受这种控制的影响。所以提取在本质上可能更具自动性[5]。以上实验证据均表明,提取过程与编码过程相比更具自动性。

Craik 等研究结果从分散注意对编码、提取阶段的影响而言,由于分散注意对编码阶段的影响较大,而对提取阶段的影响很小或几乎没有,因而得出,记忆提取似乎更具自动性。但若同时还考虑到,在提取阶段上分散注意对记忆提取的反应时的增加比编码阶段上更多,那么也可说明,记忆提取也占有一定的注意资源,从而导致提取阶段的反应时增加[6]。另外,在实验中还发现,采用什么方式(再认或自由回忆)进行测试,对记忆提取的影响也不同。因此,Craik 等提出,记忆编码实质上是对外部事件的知觉和理解的过程。这种过程需要意识参与,并需要注意资源进一步扩展编码操作,使其更精确化。而对于提取阶段,当一种记忆测验(如再认)已提供了充分可靠的提取线索时,那么此时对记忆痕迹的激活或提取是自动化的,不受资源限制;当采用另一种记忆测验(如自由回忆),由于所提供的线索很少,此时提取以一种需要意志努力的方式进行的,因而提取是控制性的,需要注意资源的[5, 7]。

3　记忆提取过程与材料熟悉程度关系

　　Anderson 在研究长时记忆信息提取的实验中研究了材料提取练习次数与提取时间的关系,采用无 Fan 数和有 Fan 数两种提取材料,研究结果表明随着练习次数的增加,提取时间按幂指数关系递减,无 Fan 数和有 Fan 数两种提取材料最终趋于同一个提取时间0.36 s,最终材料的 Fan 效应消失[8]。

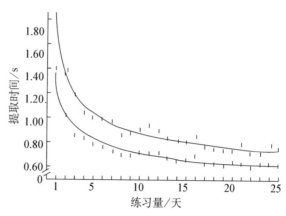

图 1　提取时间与练习量和材料 Fan 数的关系[8]

　　在 Anderson 的实验中没有研究"分散注意"条件对提取过程自动性和控制性成分的影响,作者基于 Fan 数实验范式并设置"分散注意"实验条件,研究上述影响规律。

3.1　实验材料

　　设计如表 1 所示实验材料,形如"故障-系统"的材料信息共 26 句(由 16 个故障和 16 个系统组成),材料形成了三种 Fan 数级别。

表 1　实验材料

Fan：1-1	Fan：2-1	Fan：3-1
襟翼失效-操纵系统	风扇故障-盥洗室	探测烟雾-设备舱
GPS 故障-导航系统	舱门故障-起落架	温度过热-刹车系统
电瓶故障-电气系统	油箱过热-燃油系统	压力过低-增压系统
Fan：1-2	Fan：2-2	Fan：3-2
启动故障-辅助动力	舱门故障-货舱	探测烟雾-货舱
加温故障-客舱	自动停车-辅助动力	温度过热-空调系统
引气故障-空调系统		压力过低-客舱
Fan：1-3	Fan：2-3	Fan：3-3
照明故障-厨房	风扇故障-厨房	探测烟雾-厨房
双发失效-发动机	自动停车-发动机	温度过热-发动机
压差过大-液压系统	油箱过热-液压系统	压力过低-液压系统

3.2　实验步骤

　　整个实验将在计算机屏幕上显示句子,被试通过按键做出反应。实验结果数据是屏幕输出告警信息到被试手动输出判断操作的时间间隔。实验包含三个阶段,第一阶段为自主学习阶段,被试学习以"故障-系统"的形式呈现的 26 个目标句,被试主观认为已经记住所有目标句后进入第二阶段。第二阶段为训练学习阶段,以随机顺序呈现给被试所有可能的形式为"警告信息-?"和"? -所属系统"的问题,要求被试分别报告与某个故障相关的所有系统,或与某个系统相关的所有故障。如果被试能够全部正确回忆所有答案,可认为该问题已被准确记忆。在被试能够完全回忆所有问题的答案后进入第三阶段。第三阶段为反应时测试阶段,如图 2 所示。此阶段

图 2　测试阶段实验界面

分为两种模式,一种是无分散注意影响反应时测试,一种是有分散注意影响反应时测试,分散注意机制采用图 2 中圆圈符号和十字符号的关系控制,圆圈位置随时间在屏幕内随机移动,被试在测试的同时需要操作十字符号位置以保证始终位于圆圈范围内。测试材料由 26 个目标句中随机挑选出 9 个正确句以及由这 9 个正确句重组出的 18 个错误句组成。

3.3　实验结果

实验采用组内设计的实验方法,被试为西北工业大学航空学院硕士研究生6 名。实验结果如图 3 所示,图中提取时间是屏幕输出警告信息到被试手动做出判断操作的时间间隔。由图中数据可以看出开始几次实验分散注意对提取过程具有一定的影响,表现为提取时间的增加。随着实验次数的增加,分散注意对提取过程影响逐步减弱,最终趋于同一水平。

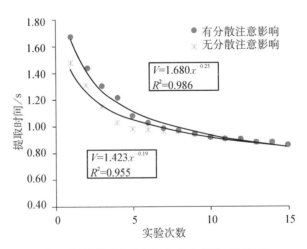

图 3　材料熟悉程度与记忆信息提取时间关系

4 结论

实验结果表明不同材料熟悉程度的被试在进行记忆提取时表现为不同的行为，开始阶段分散注意影响了记忆的提取，表现在反应时的增加，此时记忆提取主要为控制性的；随着练习的增加，进而对材料越来越熟悉，分散注意对记忆提取的影响逐步减弱，直至消失，此时记忆提取主要为自动性的。

参考文献

[1] Koler P A. Remembering operations [J]. Memory and Cognition，1973，(1)：347 - 355.

[2] Craik F I M. On the transfer of information from temporary to permanent memory [J]. Philosophical Transactions of the Royal Society of London，1983，Series B，302：341 - 359.

[3] Baddeley A D, Lewis V, Eldridege M, et al. Attention and retrieval from long-term memory [J]. Journal of Experimental Psychology：General，1984，13：518 - 540.

[4] Conway A R A, Engle R W. Working memory and retrieval：A resource-dependent inhibition models [J]. Journal of Experimental Psychology：General，1994，123(4)：354 - 373.

[5] Craik F I M, Govoni R, Naveh-Benjamin M, et al. The effects of divided attention on encoding and retrieval processes in human memory [J]. Journal of Experimental Psychology：General，1996，125(2)：159 - 180.

[6] Naveh B M. Asymmetry between encoding and retrieval processes：Evidence from divided attention and a calibration analysis [J]. Memory & Cognition，2000，28(6)：965 - 976.

[7] 张明，陈骐. 记忆提取研究的新进展[J]. 心理科学进展，2002，10(2)：133 - 146.

[8] Anderson J R. Retrieval of information from long-term memory [J]. Science，1983，220：25 - 30.

基于飞机驾驶舱警告信息的 Fan 效应研究

李云锋，薛红军，张玉刚

西北工业大学航空学院，西安 710072

摘要 本文针对"故障-系统"范式下的 Fan 效应是否显著展开了研究，采用组内设计的实验方法，获得 5 位被试在各 Fan 展开等级上的反应时间。实验结果显示基于故障概念的 Fan 效应显著，而基于系统概念的 Fan 效应不显著。结果表明系统概念是该研究范式下唯一的维度信息，该空间维度信息将作为前景维度信息，被试选择以系统概念为信息聚合点来整合信息，建构基于系统概念的情景模型。

关键词 警告信息，Fan 效应，前景-背景维度信息，情景模型

Analysis of Fan Effect Based on Aircraft Cockpit Alarm Information

Yunfeng Li，Hongjun Xue，Yugang Zhang

College of Aeronautics，Northwestern Polytechnical University，Xi'an 710072

Abstract The article researched on whether the fan effect is significant under the paradigm of "failure – system". Through the method of Within-Subject Experimental Design，an experiment was taken to attain the reaction time of 5 subjects about each fan grade. The experimental result indicates that the fan effect based on the concepts of failure rather than system is significant. As the unique dimension information under this paradigm，system concepts become the foreground and spatial information. The subjects choose system concepts as the convergent point of information organization. As a result，the subjects construct situation models based on system concepts.

Keywords alarm information，fan effect，foreground-background dimension information，situation model

1 引言

Fan 效应的概念是 Anderson 在长时记忆提取研究中首次提出的。所谓 Fan 效应是指,如果若干事件与某一概念有关,当其中任意一个事件作为再认探测事件出现时,被试对它的反应时或错误率就会随着事件数量的增多而出现反应时的延长或错误率的提高。简单地说,与某一概念有关的事件数量(Fan)越多,提取其中任意事件所花的时间越长[1]。Anderson 在 1974 年的实验一中,采用"人物-地点"范式,表现出了清晰显著的基于人物和地点概念的 Fan 效应。而在 Radvansky 等[2]1993 年的实验中,采用"物体-地点"范式,基于物体概念的 Fan 效应(多场所条件)清晰显著,而基于地点概念的 Fan 效应(单场所条件)不显著,即在不同条件下出现了不同模式的 Fan 效应。

在飞机驾驶舱警告信息中,涉及为数众多的故障和系统,某一故障可能出现在多个系统中,同样某一系统可能出现多种故障[3]。飞行员需要经过系统的知识培训才能充分熟练地掌握这些警告信息,并在紧急情况下及时、准确地做出判断,为作动赢得宝贵时间,保障飞行安全。因此,以飞机驾驶舱警告信息为学习材料,研究"故障-系统"范式下的 Fan 效应对飞行员的知识培训及认知行为建模仿真有重要的指导和参考意义。

2 Fan 效应的解释:ACT - R 理论和情景理论

目前对 Fan 效应的研究主要有两类,一类是以 Anderson 为代表的传统 Fan 效应研究,Anderson 等[1, 4—6]学者自 1974 年以来运用 Fan 效应技术进行了大量激活扩散方面的研究,为建构其 ACT - R 系列理论框架服务;另一类是 Radvansky 等[2, 7, 8]学者的研究,他们对 Fan 效应的内部机制进行了深入的研究,发现了不同模式的 Fan 效应现象,并将 Fan 效应技术与情境模型结合在一起。对于 Fan 效应的解释问题,Anderson 的 ACT - R 理论观点与 Radvansky 的情景模型观点所强调的侧重点是不同的。Anderson 的 ACT - R 观点采用"人物-地点"的研究范式,侧重于从提取加工的角度寻找对各种 Fan 效应的解释,认为提取时不同的注意权重会导致不同大小的 Fan 效应的产生。Radvansky 的情景模型观点则采用"物体-地点"和"人物-地点"研究范式,侧重于从信息内部表征的角度来寻求对 Fan 效应的解释,认为信息在记忆中是否被表征在同一个情景模型中是决定 Fan 效应产生与否的原因[9]。

关于情景模型的建构,Zwaan[10]的事件线索模型理论认为,人们在建构情景模型时至少要受到五个维度信息(时间、空间、主人公、意向和因果等抽象关系)的影响。韩迎春[11]的前景-背景维度假设将这五个维度信息分为两种:前景维度信息和背景维度信息。前景维度信息对情景模型的建构作用主要是影响人们对信息聚合点的选择(即选择哪一个共享概念来整合信息),而背景维度信息对信息聚合点的选择没有影响。至于处于活跃地位的非维度信息,更多的是作为情景建构的内容或实体而存在,对信息能否被整合或信息聚合的方向不产生影响;此外,信息整合的各个方面都要受到个体背景知识的影响。

3 实验设计

本实验在 Anderson 实验一[1]的基础上,保持实验设置基本一致,研究"故障-系统"范式下的 Fan 效应是否显著。Anderson 的 ACT-R 理论将预测到基于故障和系统概念的 Fan 效应均显著。而前景-背景维度假设的预测为,系统概念(如客舱、盥洗室等具有强烈的空间信息感)将被作为空间维度信息,并作为前景维度信息决定情景模型建构过程中信息聚合点的选择,被试将根据系统概念来整合信息,建构基于系统概念的情景模型,因此,基于故障概念的 Fan 效应显著,而基于系统概念的 Fan 效应不显著。

3.1 实验材料

以"故障-系统"范式的 26 个飞机驾驶舱警告信息(由 16 个故障和 16 个系统组成)[3]为学习材料,按照 Fan 展开等级可分为 9 类,数字组合分别表示对应项在句子材料中的联结数(Fan)。实验材料如表 1 所示。

表 1   实验材料

1-1	2-1	3-1
襟翼失效-操纵系统	风扇故障-盥洗室	探测烟雾-设备舱
GPS 故障-导航系统	舱门故障-起落架	温度过热-刹车系统
电瓶故障-电气系统	油箱过热-燃油系统	压力过低-增压系统
1-2	**2-2**	**3-2**
启动故障-辅助动力	舱门故障-货舱	探测烟雾-货舱
加温故障-客舱	自动停车-辅助动力	温度过热-空调系统
引气故障-空调系统		压力过低-客舱
1-3	**2-3**	**3-3**
照明故障-厨房	风扇故障-厨房	探测烟雾-厨房
双发失效-发动机	自动停车-发动机	温度过热-发动机
压差过大-液压系统	油箱过热-液压系统	压力过低-液压系统

3.2 实验步骤

整个实验将在计算机屏幕上显示句子,被试通过按键做出反应。实验包含三个阶段,第一阶段为自主学习阶段,被试主观认为已经记住 26 个正确句后进入第二阶段。第二阶段为训练学习阶段,以随机顺序呈现给被试所有可能的形式为"故障-?"和"?-系统"的问题,要求被试报告缺失部分的所有答案,该学习阶段要完成 5 轮无错误回答。第三阶段为反应时测试阶段。测试材料为随机挑选的故障和系统都不相同的 9 个正确句以及由这 9 个正确句重组出的 18 个错误句,随机挑选可以保证在整个测试阶段对所有实验材料的记忆。反应时间测试包含 9 组测验,每组测验 36 次测试,共 324 次测试。在每组 36 次测试中,9 个正确句每句测试两次,18 个错误句每句测试 1 次,以保证 9 个故障和 9 个系统出现次数相同(4 次),避免测试频率不同对反应时间的影响,句子的呈现顺序要控制启动效应[12]的影响到最小。每 3 组实验之间安排 5 分钟休息时间。句子输出后,被试通过快速点击键盘 D 键(正确)或 K 键(错误)来判断句子的正确性。

3.3　实验被试及仪器

被试为西北工业大学航空学院硕士研究生 5 名;所需仪器包括计算机及 C++ 平台上开发的"Fan 效应测试系统"。

3.4　实验结果

实验结果数据是屏幕输出警告信息到被试手动做出判断操作的时间间隔,实验数据如表 2 所示。

表 2　5 名被试在各实验条件下的测试值(空白处为坏点)

	Fan 展开等级																	
	正确句									错误句								
故障 fan	1			2			3			1			2			3		
系统 fan	1	2	3	1	2	3	1	2	3	1	2	3	1	2	3	1	2	3
编号	反应时间/ms																	
1	782	1,075	908	1,047	1,032	989	970		1,039	846		915	1,085	1,145	1,019	1,008	1,160	1,108
2	812	1,007	880	911	939		973	1,054	1,003	868	1,056	964	1,018	1,105	940	984	1,065	1,140
3	923	949	816	1,034	926	955		1,049	879	938	987	877	979	940	1,034	983	1,147	1,016
4	755	1,015	877	852	858	928	976	987	903	946	954	874	945	877	956	917	1,069	940
5	843	1,082	834	967	948	960	975	1,009	939	894	1,053	890	1,011	964	963	967	1,035	939

4　结果分析及结论

故障概念的 fan 效应值如图 1 所示,正确句为 1-fan(904),2-fan(954),3-fan(984);错误句为 1-fan(938),2-fan(999),3-fan(1,032);整体平均值 1-fan(921),2-fan(976),3-fan(1,008),基于故障概念的 Fan 效应清晰显著;系统概念的 Fan 效应值如图 2 所示,正确句为 1-fan(920),2-fan(997),3-fan(925);错误句为 1-fan(959),2-fan(1,038),3-fan(972);整体平均值 1-fan(939),2-fan(1,017),3-fan(948),基于系统概念的 Fan 效应不显著,近似呈现为"中间高两头低,两头值近似相等"的态势。

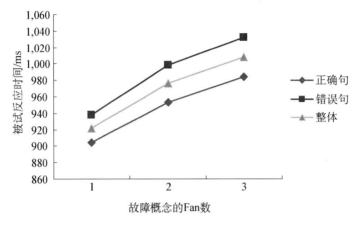

图 1 故障概念的 fan 效应图

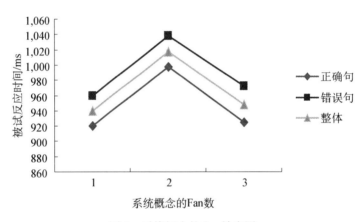

图 2 系统概念的 fan 效应图

上述实验结果与 Anderson 的 ACT－R 理论的预测不相符,故障和系统的 fan 效应并没有同时显现出来,这是 ACT－R 理论所无法解释的。相反,实验结果与前景－背景维度假设的预测大体相一致。在"故障－系统"范式下,被试在对学习材料进行编码的过程中,系统概念是唯一的维度信息,该空间维度信息将作为前景维度影响被试选择以系统概念为信息聚合点来整合信息,建构基于系统概念的情景模型,导致了基于系统概念的 fan 效应不明显,而基于故障概念的 fan 效应显著。至于同处于活跃地位的故障概念,作为非维度信息,扮演情景模型的实体或内容角色,对信息是否被整合及信息聚合方向不产生影响。针对系统概念所表现出的"中间高两头低,两头值近似相等"的态势,通过对 2－fan 系统概念材料进行分析,可以看出空间感强烈的"客舱"和"货舱"之间存在强烈的干扰,导致被试在记忆提取时要花费额外的时间对判断材料进行辨认,增加了反应时间。另外,"辅助动力"和"空调系统"的空间感较弱,情景模型在建立时难以有可以依托的信息维度,很难建立基于这两个概念的情景模型,这样也会增加反应时间。

参考文献

［1］ Anderson J R, Retrieval of propositional information from long-term memory ［J］. Cognitive psychology, 1974,6:451－474.

［2］ Radvansky G A, Spieler D H, Zacks R T. Mental model organization ［J］. Journal of Experimental Psychology: Learning, Memory, and Cognition, 1993,19:95－144.

［3］ 中国南方航空公司. A318/A319/A320/A320 飞行手册［M］.

［4］ Anderson J R, Reder L M. The fan effect-new result and new theories. Journal of Experimental ［J］. Psychology: General, 1999,128:186－197.

［5］ Anderson J R. Retrieval of information from long-term memory ［J］. Science, 1983,220: 25－30.

［6］ Sohn M S, Anderson J R, Reder L M, et al. Differential fan effect and attentional focus ［J］. Psychonomic Bulletin & Review, 2004,11,4:729－734.

［7］ Radvansky G A, Zacks R T. Mental models and fact retrieval ［J］. Journal of Experimental Psychology: Learning, Memory, and Cognition, 1991,17:940－953.

［8］ Radvansky G A, Zacks R T, Hasher L. Fact retrieval in younger and older adults: The role of mental models ［J］. Psychology and Aging, 1996,11,2:258－271.

［9］ 韩迎春,迟毓凯,邢强. Fan 效应与情境模型的研究［J］. 心理科学进展,2001,9(2):120－123.

［10］ Zwaan R A, Langston M C, Graesser A C. The construction of situation in narrative compression: An event-indexing model ［J］. Psychological Science, 1995,6:292－297.

［11］ 韩迎春. 前景-背景维度假设与情境模型［D］. 广州:华南师范大学,2002.

［12］ 游旭群,苟雅宏. 内隐记忆的启动效应［J］. 心理科学进展,2006,14(6):829－836.

Target-Pointing Movement Performance for HCI Design

Zhang Xiaoyan[1], Xue Hongjun[1], Zhou Lin[2], Zhao Junwei[2]

1. School of Aeronautics, Northwestern Polytechnical University, Shanxi Xi'an 710072

2. Commercial Aircraft Corporation of China, Ltd. Shanghai 200232

Abstract   One of the most important factors for human-computer-matching is HCI design, which is based on the characters of human target-pointing movement. The writer investigates the target-pointing movement where target width and index of movement difficulty varies as does the approach angle. Through the study of reaction time and accuracy of subjects there are some conclusions: approach angel significantly affects the character of target-pointing movement, with 0° approach angle, subjects have worst movement time and throughput, while with 60° approach angle, subjects have shortest movement time and with 90° approach angle, subjects have best throughput which is about the same as 60° angle, different muscle combinations and different effective widths would be responsible for the differences; the width has the same influence on the characters of human target-pointing movement at every approach angle, and the movement time will be more sensitive to width with the bigger index of difficulty. The conclusions can be used to instruct HCI design.

Keywords   target-pointing movement, Fitts' law, approach angle, HCI design, index of movement difficulty, effective widths

1   Introduction

Human computer interface (HCI) is always composed with windows, menus, buttons and other icons and so on. People often operate the equipment through target-pointing movement by the mouse, specialized buttons in keyboard and other input facilities. Target-pointing movement is the most important movement for the evaluation and design of HCI and interaction equipments for example buttons' shape, size and other geometrical characters [1 - 3]. The movement has been sudied for many years range from equipments' design and evaluation and human locomotion characters and throughput of HCI [5 - 6]. There are also different models trying to reflect the movement characters [7 - 9]. Fitts had built the initialized model in 20C 50G from information theory, and also validated as a more successful model. At present the general form of Fitts' Law is: $MT = a + b\log_2(W/D+1)$ [9], for which MT means movement time, W means target size or the target width, D means target distance, and a, b is the coefficient, and the value of logarithm represents the index of difficulty (ID).

The researchers have already built some models which are functions of target width and distance to try to discover the essence of target-pointing movement, but the movement characters are not just related with width and distance but approach angles or other factors. The paper tries to describe the movement from the intercross effect of approach angles, width and distance. The conclusions are useful for HCI and elements design.

2 Experiment

2.1 Subjects

、   There are seven undergraduate subjects from which three are girls. Right hand is the advantage hand of all subjects. The cognitive state is as good as usual and there is no fatigue or other uncomfort. The experiment desires the subjects react to the target as fast as he can but make sure the reaction is right.

2.2 Apparatus

The experiment needs computers and touch screen. The 19' colorful screen's refresh frequency is 60 Hz, and the display resolution is $1,600 \times 1,200$. The screen is put horizontally to make sure the experiment is two-dimension.

2.3 Design and Procedure

The experiment is in-subject design, and the variable level consists target widths (0.5 cm, 1.0 cm, 1.2 cm), ID (2, 3, 4, 5) [10] and approach angles (0°, 30°, 60°, 90°). For each variable group subject should repeat ten times, so there will be $3 \times 4 \times 4 \times 10 = 480$ experiments for each subject. Initial position is in the down left corner of the screen represented by a 10 mm × 10 mm yellow square. The subject should react to the target on the screen and reaction will be timed with the index figure leaving the initial position until the figure touched the target. If subject reacts to the wrong target there would be a warning by computer. The reaction time and accuracy will be saved.

3　Experiment results

3.1　Movement time of different approach angles

The movement time (MT) is shown in Fig. 1. The slowest movement is in the horizontal direction from left to right (approach angle is $0°$), and the fastest is from down left to top right (the approach angle is $60°$), the trend of the result is just the same as the work done by Hancock [11] shown in Fig. 2. $0°$ is different from the $90°$ while the ID is less than 5, but when the ID is 5, all the data almost

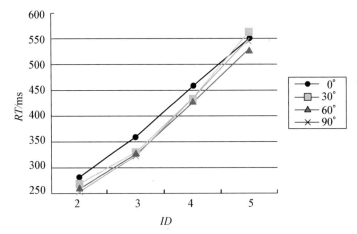

Fig. 1　MT of different angles

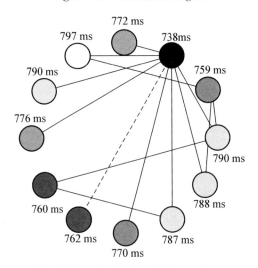

Fig. 2　Result of Hancock' experiment

become the same except $60°$ orientation. The ANOVA analysis revealed a significant main effect of different angles while $F = 11.199 > F_{0.05}(3, 96) \approx 2.7$.

3.2 Throughput of different approach angles

Throughput means index of performance and the unit is bits/second. According the research of Soukoreff, the calculate method is: $TP = \dfrac{1}{y} \sum\limits_{i=1}^{7} \left(\dfrac{1}{x} \sum\limits_{j=1}^{4} \dfrac{ID_{ij}}{MT_{ij}} \right)$. From which, y means the number of subject, x means the number of experiment variable. In this experiment, $x = 4$, $y = 7$, $ID = \{2, 3, 4, 5\}$. The throughput of each approach angle is shown in Fig. 3. The throughput increases with the approach angle increasing anticlockwise and the throughput is best at $90°$, a little better than $60°$ orientation.

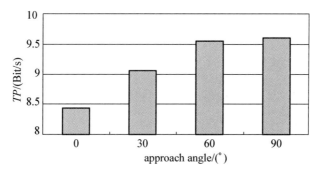

Fig. 3   Throughput of different approach angles

3.3 The effect of the size in different approach angles

The effect of the target size in different approach angles is shown in Fig. 4. The width of different angles affects the movement performance in almost the same way. While the ID is smaller, less than 4, the width has almost no effect on MT, but while the ID is bigger than 4, the width affects performance gradually, but when the width ranges from $1 - 1.2$ cm, the MT increases not so significantly. Through the ANOVA analysis, the width has a significant effect on MT.

4 Conclusions

4.1 Movement performance of different approach angles

The movement performance is different among different approach angles. The shortest MT (movement time) is $60°$ approach angle, and the opposite situation is $0°$ angle. But while the ID increases the discrepancy of MT is gradually smaller and smaller, especially when the ID is 5, $0°$ and $90°$ angle has almost the same MT. The throughput increases with the approach angle increases anticlockwisly,

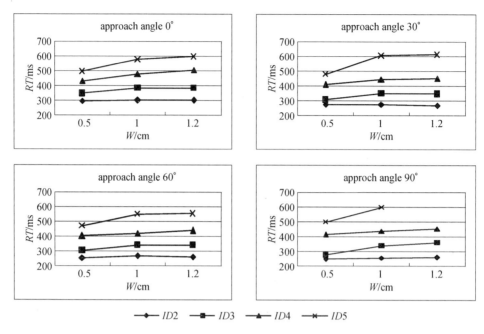

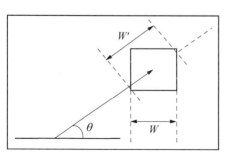

Fig. 4　Effect of width of different angles on *MT*

and the throughput is best at 90°. While the angle bigger than 60° *TP* increases gently.

The research of Hancock [11] indicated that *MT* was related with muscle group employed. The significant different performance of horizontal and vertical orientation is induced by different muscle groups when *ID* is less than 5. The horizontal movement employs figure joints and wrist joints while the vertical only need figure joints and the vertical movement fits the movement rule of hand joints much more. When *ID* is bigger, and distance is longer, horizontal movement and vertical movement both need employ wrist joints, elbow joints even shoulder joints, and then the movement time has no significant difference. And the performance of target-point movement is also related with the operation habits of human, when right hand is advantage, the subjects used to point from the down left to the up right which fits comfortable movement rule and so the shortest *MT* is in 60° direction. For the depth analysis there is different effective width in different approach angles. The effective width is defined as the width which is consistent with the approach direction shown in Fig. 5.

Fig. 5　effective width

According to Fitts' law the effective width $W' = W/\cos\theta$. When the $\theta = 60°$, the effective width is the biggest and ID is the smallest and the MT is also the shortest.

4.2   The effect of width in different approach angles

The width has a significant effect on the performance in different angles, and effects the movement performance in almost the same way. The former research of horizontal movement indicated that when the ID is smaller (2, 3) the width has no significant effect on MT, but when the ID is bigger, the situation is opposite. The error rata is higher while the size is smaller, the result indicates that when $ID > 3$, $D > 3\,cm$, and $W = 0.5\,cm$ the error rate is gradually increase, when $ID = 5$, the error rate is highest; when ID is bigger, MT is sensitive to the change of size. Horizon movement has the same performance as the other approach angles.

But when the approach angle increases (bigger than $30°$), the width is not sensitive to ID any more, when ID is less than 5, the width has no effect on MT, and the effect is mainly on error rate, bigger width, less error rate.

4.3   Suggestions for Design

HCI design not just relates with the movement time but also the error rate. Error rate is mainly related with target size and movement distance, movement time is mainly related with movement distance and approach angles. The research indicates that layout of control panel should consider the initial position of operator hand, and the layout should be on the same level; the buttons on the panel should allocated with the movement performance of hand, the emergency button should lay on the top right $60°$ position, and the effective size should be 0.8—$1.0\,cm$.

The HCI design is also related with the cognitive performance of human such as eyemovement, the integrated human performance model should be built to instruct HCI design and arrive at perfect man-machine match.

References

[1]   Christian S. An overview and evaluation of modern human interface devices January 27, 2010. http://academic.research.microsoft.com/Publication/13367837/

[2]   Ravin Balakrishnan. "Beating" Fitts' law: Virtual enhancements for pointing facilitation [J]. International Journal of Human-computer Studies, 2004(61):857 – 874.

[3]   M. Risto. Scientific user interface testing: exploring learning effects and Fitts' law [D]. University of twente, 2009.

[4]   Jacob O. Wobbrock, Brad A. Myers, Htet Aung. The performance of hand posture in front-and back-of-device interaction for mobile computing [J]. International Journal of

Human-computer Studies, 2008(66):857 - 875.

[5] Huang zhiqi, Chen dongyi, Wang houjun. Experim ent of mthe Twiddler Tapping Performance Model Based on Fitts' Law [J]. Journal of electronic measurement and instrument. Vol. 22 No. 5:33 - 37 [in Chinese].

[6] Xu ronglong, Liu zhengjie. Establishment and Verification of Menu Selection Performance Model [J]. Computer Engineering. Vol. 36 No. 19:256 - 260 [in Chinese].

[7] Michael J. Mcguffin, Ravin Balakrishnan. Fitts' law and expanding targets: experimental studies and designs for user interfaces ACM Trans. Comput. Hum. Interact, Vol 12, P:388 - 422,2005.

[8] Johnny Accot, Shumin Zhai. Beyond Fitts Law: Models for Trajectory-Based HCI Tasks [C]. In Proceedings of ACM CHI'97 Conference on Human Factors in Computing Systems, 1997:295 - 302.

[9] I. Scott MacKenzie, William Buxton. Extending Fitts' law to a two-dimensional tasks [C]. Proceedings of the CHI'92 Conference on Human Factors in Computing Systems, 219 - 226. New York: ACM.

[10] R. William Soukoreff, I. Scott MacKenzie. Towards a standard for pointing device evaluation, perspective on 27years of Fitts' law research in HCI [J]. International Journal of Human-Computer Studies, 2004(61):751 - 789.

[11] Hancock, M. , Booth, k. Improving menu placement strategies for pen input [C]. In Proceeding of Graphics interface, Canadian Human Machine Communications Society. 2004, Canada, 221 - 230.

Usability of Using Polarized Filter for Reading Lamp to Reduce Disturbing Reflected Glare

Yandan Lin, Qianyun Du, Liqing Tong, Yaojie Sun

Institute for Electric Light Sources, Fudan University; Engineering Research Center
of Advanced Lighting Technology, Ministry of Education; Shanghai 200433

Abstract　In this paper the usability of using polarized filters for reading lamp to reduce disturbing reflected glare was tested based on a within-subject design ergonomic research. Results show that with the increasing polarization degree the subjective rating score decreases, which means that increasing polarization of the optical film reduces the reflected glare. Statistical analysis shows that there is high correlation (0.05 level, 2 – tailed) between the subjective rating score on reflected glare and the degree of polarization. An univariate analysis of variance proves that the different evaluation caused by the optical films with different polarization degrees has statistically significant meaning ($F = 108.4$, $p < 0.001$). The results of this research is very useful for optical design of reading lamp, which occupies a big market in China, affecting the visual comfort of huge numbers of readers, such as students and office people.

Keywords　Polarization, Glare, Reading lamp, Visual comfort, Veiling reflection

1　Introduction

In China, many students and officers take more than 8 hours working on the table. In such long working time, desktop lamps are the most important light source, whose lighting parameters are directly relevant to visual performance and visual comfort. The desktop lamp which has higher visual comfort can decrease the degree of asthenopia and the rate of myopia in young students . Also, optimizing the lighting conditions increases the task performance, workers' efficiency and accuracy [1].

Previous research on visual tasks shows that illuminance, disability and discomfort glare, luminance distribution and colour have a complex interaction on visual comfort. The differences of individual visual ability and the observers' preference also impact their assessment [2]. Among these factors glare is one of the important parameters that greatly affect the visual comfort of lamp, including direct glare and reflected glare. A scientific normal optical design can avoid the direct glare effectively, but it can only decrease the degree of the reflected glare. So the reflected glare, the highlight reflected image of the light source on the table between the lamp and the user , is the main glare source, for most of the reading tasks. Usually, we call it as veiling reflection. Veiling reflection can change the luminance of target and backgroud, leading to the reduction of contrast rending and visibility, causing discomfort like phototropism and fusion stimulus [3 - 5].

Light vibrations are usually divided into two principal directions — parallel component (p-polarized) and perpendicular component (s-polarized). According to Brewster's law, if the reflection occurs on a glossy surface which is nonmetal and transparent such as glass, the reflected light is partially polarized as larger percentage of the perpendicular component is reflected than the parallel one. If the angle of incidence is equal with Brewster's angle, the reflected light contains only the s-polarized component. So we can decrease the luminance of reflected image on glossy surface through adding the percentage of the p-polarized light of the luminaire [6]. This principle has been widely used in many applications. The polarized sunglasses are used to make the situation under flashing water clearly. Application of polarization instruments also can be found in camera and lighting system of motor vehicles, and automotive polarizer screen shade is made. At the same time, the feasibility of using polarized light to reduce the veiling glare has been also discussed. E. E. Richman evaluate the potential benefits of polarized light by installing the polarized lighting lenses to the office illuminating system, but because the notable dimming after installment, the subjective value is not satisfied. Because that the experiment did not control other paramenters in a comparable range, the test doesn't provide any concrete conclusion about the effectiveness of the polarized lighting [7]. Mark S. Rea proved that polarization leads to a significant changes in performance even at $25°$ viewing angle while the difference between target and back groud is small [8]. And his later research found that polarized light made illuminance different with unpolarized light as the setup is composed of preferential aligned birefringent collagen fibers, it gives out a model to predict direct estimates of the attention by Haidinger's brushes for cool white fluorescent under ideal conditions [9]. Obviously, the subjective assessment about polarized lighting is not similar with unpolarized light. Boyce. P. R found the polarising luminaires contain a multi-layer polarising material can't change the value of CRF (Contrast Rendering Factor) significantly [10]. But it is compared with changes in position or between tasks, and the percentage of vertical polarised light is not high.

Actually, as the vertical polarized light can reduce the percentage of light reflected at the smooth surface of reading material and increase the refracted scatterd image light, polarizing luminaires can decrease the veiling reflection effectively if the glare between observers and lamps has the angle of incidence that is equal to or nearly Brewster's angle. Poralizing luminaires have an practical application based on the early conclusion that the viewing angles most encountered in offices range between 20° and 45° [11]. Thus a polarising desktop lamp is possible to maximumly reduce the veiling reflection by reducing vertical polarised light, if the illuminance and other photometric data is controlled in the same level. Due to the high requirement of the transmittance of the polarized film, few report was found by using this optical film on desktop lamp.

The goal of this research is to test the usability of using polarized filters for reading lamp to reduce disturbing reflection glare, therefore to offer evidence of using polaroid in design to reduce veiling reflections and improve its visual comfort.

2　Methods

2.1　Optical parameters selected

In this experiment, desktop lamps with different optical filters are used to associate the subjective evaluation given by the observers under different polarization. Here the CCT (Correlated Color Temperature) and polarization degree P of the desktop lamps are changed as independent variables. Six reading lamps with the same structure except the optical film CCT are selected for the experiment. Half of the luminaires are with the light sources of CCT of 6,500 K and the others are 4,100 K. The light source is the 27 W fluorescent bulb. In order to gain different degree of polarization, 3 optical materials are adhered to the emission surface of the luminaires. They are 3M DBEF Film, fosted glass and DRPF film, expected to have a high, middle and low polarization degree. The parameters and number of each desktop lamp is shown in Tab. 1.

Tab. 1　parameters and number of desktop lamps

Number	1	2	3	4	5	6
CCT/K	6,500	6,500	6,500	4,100	4,100	4,100
Material	DBEF	Frosted glass	DRPF	DBEF	Frosted glass	DRPF

Otherwise, the other light parameters which have a potent effect on visual comfort such as illuminance and light distribution are controlled within comparable range.

2.2　Measurement of polarization

As the light ray goes through several processes of reflection and refraction after it emits from the lamp, the original polarization of the emission light is changed in this process. As a result, the polarization of the light at the direction of the observation by the eyes is different from the light at the emission surface. In order to measure the polarization of the light exits from the lamps, an illuminance meter whose sensor is covered by a linear polaroid is used to measure the intensity of S-polarized light and P-polarized light of the lamp. Thus the polarization degree P of the material can be calculated based on equation 1 [12].

$$P(\%) = |(E_p + E_s)/(E_p - E_s)| \times 100 \tag{1}$$

where E_p and E_s are the illuminance of the parallel component and perpendicular component, at the given angle.

2.3　Subjective evaluation

For the subjective rating design, we adopted one of the classic measuring scale named 5-score rating method [5, 13]. Under each lamp, observers are asked to give his subjective evaluation about the reflected glare based on the veiling reflection intensity rating (See Tab. 2).

Tab. 2　Scale to assess the degree of veiling glare

1	2	3	4	5
Not disturbing		Just disturbing		Strongly disturbing

6 males and 6 females aged $20 \sim 23$ are selected as the observers. All 12 subjects have no difficulty in discriminating colors and have normal or corrected-to-normal visual acuity. To minimize the influence of gender difference and eliminate random error, the subjects experienced all 6 (2 CCT $\times$ 3 optical materials) experimental combinations in random order. In the experiment, the eyes of the observers are fixed with the vertical distance from the emission surface of the

luminaire of 400 mm and horizontal distance of 600 mm, required by the Chinese standard *Performance Requirement for Table Lamps for Paper Task* (GB/T 9473—2008) [14]. A diagrammatic sketch about the position of observer's eyes and the lamp is shown in Fig. 1. A coated paper with printed letters is used for being rated, and put at the point "A" shown in Fig. 1 where the veiling reflection is most obvious — both of these are to magnify the difference of reflected glare between the lamps.

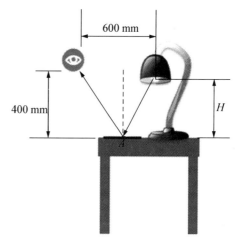

Fig. 1　The relative location of the lamp and the observers

Observers are ensured to well understand the target of their subjective assessment, giving an rating for the degree of veiling reflections. The formal experiment started after several minutes of adaptation and a preliminary experiment. Fig. 2 are the photos of the real site when the experiments were being carried out.

Fig. 2　Photos of the real site during the experiment

3　Results

3.1　Degree of Polarization

The polarization degree P of the 3 films were measured twice at the position "A" shown in Fig. 1, according to the standard observing position. Average results of P of the 3 films are respectively 84.45%, 9.11% and 7.24% with the standard error of 0.42%, 0.37% and 0.65%, respectively.

3.2　The Effect of Different Optical Films

The average subjective rating scores on the disturbing reflection glare caused by different optical films were shown in the Fig. 3. It is clear that with the increasing P the average subjective rating scores decrease, which means that increasing polarization of the optical film reduces the reflection glare effect. The ANOVA result (See Tab. 3) shows that the rating difference caused by different optical films is statistically significant ($p < 0.05$), however there is no significant difference between different CCT, neither is there interactive effect between CCT and Film. A further statistical analysis shows that correlation between the degree of polarization and rating score is significant at the 0.05 level (See Tab. 4).

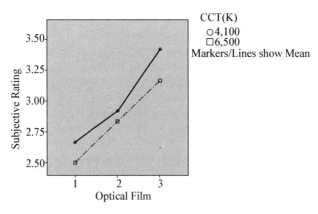

Fig. 3　The average subjective rating scores under different optical films (Polarization degrees P of Film No. 1 to No. 3 are respectively 84.45%; 9.11% and 7.24%)

Tab. 3　Analysis on the variance of the rating scores
Tests of Between-Subjects Effects

Dependent Variable: Subjective Rating

Source	Type III Sum of Squares	df	Mean Square	F	$Sig.$
Model	619.167[a]	6	103.194	108.395	0.000
Film	6.083	2	3.042	3.195	0.047
CCT	0.500	1	0.500	0.525	0.471
Film * CCT	0.083	2	0.042	0.044	0.957
Error	62.833	66	0.952		
Total	682.000	72			

a. R $Squared = 0.908$ (Adjusted R $Squared = 0.899$).

Tab. 4   Analysis on the correlation between degree of polarization and rating scores Correlations

		Subjective Rating	Degree of Polarization $P/\%$
Subjective Rating	Pearson Correlation	1	−0.236*
	Sig. (2 - tailed)		0.046
	Sum of Squares and Cross-products	69.500	−600.850
	Covariance	0.979	−8.463
	N	72	72

*. Correlation is significant at the 0.05 level (2 - tailed).

4   Summary

The subjective evaluation proves that polarized luminaires can decrease the degree of veiling reflection when the illuminance of lamps are similar. There is a significant difference of visual comfort between the 3 films. But no palpable change has been found between different CCTs, and an interaction between films and CCTs doesn't exist, either.

The experiment turns out that through a scientific optical design, we can effectively use the vertical polarized light to make the lighting more comfortable.

What's more, more experiments under different conditions with other kinds of visual tasks are needed for all-sided evaluation.

Acknowledgement

This work is supported by the National Basic Research Program of China (973 Program No. 2010CB734102); Ministry of Science and Technology of China (Project No. 2009GJC00008); Science and Technology Commission of Shanghai Municipality (Project No. 09DZ1141004). The authors would like to thank 3M group for offering the polarization technology, and thank the group members' help during the experiments. They are Ms. Wenting Cheng, Ms. Jinjin Qiu and Ms. Yihong Liu.

References

[1]   Veitch J A, Newsham G R, Boyce P R, et al. Office lighting appraisal, performance, and well-being: a linked mechanisms map [Z]. Beijing, China: 2007,1 - 4.

[2]   Boyce P R. Current knowledge of visual performance [J]. Lighting Research and

Technology, 1973(5):204.

[3] Boyce P R, Slater A I. The application of CRF to office lighting design [J]. Lighting Research and Technology. 1981,13:65.

[4] Reitmaier J. Some effects of veiling reflections in papers [J]. Lighting Research & Technology. 1979,11(4):204 - 209.

[5] Uitterhoeve W L, Kebschull W. Visual performance and visual comfort in offices [J]. 2nd European Lighting Congress. 1973.

[6] Born M, Wolf E. Principle of Optics [M]. 1975.

[7] Richman E E. Results of an attempted field test of multi-layer light polarizing panels in an office space [J]. Prepared for the U. S. Department of Energy, 2001,4.

[8] Rea M S. Visual performance with realistic methods of changing contrast [J]. Journal of the Illuminating Engineering Society, 1981,10(3):164 - 177.

[9] Rea M. Photometry and visual assessment of polarized light under realistic conditions [J]. Journal of the Illuminating Engineering Society, 1982,11(3):135 - 139.

[10] Boyce P R, Slater A I. The application of CRF to office lighting design [J]. Lighting Research and Technology, 1981,13:65.

[11] Japuntich D A. Polarized task lighting to reduce reflective glare in open-plan office cubicles [J]. Applied Ergonomics. 2001(32):485 - 499.

[12] Rea M. The Iesna Lighting Handbook [M]. 9th Edition. The Illuminating Engineering Society of North America, 2000.

[13] De Boer J B. Performance and comfort in the presence of veiling reflections [J]. Lighting Research and Technology, 1977,9(169).

[14] GB/T 9473—2008. Performance Requirement for Table Lamps for Paper Task [S], 2008.

The Effect of LED Lighting on Color Discrimination and Preference of the Elderly

Wenting Cheng, Jiaqi Ju, Yaojie Sun, Yandan Lin

Institute for Electric Light Sources; Engineering Research Center of Advanced Lighting Technology, Ministry of Education, Fudan University, Shanghai 200433

Abstract The purpose of this study is to find out the effects of LED lighting on elderly people's color discrimination and preference, for an improvement of lighting environment set for the elderly. In this study, experiments were conducted under LED lighting with two different spectra (with CCT of 2,800 K and 6,000 K), and three different illuminance levels (30 lx; 100 lx; 1,000 lx). 6 Elderly observers (aged 55~65) participated in the experiments, and were exposed to each lighting condition to finish one trial of color discrimination test (Farnsworth-Munsell 100 - hue test) and a 7 - scale, 6 - item preference evaluation test. It comes to the conclusion that the elderly have a better performance of color discrimination with higher CCT of LED light sources, which compensates their decreased lens transmission at short wavelength. Their performance also increases with higher illuminance (30~1,000 lx) of LED lighting. Meanwhile, they prefer a higher illuminance, which makes them feel the lighting environment more comfortable, brighter and better for reading, but for CCT, although they feel a higher CCT is better for reading, they still have different tastes on CCT of light sources.

Keywords the elderly, color discrimination, preference, LED lighting

1 Introduction

With increasing age, changes occur in the visual system of human eyes, including optical and neural characteristics, which have an impact on the people's visual capabilities. A consideration of the effects of age on threshold performance reveals that the effect of increasing age is almost always negative, in the sense that the visual system becomes less discriminating, and more sensitive to adverse conditions. Specifically, older people tend to show reduced visual field size, increased absolute threshold luminance, reduced visual acuity, reduced contrast sensitivity, increased sensitivity to glare, and poorer color discrimination [1].

As a result, people's demands of light change with their ages. However, most of the lighting applications in daily life are set according to the experience of young adults, so to create a favorable lighting condition for the elderly poses a special set of problems, for examples, in the lighting design of nursing home for the elderly, or in the development of reading light especially for the elderly. For that purpose, two aspects are considered. One is the effect of light on their visual performance; the other is that on their perception and mood.

For visual performance measurements, abstract visual tasks are widely used for laboratory study instead of real tasks, to avoid the influence of people's cognitive level and motor level, and to provide a simple measurement of visual performance through task performance. Hue discrimination task is a common-used task. As early as 1970s, Boyce et al. [2] firstly used FM 100 - Hue Test to study the hue discrimination of light sources and found out that lamp type (spectrum) and illuminance were important for the older age group of people. Knoblauch et al. [3] examined the effect of age and illuminance on the FM - 100 hue test using the MacBeth Executive Daylight source. The result showed that the number of errors committed on the test increased both with increasing age and with decreasing illuminance level. In Boyce's later study [4], FM - 100 test was used again to study the effect of CCT on color perception and preference. Performance significantly increased with increasing illuminance under the 2,700 K and 2,500 K lamps. Darcie. A. O'Connor and Robert G. Davis [5] used a similar color discrimination test, L'Anthony Dichotomous 15 - Hue test [6], as a visual task to study the effect of light source type and illuminance of lighting and studied halogen lamp and compact fluorescent lamp. Recent study of Mark S. Rea et al. [7], Ferenc Szabo [8] and Elodie Mahler et al. [9] also used color discrimination as a method to evaluate quality of LED lighting.

Subjective methods are also commonly used for ergonomics studies [10 - 13]. For evaluation of lighting conditions, semantic differential rating scales is widely used by researchers [14 - 16]. With some of them focused on the elderly [17, 18], most invited subjects with young ages. So preference of elderly people need further study.

Despite the achievements that have been made previously, performance of the elderly on a color discrimination task and their preference under different lighting condition still needs further investigation. Specifically, the effects of Correlated Color Temperature (CCT) and illuminance of LED reading light on the elderly should be tested. Therefore, this study focused on LED lighting, with two different spectra, and the illuminance levels covered a wide range for indoor lighting, and for a precise result, the color discrimination test used in the experiments was the original version with a complete series of color samples. Moreover, with the increasing age, reading ability and enjoyment can be increasingly difficult, so to provide a favorable reading lighting for the elderly is very important. For this consideration, the experiments set subjective evaluation questions based on the reading application of light. Research in this field may enrich our knowledge of the needs of lighting for the elderly; hence help propose better lighting conditions for reading environment.

2　Methods

2.1　Experiment set-ups

Experiments were done in a chamber (See Fig. 1). The chamber is formed by putting a compartment in a big dark room. Dimension of the compartment is 1.8 m (length) × 1.0 m (width) × 2.8 m (height). On the ceiling of the compartment, there mounts an LED panel (Fig. 2). The light output of this LED panel such as illuminance and CCT can be adjusted. The subjects first underwent dark adaptation for 20 minutes before the test starts. For each subject, the visual task was done binocularly. The important parameters used in the experiments and their values are summarized in Tab. 1.

Fig. 1　Experiment chamber

Fig. 2　LED panel

Tab. 1　Values of the parameters used in experiments

Parameters		Values
Fixed parameters	Adaptation	20 min
	Viewing Status	Binocular
	Light Source	LED
Variable parameters	Mean Illuminance on Task(lx)	30; 100; 1,000
	CCT of Lamp(K)	2,800; 6,000

2.2　Observers

A total of 6 observers participated in the experiments, aged 55 to 65, 4 were female and 2 were male. They all had normal color vision, without any vision diseases, and were corrected to normal visual acuity.

Ishihara-color-test plates were used in the experiment to screen the subjects, to ensure that the test results were influenced by people's deficit in color vision.

2.3　Color discrimination and preference test

The Farnsworth-Munsell 100 – Hue Test was used in the experiments. It consists of 85 color discs arranged in four series and is used widely as a test of color vision (see Fig. 3). It has also been used to evaluate the color quality of different light sources. The color discs are randomly arranged beforehand. The subjects should put all discs in proper sequence. An error score is introduced to evaluate how the visual task is accomplished.

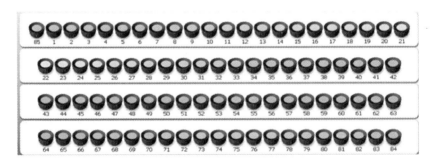

Fig. 3　Caps of Farnsworth-Munsell 100 – Hue Test

For the preference test, the method of semantic differential was adopted, and a questionnaire was designed with 7 scales, 5 pairs of adjectives and an extra question (See Tab. 2). Subjects were asked to rating on the questionnaire according to their perception of the lighting condition.

Tab. 2   questionnaire for preference test

	7	6	5	4	3	2	1	
	How do you like the lighting condition?							
Like								Dislike
Comfortable								Uncomfortable
Bright								Dark
Good for reading								Not good for reading
Good color								Bad color
	Are you willing to choose this as your daily lighting condition for reading?							
Yes								No

2.4 Procedure

Subjects were instructed as to the purpose of the study at the beginning of each experiment. A few trial presentations were conducted to aid in clarifying the procedure, until the subjects were familiar with the test, to avoid the effect of experience.

After trial experiments, subjects were asked to do the FM 100 – Hue test under several lighting condition one by one (See Fig. 4). Half of the subjects did the tests with one order; the others did them with the reverse order. There was no time limitation for the test, but the duration of each test was recorded. Subjects finished a questionnaire of preference test after doing the FM 100 – Hue test.

Fig. 4   One of the subjects doing FM 100 – Hue test

Before each test under one lighting condition, 15 minutes were given for each subject to adapt to the lighting condition; enough relaxation time was also given between the tests to avoid eye fatigue.

3   Results

The average error distributions of the subjects under different lighting conditions are shown in Fig. 5. The errors decrease with increasing illuminance and increasing CCT, and errors are not randomly distributed along the hue circle, but oriented mostly along a tritan axis.

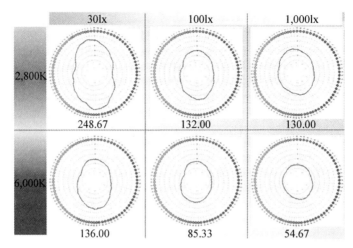

Fig. 5　Mean error score distribution under different CCTs and illuminances

The average error scores under different lighting conditions are shown in Fig. 6. It also shows a decrease trend of error scores with increasing illuminance and CCT, which means better performances. For preference tests, rating of each item (i. e. each pair of adjectives) is presented separately. For the item Like/Dislike, there is an obvious increase of average rating scores with increasing illuminance under both CCT (See Fig. 7),

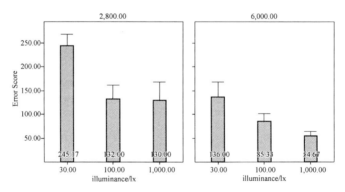

Fig. 6　Mean error scores under different illuminances (CCTs split)

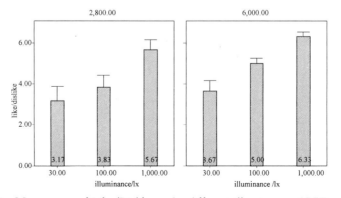

Fig. 7　Mean scores of Like/Dislike under different illuminances (CCTs split)

and lighting conditions with CCT of 6,000 K get higher rating scores with the same illuminance level.

The results of the other items are shown in Fig. 8 ~ 12. They are Comfortable/Uncomfortable (See Fig. 8), Bright/Dark (See Fig. 9), Good for reading/Not good for reading (See Fig. 10), Good color/Bad color (See Fig. 11),

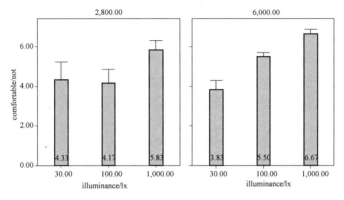

Fig. 8   Mean scores of Comfortable-uncomfortable
under different illuminances (CCTs split)

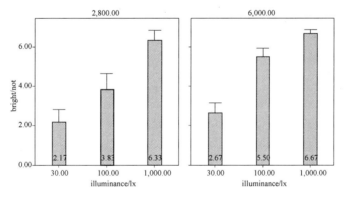

Fig. 9   Mean scores of Bright/Dark under different illuminances (CCTs split)

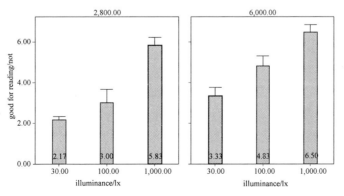

Fig. 10   Mean scores of Good for reading/Not good for reading
under different illuminances (CCTs split)

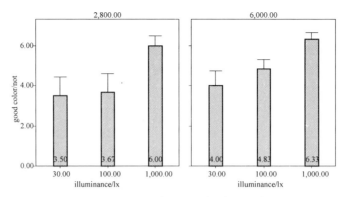

Fig. 11　Mean scores of Good color/Bad color under different illuminances (CCTs split)

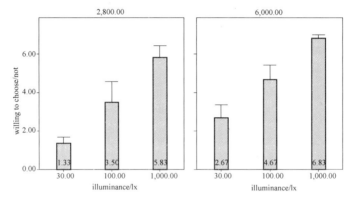

Fig. 12　Mean scores of Willing to choose for daily life/Not under different illuminances (CCTs split)

Willing to choose/Not willing to choose (See Fig. 12). In most of the items，the trends are similar to that in the item Like/Dislike.

4　Discussion

Illuminance and CCT are used as two variables in the experiments. Analysis of Variance (ANOVA) is conducted for all the subjective evaluation items as well as the error scores.

Tab. 3 shows the effects of CCT from the result of ANOVA.

Tab. 3　the effects of CCTs from ANOVA

	F	$Sig.$
Error score	25. 614	0. 004
Like/Dislike	5. 976	0. 058
Comfortable/Uncomfortable	0. 954	0. 374

(continued)

	F	Sig.
Bright/Dark	2. 419	0. 181
Good for reading/Not good for reading	13. 75	0. 014
Good color/Bad color	2. 222	0. 196
Willing to choose/Not willing to choose	5. 288	0. 07

Significant difference is found between two CCT levels for error scores ($P = 0.004$) and the "good for reading/not" item in the preference test ($P = 0.014$), but not in other items. It comes to the conclusion about CCT that:

Firstly, visual performance of color discrimination of the elderly is influenced by the spectrum of light source. A higher CCT can improve the visual performance.

The possible reason is that it is the light of short wavelength that matters. Study [19] showed that transmission of lens decrease with increasing age, and most prominently at shorter wavelengths. That can explain why the error distribution oriented mostly along a tritan axis (like a tendency of s-cone deficiency). High CCT light sources, with more proportion of short wavelength light, somehow compensate the low transmission of blue light, and can benefit the color discrimination performance of the elderly.

Secondly, the elderly think a higher CCT light source is more suitable for reading. But during the experiments, subjects reported orally that, they had different needs for reading lamp. For example, they thought a cooler light is good for reading because it made them feel alert, but sometimes they would rather choose a warmer one to make them comfortable and dozy, especially for bed-time reading.

Tab. 4 shows the effects of illuminance from the result of ANOVA.

Tab. 4   the effects of illuminanc from ANOVA

	F	Sig.
Error score	66. 147	0. 001
Like/Dislike	21. 69	0. 007
Comfortable/Uncomfortable	8. 375	0. 037
Bright/Dark	24. 189	0. 006
Good for reading/Not good for reading	33. 299	0. 003
Good color/Bad color	3. 348	0. 14
Willing to choose/Not willing to choose	413. 372	0. 00

The analysis shows that the three illuminance levels significantly differ for error score ($P = 0.001$), Like/Dislike ($P = 0.007$), Comfortable/Uncomfortable ($P = 0.037$), Bright/Dark ($P = 0.006$), Good for reading/Not good for reading ($P = 0.003$), Willing to choose/Not willing to choose ($P = 0.00$), but not for the Good color/Not item.

It shows the effects of illuminance on the visual performance of the elderly, which is in accord with the previous study result using traditional light sources, that is, visual performance improves with increasing illuminance under LED lighting with both CCT. Illuminance also has a significant effect on the six items of subjective ratings, except the Good color/Not one that related closer to spectrum instead of amount of light. Subjects consider, as illuminance increases, their preference increase, and they feel more comfortable, brighter, easier for reading, and they preferred to choose lighting conditions with higher illuminance.

5　Conclusion

It comes to the conclusion that the elderly have a better performance of color discrimination with a higher CCT of LED light sources, which compensate their decreasing lens transmission at short wavelength. Their performance also increases with illuminance (30—1,000lx) of LED lighting. Meanwhile, they prefer a higher illuminance, which makes them feel the lighting environment more comfortable, brighter and better for reading, but for CCT, although they feel a higher CCT is better for reading, they still have different tastes on CCT of light sources. Therefore, in order to provide a suitable reading environment for the elderly, in consideration of both performance and preference, besides the consideration of illuminance, we should also pay attention to the spectra of light sources, to guarantee enough blue light to compensate the lens transmission of the elderly decrease, in addition we should take into account different applications and people's different tastes of CCT.

Acknowledgement

The work is supported by the National Basic Research Program of China (973 Program No. 2010CB734102).

References

[1]　Boyce P R. Human factor in lighting London Taylor & Francis [J]. Coloration Technology,

2011,127:101 - 103.

[2] Boyce P R, Simons R H. Hue Discrimination and Light Sources [J]. Light Res. &.
Technol, 1977,9(3):125 - 140.

[3] Kenneth K, et al. Age and illuminance effects in the Farnsworth-Munsell 100 - Hue test
[J]. Applied Optics, 1987,26(8):1441 - 1448.

[4] Boyce P R, Cuttle C. Effect of Correlated Colour Temperature on the Perception of Interiors
and Colour Discrimination Performance [J]. Light Research and Technology, 1990,22(1):
19 - 36.

[5] Darcie A O C, Robert G. Davis Lighting for the Elderly: The Effects of Light Sources
Spectrum and Illumiance on Color Discrimination and Preference [J]. LEUKOS, 2005,2
(2):123 - 132.

[6] Lanthony P. The Desaturated Panel D - 15 [J]. Documenta Ophthalmologica, 1978,46(1):
185 - 189.

[7] Mark S R, Jean P. Freyssinier-Nova Color rendering: a tale of two metrics [J]. Color
research and application, 2008,33,3:192 - 202.

[8] Szabo F. A Comparative Study of New Solid State Light Sources CIE 26TH SESSION —
BEIJING 2007 D1 18 - 21.

[9] Elodie M et al. Testing LED lighting for colour discrimination and colour rendering [J],
Color research and application 2009,34,1:8 - 17.

[10] Parsons K C. Environmental ergonomics: a review of principles, methods and models [J],
Applied Ergonomics, 2000,31:581 - 594.

[11] Ki Y C, et al. Evaluation of Radio-Frequency Identification Projects in Public Sectors of
Korea [J]. Human Factors and Ergonomics in Manufacturing &. Service Industries, 2011,21
(1):44 - 51.

[12] Kim C, et al. Affective Evaluation of User Impressions Using Virtual Product Prototyping [J],
Human Factors and Ergonomics in Manufacturing &. Service Industries, 2011,21(1):1 - 13.

[13] Lin C J, Yenn T C, Yang C W. Evaluation of Operators' Performance for Automation
Design in the Fully Digital Control Room of Nuclear Power Plants [J]. Human Factors and
Ergonomics in Manufacturing &. Service Industries, 2010,20(1):10 - 23.

[14] Hajimu N, Yoshinori K. Relationship between illuminance/color temperature and preference
of atmosphere [J]. J. Light &. Vis. Env. 1999,23,1:29 - 38.

[15] Naoyuki O, Hironobu T. Preferred combinations between illuminance and color temperature
in several settings for daily living activities [J], CIE 26th Session-Beijing, 2007.

[16] Insiya S, Nadarajah N. White LEDs in landscape lighting application [C]. Solid state
lighting: Proceedings of SPIE, 2002.

[17] Park N K, Cheryl A. Farr, Retail store lighting for elderly consumers: an experimental
approach [J]. Family and consumer sciences research journal, 2007;35;316,DOI: 10. 1177/
1077727X07300096.

[18] Robert G, Davis, Antonio G. Task lighting for the elderly [J]. Journal of the illuminating
engineering society, 2002,31,1:20 - 32.

[19] Line K, et al. Age-related changes in the transmission properties of the human lens and their
relevance to circadian entrainment [J]. Journal of Cataract &. Refractive Surgery, 2010,36,
2:308 - 312.

Effects of Correlated Color Temperature on Spatial Brightness Perception

Jiaqi Ju, Dahua Chen, Yandan Lin

Department of illuminating engineering and light sources, Engineering Research Center
of Advanced Lighting Technology, Ministry of Education,
Fudan University, Shanghai 200433

Abstract　The relationship between lamp color characteristics and brightness perception is not well known. In this study, nine lighting environment with correlated color temperature (3,000, 5,000, 8,000 K) and illuminance (1,000, 300, 100 lx) were created. Both the side by side visual matching and spatial brightness scaling experiments are designed to verify the effects of correlated color temperature on spatial brightness perception. The results of the study show that lighting with high correlated color temperature will have stronger spatial brightness perception than lower ones.

Keywords　correlated color temperature, brightness perception, lighting environment

1　Introduction

The relationship of correlated color temperature and illuminance has been studied for more than half century to find good lighting solution which enjoy both visual comfort and energy saving. The Kruithof [1] curve relates the illuminance and correlated color temperature of visually-pleasing light sources in 1941. With 2 example light sources, Kruithof created a pleasing light region for people. According to the Kruithof curve, the observer prefers lower correlated color temperature lighting when the light level is lower and prefers a higher correlated color temperature when the light level is higher. Early to 1954, Harrington [2] found that high correlated color temperature lamp reach the same apparent brightness as lower correlated color temperature lamp, but with less photopic illuminance.

High correlated color temperature with high illuminance has been introduced to many international and national lighting design guides. However, in 1990, Boyce and Cuttle [3] found that the correlated color temperature of the lamp has little effect on people's impression of the lighting in the room. Other researches show different conclusions [4, 5].

Spatial brightness is the sense of overall brightness perceived when the observer is immersed in the large space with a full visual field [6]. Both side by side visual matching tests and spatial brightness scaling experiments are designed to verify the effects of correlated color temperature on spatial brightness perception in this study. Also, the equivalent luminous flux modified by 10 degree $V(\lambda)$ is calculated from the spectral power distribution of the lamps to compare the results with experiments study.

2   Research methods

Ten subjects without vision problems take part in the experiments, 5 male and 5 female, aged from 18 to 25. After the procedures of the experiments were explained, the subjects were asked to wait in a dark room for 20 minutes to remove any before effects on the experiment. Nine lighting environments will be created by fluorescent lamps in a small partition room for the subjects' scaling. The nine lighting environments are randomly listed in Tab. 1. The partition room is 1. 45 m in length, 1. 40 m in width, and 2. 15 m in height. The walls of the partition room are painted with white diffuse materials. All the lamps are with high color rendering index over 80, and every lighting environment has more than 0. 7 in uniformity (minimum illuminance / average illuminance) on work plane for visual evaluation. Seven-point scaling method was used in the subjective evaluation, with the range 1 meaning very dim to 7 meaning very bright. Five minutes adaptation of each lighting condition was given to the subjects before they scaled. The first subject did the experiment from lighting No. 1 to No. 9, and the second one from No. 9 to No. 1. The third, fifth, seventh, ninth subject had the same sequence of the first subject. And the others had the same sequence of the second subject to remove the sequence effects of the experiments.

Tab. 1   Nine lighting environments for spatial brightness scaling The illuminance is the average horizontal value on the work plane.

	3,000 K	5,000 K	8,000 K
100 lx	1	4	7
300 lx	2	5	8
1,000 lx	3	6	9

Besides the spatial brightness scaling test, side by side visual matching test was also done by all the subjects. Another partition room which is the same as used in scaling test was set beside the former partition room (See Fig. 1). Every one of the nine lighting environments was set in the left room as the reference lighting condition, the subject was asked to adjust the lighting in the second room by themselves with different correlated color temperature to reach the same subjective brightness of the work plane. Every matching test was done twice. In the first trial, the subjects adjusted the lighting from very dim to where they feel the same as reference lighting condition. In the second trial, they adjusted the lighting from very bright to where they feel the same as reference lighting condition. After every trial, the average horizontal illuminance on the work plane was recorded. The average value of every two trials was taken as the result of each test.

Fig. 1　The real environment for experiments

The spectra of the three different correlated color temperatures are tested. (Fig. 2) As common, the luminance Φ can be calculated by the formula below (Eq. (1)).

$$\Phi = K_{\mathrm{m}} \int_{380}^{780} P(\lambda)V(\lambda)\mathrm{d}\lambda = K_{\mathrm{m}}\alpha \int_{380}^{780} P'(\lambda)V(\lambda)\mathrm{d}\lambda \tag{1}$$

Where $V(\lambda)$ is luminous efficacy function for photopic vision. K_{m} is maximum spectral luminous efficacy, for photopic vision with $V(\lambda)$, it is 683 lm/W. $P(\lambda)$ is the absolute spectral power distribution; $P'(\lambda)$ is relative spectral power distribution, and α is a constant. The brightness perception can be evaluated by Fechner's law (Eq. (2)).

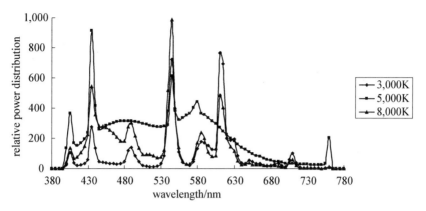

Fig. 2   the SPD of the three CCT lighting environment

$$S = K \lg I \tag{2}$$

Where S refers subjective feeling. I is a physical parameter and K is a constant. For this experiment, the I refers luminance and S means subjective brightness rating. So the Fechner's law can be changed to Eq. (3).

$$S_{BR} = K \lg L \tag{3}$$

Because the luminance L and illuminance E can be connected in Eq. (4) and the two partition rooms are the same, the Fechner's law can be expressed as Eq. (5).

$$\rho E = \pi L \tag{4}$$

$$S_{BR} = K \lg E \tag{5}$$

3   Results

The results of subjective spatial brightness scaling are shown in Fig. 3 and Fig. 4. For each color temperature environment, the higher the illuminance is, the higher the rating score is. It is really true for our common sense that with more light, you get more brightness. However, for each light level, the 8,000 K and 5,000 K will have higher scaling value than 3,000 K. The results show that 5,000 K with 1,000 lx gets the highest score of 5. 8, while 3,000 K with 1,000 lx gets 4. 9. In each light level, lighting environment with high color temperature always gets higher scores than the same level for 3,000 K.

According the Eq. (5), the relationship between illuminance and brightness scaling can be checked by Fechner's law. The results show that the three CCT lighting environments do not fit well, especially for high light level (Fig. 5). For

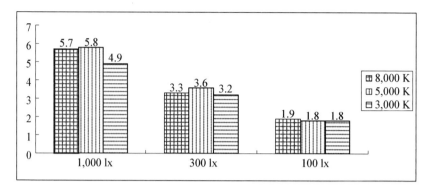

Fig. 3   The subjective spatial brightness scaling for different illuminance

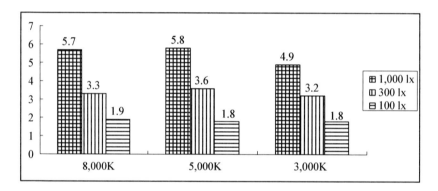

Fig. 4   The subjective spatial brightness scaling for different CCT

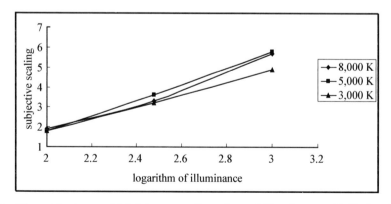

Fig. 5   The subjective spatial brightness scaling of tested illuminance with Fechner's law

spatial brightness perception, the traditional 2 - degree $V(\lambda)$ which is applied to the measurement sensor is inappropriate because both the cones and rods will function for this situation. The 10 - degree $V(\lambda)$ is used to modified the tested illuminance for verifying the experiment results. The modified illuminance is listed in Tab. 2. The modified illuminance of the three CCT fit better than the 2 – degree $V(\lambda)$ tested illuminance (Fig. 6).

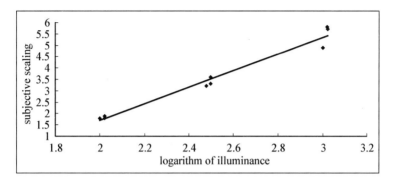

Fig. 6   The subjective spatial brightness scaling of modified
illuminance by 10 – degree $V(\lambda)$

Tab. 2   The modified illuminance by 10 – degree $V(\lambda)$

	3,000 K	5,000 K	8,000 K
100 lx	100 lx	105 lx	105 lx
300 lx	301 lx	315 lx	316 lx
1,000 lx	1,002 lx	1,049 lx	1,052 lx

For side by side visual matching test, taking 8,000 K lighting environment as reference, the 3,000 K has to be adjusted to 1,117 lx, 348 lx, 139 lx to have the same brightness perception of 1,000 lx, 300 lx, 100 lx of 8,000 K (See Tab. 3). If 5,000 K lighting environment is taken as reference, the 3,000 K needs to be adjusted to 1,113 lx, 414 lx, 145 lx (See Tab. 4). In every lighting condition, lighting environment with high color temperature needs less tested illuminance than 3,000 K to reach the same brightness. The 10 – degree modified illuminance is also used to check the difference between different CCT (See Tab. 5. and Tab. 6). The difference for 100 lx is most distinctive, reached 40% for 5,000 K.

Tab. 3   The illuminance for 3,000 K to reach the same brightness as 8,000 K

8,000 K	3,000 K	Difference
1,000 lx	1,117 lx	11.7%
300 lx	348 lx	16%
100 lx	139 lx	39%

Tab. 4   The illuminance for 3,000 K to reach the same brightness as 5,000 K

5,000 K	3,000 K	Difference
1,000 lx	1,113 lx	11.3%
300 lx	414 lx	38%
100 lx	145 lx	45%

Tab. 5   The modified illuminance for 3,000 K to reach the same brightness as 8,000 K

8,000 K	3,000 K	Difference
1,052 lx	1,119 lx	6.4%
316 lx	349 lx	10.8%
105 lx	139 lx	32.4%

Tab. 6   The modified illuminance for 3,000 K to reach the same brightness as 5,000 K

5,000 K	3,000 K	Difference
1,049 lx	1,115 lx	6.3%
315 lx	415 lx	31.7%
105 lx	145 lx	40%

4   Discussion

Both the subjective brightness scaling test and the side by side visual match test show that the lighting environment with high CCT shows more brightness perception. From classic Fechner's law of psychophysical scaling, the relation between psychological and physical scales can be expressed as to reach the same brightness perception, the same physical stimuli is required. It shows different tested illuminance or 10 - degree modified illuminance to reach the same brightness perception for different CCT. Higher CCT needs less illuminance. It means the $V(\lambda)$ can not evaluated the light in short wavelength because the lighting environment with high color temperature means more blue light components. The 10 - degree $V(\lambda)$ is still need to be modified to give more reliable results for spatial brightness perception. For each light level, the 3,000 K and 8,000 K need to have the same photometric stimuli to reach the same brightness perception as Eq. (6).

$$C(\lambda)P(\lambda)_{8,000} = C(\lambda)P(\lambda)_{3,000} \tag{6}$$

The $C(\lambda)$ is a modified spectral sensitivity function for spatial brightness perception. The vision studies of photopic and scotopic use S/P to evaluate a light source performance in low light level.

$$S/P = \frac{1,699 \cdot \int_{380}^{780} P(\lambda)V'(\lambda)d\lambda}{683 \cdot \int_{380}^{780} P(\lambda)V(\lambda)d\lambda} = \frac{1,699 \cdot \sum_{380}^{780} P'(\lambda)V'(\lambda)\Delta\lambda}{683 \cdot \sum_{380}^{780} P'(\lambda)V(\lambda)\Delta\lambda}$$

$P'(\lambda)$ is relative power distribution. The C/P can be defined as the similar style.

$$C/P = \frac{Kc \cdot \int_{380}^{780} P(\lambda)C(\lambda)d\lambda}{683 \cdot \int_{380}^{780} P(\lambda)V(\lambda)d\lambda} = \frac{Kc \cdot \sum_{380}^{780} P'(\lambda)C'(\lambda)\Delta\lambda}{683 \cdot \sum_{380}^{780} P'(\lambda)V(\lambda)\Delta\lambda}$$

Then for each light level, the Eq. (6) can be changed to Eq. (7) below.

$$\left(\frac{C}{P}\right)_{8,000} E_{P8,000} = \left(\frac{C}{P}\right)_{3,000} E_{P3,000} \tag{7}$$

Ep. (6) is the known tested illuminance in Tab. 3. And Eq. (7) can be translated to Eq. (8).

$$\frac{\left(\dfrac{C}{P}\right)_{8,000}}{\left(\dfrac{C}{P}\right)_{3,000}} = \frac{E_{P3,000}}{E_{P8,000}} = 1.16 \tag{8}$$

The Eq. (8) can be extended to the equation below which is only related to the relative power distribution.

$$\frac{\left(\dfrac{C}{P}\right)_{8,000}}{\left(\dfrac{C}{P}\right)_{3,000}} = \frac{\dfrac{K_B \cdot \sum_{380}^{780} P'_{8,000}(\lambda)C(\lambda)\Delta\lambda}{683 \cdot \sum_{380}^{780} P'_{8,000}(\lambda)V(\lambda)\Delta\lambda}}{\dfrac{K_B \cdot \sum_{380}^{780} P'_{3,000}(\lambda)C(\lambda)\Delta\lambda}{683 \cdot \sum_{380}^{780} P'_{3,000}(\lambda)V(\lambda)\Delta\lambda}} = \frac{\sum_{380}^{780} P'_{8,000}(\lambda)C(\lambda)\sum_{380}^{780} P'_{3,000}(\lambda)V(\lambda)}{\sum_{380}^{780} P'_{3,000}(\lambda)C(\lambda)\sum_{380}^{780} P'_{8,000}(\lambda)V(\lambda)} = 1.16$$

The similar method can be used to get the relationship between 5,000 K and 3,000 K. Three characteristic spectral lines, 435 nm, 545 nm, and 610 nm are applied to evaluate the three-band fluorescent lamp. And finally two relative values can be calculated as below.

$$\frac{C(435)}{C(545)} = 0.073,3 \qquad \frac{C(435)}{C(610)} = 0.457,1$$

A target Gaussian distribution function is applied to describe the $C(\lambda)$, which is related to the maximum spectral sensitivity wavelength and a constant A.

$$C(\lambda) = e^{-2\left(\frac{\lambda-\lambda_{max}}{A}\right)^2}$$

For $V(\lambda)$, $\lambda_{max} = 555$, $A = 84.56$ ($R^2 = 0.99$); For $V'(\lambda)$, $\lambda_{max} = 507$, $A = 78.41$ ($R^2 = 0.99$). The function is effective to evaluate the vision spectral sensitivity (Checked by Origin 7.5).

Use the relative value of $C(435)$, $C(545)$, and $C(610)$, the $C(\lambda)$ can be calculated. And the results are $\lambda_{max} = 530$, $A = 82.06$ (See Fig. 7).

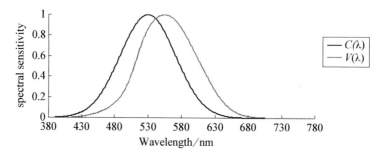

Fig. 7　the $C(\lambda)$ spectral response sensitivity function

5　Conclusion

The discovery of the vision science showed that the pupil size of our eyes changes according to different light level and different spectrum [7, 8]. The change of pupil size can change the light amount into the eyes, thus change the brightness. Details of the mechanism behind these effects are still being researched. In this study, both subjective spatial brightness scaling and side by side visual matching test are set up to verify the effects of correlated color temperature on spatial brightness perception. The high correlated color temperature does have more effects on spatial brightness perception. Comparing with the $V(\lambda)$, the calculated spectral sensitivity model $C(\lambda)$ has peak wavelength at 530 nm. This does not mean that our eye is most sensitive to 530 nm, but it is a synthetic model to evaluate real lighting environment. It may take account of the effects of pupil size variation which is overlooked in $V(\lambda)$.

References

[1]　Kruithof A A. Tubular luminescence lamps for general illumination [J]. Philips technical

review, 1941;6:65 – 96.

[2]   Harrington R E. Effects of color temperature on apparent brightness [J]. Journal of the optical society of America, 1954;44(2):113 – 116.

[3]   Boyce P R, Cuttle C. Effect of correlated color temperature on the perception of interiors and color discrimination performance [J]. Lighting Research & Technology, 1990,22(1):19 – 36.

[4]   Berman S M, Jewett D L, Fein G, et al. Photopic luminance does not always predict perceived room brightness [J]. Lighting Research & Technology, 1990,22(1):37 – 41.

[5]   Akashi Y, Boyce P R. A field study of illuminance reduction. Energy and Buildings, 2005,38 (6):588 – 599.

[6]   Fotios S A. Lamp colour properties and apparent brightness: a review [J]. Lighting Research & Technology, 2001,33(3):163 – 181.

[7]   Berman S M, Clear R D. Past vision studies can support a novel human photoreceptor [J]. Light & Engineering, 2008,16(2):88 – 94.

[8]   Gamlin P D, McDougal D H, Pokorny J, et al. Human and macaque pupil responses driven by melanopsin containing retinal ganglion cells [J]. Vision Research, 2007,47(7):946 – 954.

[9]   Rea M S, Figuerio M G, Bullough J D, et al. A model of phototransduction by the human circadian system [J]. Brain Research Reviews, 2005,50:213 – 218.

An Integrated Experimental Platform for the Crew Workload Measurement

Dayong Dong, Shan Fu

Shanghai JiaoTong University, Shanghai 200240

Abstract   The measurement of crew work load is an important part in the evaluation of the airworthiness of human factors. In clause 25. 152,3 of air worthiness' regulation on minimum flight crew, it is explicitly required that crew work load be put into consideration. The applicant must demonstrate in a rational manner the appropriate level of the crew that is to be given certification in order to ensure the safety of the aircraft.

The evaluation of crew workload is to be done in an objective way so as to offset the uncertainty of subjective evaluations. The research of this paper constructs an integrated simulator experiment platform to collect and analyze the physiological data of pilots during the experiment, to achieve a quantitative measurement of crew work load. The integrated experiment platform mainly comprises of two parts: firstly, a flight simulation system that can simulate various flight task scenarios; the other is a pilot physiological data collection and analysis system, where objective measurement of the flight crew is achieved by recording and analyzing the relationship between physiological and flight data of the tested pilots. In this research, experiments of approaching and landing are done and through data recording and analysis, the feasibility of the integrated experiment platform is tested.

Keywords   Airworthiness, simulator, crew workload, human factors, physiological

1   Introduction

Airworthiness certification is issued for an aircraft by the national aviation authority in the state in which the aircraft is registered. The main purpose of airworthiness certification is to insure the aircraft must be in a condition for safe operation. The human factors have been widely recognized as critical to aviation safety and effectiveness (FAA, 1993). Human error has been documented as a primary contributor to more than 70 percent of commercial airplane hull-loss accidents (Boeing). As a result, in the type certification processes of air crafts, human factors must be considered as a key point of evaluation.

The current FAA (Federal Aviation Administration) and CAAC (Civil Aviation Administration of China) airworthiness standards (25. 152,3) explicit requirement on crew workload: "The minimum flight crew must be established so that it is sufficient for safe operation, considering the workload on individual crewmembers. " Measurement and evaluation of light crew workload is the main work of essential flight crew certification. There are mainly four ways of flight crew workload measurement and evaluation. (Cowin workload, 1989): subjective workload measurement, physiological workload measurement, performance workload measurement, and analytical assessment technique of Timeline Analysis (TLA). Applicant must use an appropriate workload measurement method to demonstrate that the crew workload in normal, non-normal or emergency situation and could be maintained at an acceptable level. Though subjective evaluation is still the prevalent workload measurement method. However, as subjective method relies too much on the pilots' subjective opinions, during its application inconsistency of experiment results might result from different understandings of questionnaires. Physiological measurement as an objective measurement method is becoming increasingly accepted, with many researchers trying to measure the workload of pilots with various physiological parameters (Malcolm, 2002) (Roscoe, A. H. , 1992). (Wilson, G. F. , 2001, 2002) Thus technologies of objective workload measurement will become an inevitable trend.

This research constructs an integrated simulation experiment platform that performs real time collection of physiological parameters during simulated flights and uses the parameters for workload measurement and evaluation. The platform comprises of mainly two parts. One is a flight simulation system that simulates task scenarios of certain models aircrafts. It includes aircraft dynamics simulation, virtual meters, operational and control devices and external view. Tested pilots could undergo flight task experiments of various scenarios in this simulation system.

The other part of the experiment platform is the pilot physiological data collection and analysis system, where real time collection could be performed during simulated flight experiments. The data are then integrated via TCP/IP network, then gathered and analyzed as a work station. At the current stage, the data that can be collected mainly include: eye movement data, heart rate data, pilot behavioral video data and flight performance data. The system is open, in that by expanding data ports data other than aforementioned could be collected. Through analysis of the collected data, objective measurement and evaluation of flight crew workload could be achieved. In this research, confirmatory experiments are done on the workload measurement during approaching and landing. The results show that the platform is capable of collecting and processing physiological and performance data from simulated flight tasks, and thus is a technology to achieve objective evaluation of workload.

2　System Platform Framework

Of the MOC (Means Of Compliance) for air worthiness evaluation, testing through simulator experiments is one of the means. It is also one major method used in related researches on human factors. Common processes of simulator experiment are that pilots conduct simulated flight tasks in simulators and fill in questionnaires, both in-flight and post-flight, which are referenced against rating scales, to evaluate crew workload. In order to collect physiological data of subject pilots and to measure and evaluate crew workload objectively, this research constructs an integrated experiment platform for flight simulation and physiological data collection and analysis. The platform comprises of two parts, to achieve flight simulation and collection of physiological data of subject pilots. See Fig. 1 for the framework of the system platform.

2.1　Flight Simulator

Flight simulator mainly constructs a virtual flight environment, where pilots could undergo experiments of various tasks of flight simulation. It is made up of flight deck display system, visual scenery system, flight control system, control devices, etc. The platform adopts a module design, and simulations of different scenarios could be performed by adjusting the corresponding modules.

1) Flight deck display system

Flight deck display system is the most important human-machine interface in the cockpit and its design is of critical importance to crew workload. In this research, VAPS is chosen as development tool, and flight deck display system simulation is designed successfully. The systems is made up of: Principle Flight

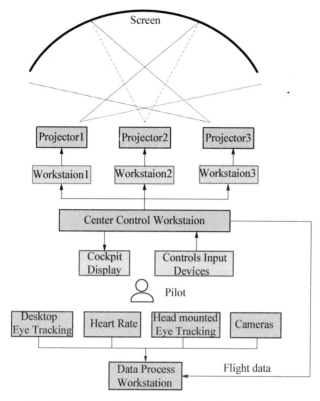

Fig. 1   framework of integrated experiment platform

Display (PFD), Multi-Function Display (MFD), Engine Indicating and Crew Alerting System (EICAS) (yet unwritten), mode control panel, overhead panel and control display unit, etc.

Among the listed, PFD is a modern aircraft instrument dedicated to flight information. It is located right in front of the pilot, provides an accurate virtual horizon, and displays angle of pitch of the plane, roll angle, and other information concerning the attitude of the plane; displays flight altitude, speed, Mach number and lifting speed and other flight parameters; also displays heading, flight track, Flight Director and other navigation information.

MFD is a screen in an aircraft that can be used to display information to the pilot in numerous configurable ways, allowing the pilot to display their navigation route, moving map, weather radar, (ground proximity warning system) GPWS, (Traffic collision avoidance system) TCAS and airport information.

Engine Indicating and Crew Alerting System (EICAS) is an integrated system to provide aircraft crew with aircraft engines and other systems instrumentation and crew annunciation. EICAS typically includes instrumentation of various engine parameters, including for example revolutions per minute, temperature values, fuel flow and quantity, oil pressure etc.

Mode Control Panel (MCP) is located at the middle of the glare shield. It is an instrument panel that controls advanced autopilot and related systems. It contains controls that allow the crew of the aircraft to select which parts of the aircraft's flight are to be controlled automatically. The MCP can be used to instruct the autopilot to hold a specific altitude, to change altitudes at a specific rate, to hold a specific heading, to turn to a new heading, to follow the directions of a flight management computer (FMC), and so on.

Overhead Panel is installed above the pilot, made up of many switches which control lighting, fire protection of the cockpit and anti-ice, etc.

CDU is the core of Flight Management System (FMs). Pilots control FMS by controlling CDU. FMS is a specialized computer system that automates a wide variety of in-flight tasks, especially in-flight management of the flight plan.

Design of the display and control system followed relevant requirements on "electronic cockpit display" issued by FAA (06/21/2007).

2) Visual scenery system

The hardware of the visual scenery system is made up of three-channel projector and large dome screen projection. Vega Prime software is used to realize combination of visual scenery projection and three channel data mixing.

Visual Scenery based on Vega Prime is mainly made up of three parts: construction of model of the visual scenery, graphic interface design of LynX Prime, and visual scenery program design. The construction of visual scenery model mainly means using Creator to build static visual scenery model pool and using Creator Terrain Studio to generate large scale terrain visual scenery model pool. Graphic interface design of LynX Prime includes settings of the basic environment, model initial location, and frequently used weather and so on. Visual scenery simulation program is designed to finish to whole process of simulation to provide users with real time visual scenery simulation results. Drawing from the model pool generated during visual scenery construction using API functions from Vega Prime, and documents generated during LynX Prime's graphic interface design, the program drives the entire simulation.

Modeling of virtual three-dimensional entities in simulated scenarios is done in Creator; and the three-dimensional models are generated into virtual scenarios in Vega Prime, and the scenarios are controlled through Vega Prime's cross functions.

3) Flight dynamics model

Flight dynamics simulation model is one of the core software systems of flight simulator. Its main function is to calculate aircraft atmospheric environment parameters, aircraft engine parameters, aerodynamic equation and kinetic equation and attitude parameters. It is the data hub of the flight simulator. It takes flight environment and pilots' operational instruction signals to levers, helm, throttle lever and cockpit equipment as input parameters, and calculates in real time the attitude, location, speed, altitude and other flight parameters, which will all be transferred to other sub-systems as driving instructions and computing input parameters, also transferred via network to visual scenery system to render external visual scenery.

In this research, the commonly used 6-degrees-of-freedom kinetic equation is adopted to solve the aircraft's kinematical equation and dynamic equation and then transfer to the plane's attitude, which is then delivered to the visual scenery system. In the development of the software, open data ports are designed for aircraft aerodynamic parameters and engine parameters, so as to achieve dynamic simulation of various types of aircrafts.

4) Control devices

Control devices mainly include steering wheel/lever, pedal, throttle, flap handle and other control input devices. To give the system better compatibility, all control devices all use standard USB ports. For cockpit layout, steering wheel style layout of Boeing and side stick style layout of Airbus are referred to, and switch between the two can be easily made, to simulate various design plans.

5) Fidelity

Flight simulation should satisfy requirements of human factor experiment fidelity. In system development processes, "General Specification for Airplane Simulator" should be followed to finish design of the simulation system. In this way the validity of data gathered from experiments can be ensured.

2.2　Physiological Data Collection

To evaluate crew workload with multi physiological parameters, multi-channel physiological data collection system is constructed. This research chooses heart rate and eye movement parameters that are effective for work load evaluation, together with pilot behavioral analysis based on videos recorded, considering both physiological parameters and behavioral features, to construct flight crew workload evaluation model.

1）Heart rate

In this research, we adopt POLAR's RS800CX multiterm movement heart rate watch as heart rate parameter collection system. In the experiments, subjects wear heart rate sensors to collect heart rate data in real time. And via a wireless transmission module, the collected heart rate data will be to the data analysis software at a terminal workstation.

2）Eye movement

According to statistics, over 60% of all information input is through vision. Eyes, as the main organ for visual information input, displays features during information obtaining corresponding to mental workload during task performing.

Pilot's workload during flying can be shown in data of eye movement. Obtaining of eye movement data is through a SmartEye desktop eye tracking system. By recording and analyzing distribution of eye focus and pupil data during the experiments, a relationship between workload and eye movement data can be determined.

3）Operation behavior

Pilot's control of the aircrafts is ultimately through the interaction between the limbs and the display control interface. So through analyzing pilots' behavioral features in flight tasks, pilot performance can be evaluated. When combined with eye movement data, pilot's eye-hand coordination modes in executing operations can be analyzed as well. Thus further studies on the relationship between information obtaining in the interaction process, information processing and information execution can be done.

For behavioral data collection, general video collection modules are used. Videos collected via camera during performing tasks will be transferred to terminal workstation for analysis with data processing software.

4）Data summarization and analysis

For data summarization and processing, a workstation is linked to all physiological collection devices to collect and store all data from experiments. And after finishing experiments, integrated data analysis software synchronizes the data for analysis.

3　Experiment

To verify the validity of the experiment platform, subject students went through many experiments involving approaching and landing. And through the simulation, various parameters were collected and processed. See Fig. 2.

Fig. 2   Experiment

Initial status of experiment task: altitude 2,000 ft, speed 250 kn, distance from runway 6 mile, flying level straight towards the runway.

Basic operations:

(1) Descend to 1,000 ft, decelerate to 200 kn;

(2) Extend gear;

(3) Descend to 700 ft, decelerate to 150 kn, extend flap to 20°;

(4) Descend to 400 ft, decelerate to 130 kn, extend flap to full;

(5) Flatter out, land gear down;

(6) Brake.

6 subjects in total, experimenting 2～3 times per day, all data recorded, and analyzed with software.

4   Results

During experiments, various parameters of the aircraft during the entire flight are recorded as well, including space coordinates (longitude, latitude, altitude), attitude, (angle of pitch, roll angle, orientation), status of various parts of the aircraft (status of engine, flaps, ailerons, elevator, rudder), etc.

Tab. 1   **Experiment Data**

Physiological and Mental Measurement Equipment	Form of Recorded Data	Content Information
Polar Heart Rate Watch	1 text file	Heart Rate
SmartEye Desktop Tracking System	1 video file	Video of scenario in front of the subject
	1 text file	Coordinates of focus
		Number of gazing
		Number of blinking
		Number of glancing

（continued）

Physiological and Mental Measurement Equipment	Form of Recorded Data	Content Information
SMI Eye Tracking System	1 video file	Video of scenario in front of the subject
	1 text file	Coordinates of focus
		Diameter of pupils
		Duration of gaze
Hand Movement Capturing Camera	1 video file	Operational behaviors of the subject
Body Movement Measurement Camera	1 video file	Range of body movement

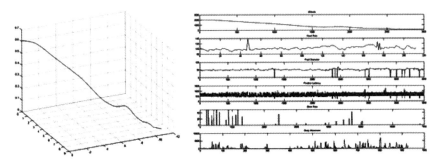

Fig. 3　Experiment Data

5　Conclusion

Through analysis of the experiment data, the following results can be gathered:

（1）During approaching, as the plane descends, the subject's heart rate fluctuates but the range of fluctuations tends to gradually lessen after gradually increasing. And heart rate rises more obviously near landing than when ascending.

（2）When plane altitude changes intensely, pilot's operation movements are more frequent, range of movements larger and amount of movement larger.

（3）Heart rate and movement range are rather sensitive to changes in plane altitude.

（4）When the simulator has just gone into initial status, the subject has a higher frequency of blinking. And when the plane is descending, frequency of blinking decreases, but rises again when approaching the ground.

（5）Relationship of pupil diameter of subjects and duration of gazing, to plane altitude is not clear.

This research constructs a rather complete human factor experiment platform that could undergo pilot-in-the-loop flight experiments. And through collection and processing of experiment data during flights, objective analysis of parameters, pilot performance and workload could be achieved. In follow-up studies to come, more experiments will be conducted using this platform, for deeper research into the methods for objective analysis of human factors in aviation.

Acknowledgments

The authors would like to acknowledge the support of 973 Program (No. 2010CB734103)

References

[1] Boeing Aero Magazine No. 8. Human factors. http://www. boeing. com/commercial/ aeromagazine/aero_08/human_textonly. html.

[2] Corwin, William H. Assessment of Crew Workload Measurement Methods [S]. Techniques and Procedures. ADA217699, WRDC – TR – 89 – 7006. Volume I and Volume II. 1989.

[3] Eggemeier, F. T. , & Wilson, G. F. Performance and subjective measures of workload in multitask environments [M]. In D. Damos (Ed.), Multiple-task performance (pp. 217 – 278). London: Taylor & Francis. 1991.

[4] FAA. 9550. 8 – Human Factors Policy [S]. 1993.

[5] Malcolm A. Bonner & Glenn F. Wilson. Heart Rate Measures of Flight Test and Evaluation [J]. The International Journal of Aviation Psychology, 2002,12:1,63 – 77.

[6] Roscoe, A. H. Assessing pilot workload: Why measure heart rate, HRV, and respiration? [J] Biological Psychology, 1992,34,259 – 288.

[7] Wilson, G. F. In-flight psychophysiological monitoring [M]. In F. Fahrenberg & M. Myrtek (Eds.), Progress in ambulatory monitoring (pp. 435 – 454). Seattle, WA: Hogrefe & Huber. 2001.

[8] Wilson, G. F. Psychophysiological test methods and procedures [M]. In S. G. Charlton & T. G. O'Brien (Eds.), Handbook of human factors testing and evaluation (2nd ed. , pp. 127 – 156). Mahwah, NJ: Lawrence Erlbaum Associates, Inc. 2002.

The Precision Analysis of the Eye Movement Measurement System

Yuezhi Huang, Shan Fu

School of Aeronautics and Astronautics, Shanghai JiaoTong University,
Shanghai 200240

Abstract　The significance of mental workload has become more important in the cockpit system designing. The aircraft's design, to meet the mental workload standards, ensures the pilots maintain safety, healthy, comfort and long-time working efficiency. Eye movement, behavior performance and heart rate are good indicators of mental workload assessment. This paper describes the pilot mental workload research of ergonomics in the civil aircraft cockpit. And then we analyze the information that the pilot need to focus on, and the analysis precision need to be achieved during the aircraft approach and landing phase. In the end, the eye movement tracking system is established in the simulated cockpit. There are many different results after a simulated flight mission, and most of the recorded data can be used to analyze and compare the expected and actual gaze positions comprehensively, so as to assess the precision of the eye tracking system in the application under the simulated cockpit conditions. The conclusions will provide a theoretical basis to calculate and measure the parameters of eye movement measurement.

Keywords　mental workload, eye movement measurement, precision analysis

1　Introduction

The mental workload plays an important role in the cockpit system design: guiding the system design and evaluation by measuring the mental workload; ensuring the information provided to pilot not overloaded; minimizing the system effect on the pilot's mental workload when busy driving. Therefore, the aircraft designed to meet the standard of mental workload, enables pilot to maintain safety, healthy, comfort and long-time working efficiency.

There are many methods to measure the mental workload, and they can be divided into three kinds [1 - 3]: performance measurement, subjective measurement and physiological measurement. Physiological measurement reflects the changes of the mental workload by measuring some of the physiological index. As the one that is a more objective measurement technical, physiological measurement has shown its strong availability and attracts more and more attention with these prominent characteristics: ① Physiological measurement, as a useful technology for the real-time measuring, can respond the change of mental workload rapidly, and the parameters can be continuously measured. ② Although the participants were equipped with specialized equipment, the main operation will not be interfered and disrupt.

Typical measurements of physiological parameters include : heart rate measurement, eye movement measurement, respiration measurement, behavior measurement and body fluids analysis and so on. Some index of the eye movement, such as blink frequency, blink duration, saccade frequency, saccade duration, saccade amplitude, pupil diameter are the result of the eye muscle stretching, which can be used as the index of participants' mental workload. Reference [4] concluded that the blink frequency and blink duration decreased with the level of mental workload increased. The study shown that when the tasks of the participant increased, the attention was affected by the mental concentration, making the eyes remain open state longer. Reference [5] concluded that saccade duration was positively correlated with the workload. So in the modern cockpit environment, capturing the pilot's situation by measuring the eye movement is an objective, directive and effective way.

Fig. 1 shows the eye movement measurement is a part of the mental workload assessment. Finally, we can integrate the eye movement measurement, behavior measurement and heart rate measurement into a comprehensive, flexible pilot monitoring system. By analyzing the multi-dimensional index, the assessment of pilot's mental workload will be more accurate, reliable and robust.

However, the procedure of eye movement measurement is complex, and it's sensitive to the equipments, environment and other factors. Therefore, the analysis of precision requirement of eye movement measurement is necessary. The article firstly collates the information needs to be focused on during approach and landing phase, analyzes the precision need to be achieved, and then Establishes an

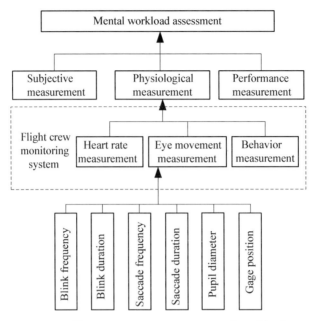

Fig. 1 Eye movement measurement and workload

initial monitoring system in civil aircraft cockpit to simulate the behavior of eye movement, using the existing eye movement measurement equipment.

2 The measuring precision of eye movement during the aircraft approach and landing phase

It's important and useful in real missions to assess the mental workload of the pilots in specific tasks scenario, hence the conclusions can be more reliable. In order to reach these purposes, we focus on analyzing the workload of pilots during the approach and landing phase.

Whether the flight is intended for personal, business or other purposes, every kind or every time of flight has a common goal that is safely landing. According to statistics, nearly 50% of the general aviation aircraft accidents occurred in the approach and landing phase [6].

However, it should be noted that both the takeoff and landing phases together represent only about 5% of the flight time, but about 54% of the accidents occurred within the two phases. Joseph T. Nall pointed out in the Air Safety Foundation's annual report: the accidents mainly occurred in a phase when the pilot was required to complete many important tasks in a relatively short flight time'. It will help to improve the safety margin and to reduce the overall general aviation accident rate if we could have modified the distribution of pilot mental workload in the two phases. Thus, assessing the pilots' mental workload during the approach and landing phase and then improving the design of cockpit are extremely important.

Referring to the pilot manual [7], we concluded that the pilot acts shown as Fig. 2 during the approach and landing phase.

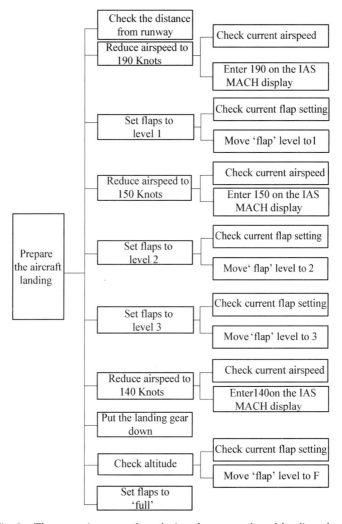

Fig. 2   The operating procedure during the approach and landing phase

During this phase, the pilot acts including: airspeed setting, flap setting, and putting the landing gear down. At the same time, the pilot also needs to pay attention to the related information. The mode of eye movements includes fixation and saccade. According to the mode, the result can be concluded that whether the attention of the pilot is focusing on a specific region or a particular reading on the display. The precision of the eye movement tracking system should identify the value of the reading focused on, when the pilot pays attention to a specific reading. Based on the analysis records, the precision of the pilot's attention can be confirmed, and it can be used to design the monitoring and controlling system. The analysis indicates that the landing phase is a high-intensity visual activity. In order to achieve the airspeed, height, and the flight attitude, and then control the flight to land successfully, the pilot needs to saccade between PFD (Primary Flight Display) and the visual scene outside. Hence, it required that the eye tracking system should locate the specific value exactly.

3 Eye movement experiment and precision assessment

3.1 Participants

There are 10 participants including both undergraduate and graduate students from Shanghai JiaoTong University in the test. They are 22 years old in average, ranging from 20 to 36. All the participants are with normal or corrected-to-normal vision.

3.2 Test environment and equipments

We conducted this test on the simulated platform of the civil aircraft cockpit in Institute of Man-Machine-Environment System Engineering Shanghai Jiao Tong University. And the eye tracking system is equipped with Smart Eye, which is the latest top production of Smart Eye Company. The system samples the eyeball position every 0.17 s (frequency of 60 Hz).

The experiment material is divided into two groups: one group is concentric circles with different radius, including 0.5, 2.5, 5.5 and 9.5 cm (shown as Fig. 3); the other group is the Primary Fight Display in the simulated civil aircraft cockpit, the display consists of airspeed indicator, altimeter, vertical speed indicator mainly (See Fig. 4).

Fig. 3　concentric circles with different radius　　　　Fig. 4　Primary Flight Display

3.3 Procedures and Results

To ensure the participants were familiar with the experiment procedures, the tasks were explained and demonstrated to the participants as follow [8]:

Place the camera: the cameras can be positioned totally physically independent from each other. However, in order to maximize the performance of the system, the issue should be taken into consideration: It is advantageous if each camera is oriented in such way that the Participant's head is at the center of the camera image, while the Participant is positioned in a neutral position. This will maximize the allowable movement in x, y and z positions, without moving out of view of the cameras.

Adjust the camera brightness and focus: To be able to use corneal reflection we need to make sure that the eyes are visible in two cameras at all times. If the image is too bright, there are many saturated pixels. The system uses IR-diodes to illuminate the face of the Participant and to minimize the effect of varying lighting conditions.

Calibrate the camera: It is very important that the positions of the cameras and their orientation in respect to each other are kept constant once the camera calibration process has been performed. Every time the cameras are moved, we have to redo the camera calibration process.

Define a coordinate system: Hold the chessboard steady in the position where you wish to define the origin of the coordinate system. When the system has calculated the pose of the chessboard, press the OK button to define the coordinate system.

In order to establish the mapping relationship between pupil and gaze coordinates, we firstly used five pre-marked points to calibrate. Based on the established coordinate, the experiment material was switched to the concentric circle. Then the participants were instructed to observe the specific location. As shown in the Fig. 5, the tracking position was better overlapped with the actual observation position.

Next, the participants observed the Primary Flight Display for eye tracking precision experiments. As shown in the Fig. 6, the red-cross cursor denoted the participant was gazing on the current airspeed value.

We chose another 9 points to perform the precision validation arbitrarily. Fig. 7 shows min., max. and mean precision for all test participants at different points. The red line illustrates the precision of the ten points ranging from 0.8 degree to 1.3 degree, and the average precision of the participants is 1.01 degree.

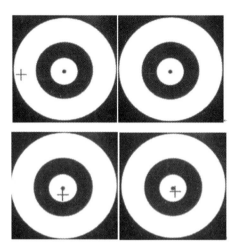

Fig. 5　Concentric circle experiment shows

Fig. 6　Read the current airspeed value

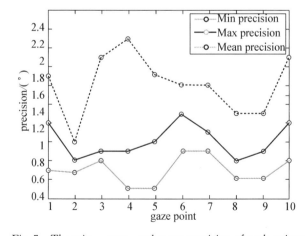

Fig. 7　The min., max. and mean precision of each points

4 Conclusion

The precision of eye-gaze tracking systems is a key factor in determining the usability of eye-gaze tracking for mental workload assessment. In this paper, we found that the gaze position, calculated out by the eye tracking system, can be accurately consistent with the display reading value, and also can distinguish the text between multiple lines, and the fixation point is relatively stable. Thus we can use this measurement system to carry out the experiment and provide more reliable data.

To be noted that, during the experiment, we found that the precision of the eye tracking system was affected by the camera position, illumination intensity, and personal profile and so on. Hence, the system should be established appropriately. Especially the position of camera need to be calibrated after the participant seated. Based on the improved measuring system and follow with the right experiment procedure, the result then can be concluded reliably.

In the future, we will focus on the research and analysis of the mental workload through all the eye movement parameters based on the system above.

Acknowledgment

This work was supported by a grant from the Major State Basic Research Development Program of China (973 Program) (No. 2010CB734103).

References

[1] Brad Cian. A Review of the Mental Workload Literature [R]. Defence Research and Development Toronto (Canada), 2007. RTO - TR - HFM - 121 - Part - II.

[2] William H. Corwin, Diane L. Sandry-Garza. Assessment of crew workload measurement, methods, techniques and procedures [J]. Process, Methods and Results. WRDC - TR - 89 - 7006. Vol. 1. 1990.

[3] Ulf Ahlstrom. Participantive Workload Ratings and Eye Movement Activity Measures [R]. Technical Report. 2005.

[4] Jeff Braun, Eric Lichtenstein, Joe Long. Human Performance-Based Measurement System [R]. Cybernet Systems Grp Ann Arbor Mi, 1999.

[5] Yanhong Xu. Assessment of workload changing on the operators using eye movement parameters [J]. Flight Surgeon. 1996,1:070 (in Chinese).

[6] Airplanes B C. Statistical Summary of Commercial Jet Airplane Accidents: Worldwide Operations 1959 - 2005 [J]. Aviation Scfety. Boeing Commeroial Airlines, Seattle, Washington. 2006.

[7] Beeing Company. Boeing 777 - 220 ⋯ Manual [M]. USA. Chicago: 2012.

[8] Smart Eye AB. Smart-Eye Pro 5. 5 User Manual [M]. Gothenburg. Sweden: Smart Eye AB, 2009.

Effects of Mental Workload on Long-latency Auditory-evoked-potential, Salivary Cortisol, and Immunoglobulin A

Ying Lean[1], Fu Shan[1], Qian Xuemei[2], Sun Xiaojiang[2]

1. Dept. of Aviation Medicine, Physiology & Human Factors, School of Aeronautics and Astronautics, Shanghai Jiao Tong University, Shanghai 200240

2. Dept. of Neurology, Shanghai No. 6 Hospital, Shanghai 200233

Abstract   This paper researches on the effects of mental workload on long-latency Auditory-evoked-potential (AEP), salivary cortisol, and Immunoglobulin A (IgA). 20 healthy subjects (11 males and 9 females) participated in the experiment voluntarily. The mental task consisted of two parts: arithmetic task and reading comprehension task. The Latencies of N1, P2, N2, P3, and Mismatch negativity (MMN) all increased significantly after the mental tasks were adopted at all of the three recording sites: Cz, Fz, and Pz ($p < 0.05$). In this experiment, changes of salivary cortisol and s – IgA levels due to mental tasks were not significant. With the introduction of mental tasks, more processing resources are allocated to the primary task (mental task), and decreased processing resources available for the secondary task (auditory task), which is reflected on the increases in the Latencies of probe-evoked AEP components.

Keywords   Mental Workload, Auditory-evoked-potential (AEP), Event-related-potential (ERP), Cortisol, Immunoglobulin A (IgA)

1   Introduction

When hearing organs are stimulated by a sound with a certain intensity or frequency, there would be a sequence of electrical activities in hearing systems from the cochlear nerve to the cortex, which are named as 'Auditory-evoked-potentials (AEPs)'. 'Long-latency AEP' contains the wave components of P1, N1, P2, N2, P3, etc. These components are used to evaluate the functions of hearing system, and are also closely related to the mental activities of cerebral cortex, such as attention, information processing, resource allocation and other cognitive activities [8, 35]. Mismatch negativity (MMN) is the negative component of Long-latency AEP by subtracting the standard wave from the deviant wave, and usually peaking at about $100\sim200$ ms after the onset of stimulus. It is an objective measurement for the accuracy of central auditory processing and auditory memory traces in human brain. Furthermore, it is also an objective index of general brain degeneration and index of the gross functional state of the brain in neurology and cognitive neuroscience [17].

The amplitudes or latencies of these AEP components, especially MMN, N1, P2, and P3, are used as indices to evaluate the speech performance and ability to process auditory information for cochlear implant subjects [10, 24, 31]. It was found that trained musicians showed more efficient neural detection and enhanced sensitivity to acoustic changes, and demonstrated superior auditory sensory-memory traces for acoustic features. Thus, the effect of music training on central auditory processing and the modulation of auditory neural system was occurred [18].

AEP with task-irrelevant auditory probes was an effective non-invasive method for mental workload evaluation in complex mental tasks [6, 21, 27]. Long-latency AEP was recorded from 10 radar operators when performing simulated radar-monitoring tasks. It was observed that the amplitudes of N1, N2, P3, and MMN decreased all with the introduction of mental tasks and decreased further when the tasks became more difficult [14].

Other similar studies had similar results and conclusions. When performing gauge monitoring and arithmetic tasks, N1 and P3 elicited by irrelevant auditory probes were recorded from 15 healthy adults at the sites of Fz, Cz, Pz and Oz. The amplitudes of N1 and P3 decreased significantly during these tasks [29]. For 14 male subjects playing computer video games, amplitudes of N1, P2, N2, P3 at Fz, Cz, and Pz electrodes decreased all when their mental workload increased [1]. Compared to the non-stressful condition, high attentional stress and mental workload led to a decrease in MMN amplitude during the IQ test. All these facts demonstrated that attentional stress and mental workload might attenuate cortical auditory processing capabilities [26].

Biochemical indices, collected from subjects blood, salivary, or urinary samples, can be used to evaluate mental workload. Blood sampling causes harms and pains to subjects. Compared to urinary sampling, salivary sampling is relatively more convenient to collect and the procedure is easier. In this research, cortisol and secretory immunoglobulin A (s – IgA) are used together with electrophysiological measurements to get a more comprehensive evaluation of mental workload.

As an indicator of hypothalamic-pituitary-adrenocortical (HPA) axis activation, cortisol awakening response is related to a number of psychosocial factors [3]. By a systematic review and meta-analysis, it was summarized that increased cortisol awakening response was generally associated with work stress, job strain, mental overload, overcommitment to work, etc. On the contrary, decreased cortisol awakening response was associated with fatigue, burnout, exhaustion, etc [3].

As one kind of immunoglobulin in human secretions, secretory immunoglobulin A (s – IgA), which could be sampled non-invasively from saliva, could be used as a measurement of mental workload and the effect of mental workload on s – IgA is bi-directional. According to an early article published on an international famous medical journal, it was observed that the secretion rate of salivary s – IgA was significantly lower in high stress than low stress periods for 64 first-year dental school students [9].

However, other similar researches had different results and conclusions. Observations among 10 second-year medical students showed that the concentration of salivary s – IgA tended to be higher on the day before academic exams and during exams and lower on the days between these exams. The increase in s – IgA before exams might be due to the anticipation of the forthcoming exams [19]. In addition, other several studies concluded that mental arithmetic task resulted in significant increases in salivary s – IgA concentration and s – IgA secretion rate, as well as increased heart rate (pulse rate) and blood pressure [22, 23, 33]. So, the previous hypothesis that mental workload and psychological stress led to increased susceptibility to respiratory infections due to decreased s – IgA might require further investigation [23]. Furthermore, s – IgA reaction to mental tasks depended not only on the external tasks, but also on the individual perceptions of the tasks [32].

While the above mentioned articles mainly focus on the changes of 'amplitudes' of N1, P2, N2, P3 and MMN, this research aims at researching on the changes of their 'latencies' as well as cortisol and s – IgA from saliva samples.

2　Materials and Methods

2.1　Subjects

20 healthy subjects (11 males and 9 females), aged from 20 to 35 years (25±3), participated in the experiment voluntarily. They were all without past or present hearing or nervous disorders and free from upper-respiratory tract diseases. Pure tone audiometry examination before the experiment showed that

their hearing levels were all within normal range. Informed consent was obtained from all subjects prior to the experiment. To assure the quality of AEP recordings, subjects were asked to wash their hair the night before the experiment day. And also, subjects were directed to sleep well and remain in a normal mood before the experiment day.

2. 2 Methods and Procedures

The mental task consisted of two parts: arithmetic task and reading comprehension task. The arithmetic task contained 20 multiplication questions. The reading comprehension task contained 4 passages, 5 questions taking the form of multiple choice for each passage. So, the total number of questions in the reading comprehension task was also 20. The reading comprehension task tested the subjects abilities of verbal scanning and understanding, reasoning, short memory, etc. All the subjects were asked to finish all the questions in 45～60 min.

This study was set up as "Self-control design". Prior to the experiment, each subject sat quietly for 15 - 20 min for a rest. Then, the experiment began. Firstly, a pre-task saliva sample was collected and AEP was recorded prior to the mental task for each subject. Secondly, the mental task began, the duration of which was about 45 - 60 min. The sequence was fixed and arithmetic task was always performed first. Finally, on completion of the task, a post-task saliva sample was collected and AEP was recorded again for each subject.

After the saliva samples were collected, they were immediately labelled and frozen at −20℃ until analysis. The levels of salivary cortisol and s - IgA were quantified by means of ELISA and conducted by a biomedical service company without knowing the experimental details.

To record AEP, after cleaning subject's skin with alcohol, three electrodes were put on the scalp at the sites of Fz, Cz, and Pz, according to the international 10～20 systems for Electroencephalogram (EEG) recordings. The reason why to choose these three electrode sites was that they were mostly used in recording AEP [1, 14, 29]. Another two reference electrodes were placed at the back of the left and right ear, respectively. The AEP was amplified and band-pass filtered (0. 1— 100 Hz).

For the long-latency AEP recordings, the repetitive standard stimuli were 500 Hz with the occurrence probability of 0. 80, and the rare deviant stimuli were 2,000 Hz with the occurrence probability of 0. 20. They were both presented at the intensity of 80 dB SPL, with the duration of 100 ms, and based on a randomized series. The stimulus interval from the end of one stimulus to the onset of the next stimulus was 500 ms. The tones were presented binaurally through the headphones. Subjects were instructed to respond to the deviant stimuli by pressing a button and ignore the standard stimuli. The experiment was conducted in a soundproof and electrically shielded room and the subjects were in a relaxed state with their eyes open.

2.3 Statistical Methods

SPSS13. 0 for Windows was used for statistical analysis. $p < 0.05$ is considered statistically significant. The test data were all normally distributed ($p > 0.05$, Shapiro-Wilk Test). For AEP components, firstly, Paired Samples t-test was used to see whether there were significant statistical differences for each AEP component and for each recording site after the mental tasks. Then, Repeated Measures ANOVA was done to see whether there existed statistical significances among the three different recording sites: Cz, Fz, and Pz. For salivary cortisol and s-IgA, Paired Samples t-test was used to see whether there were significant statistical differences for the changes of salivary cortisol and s-IgA concentrations after the mental tasks were adopted.

3 Results

3.1 AEP Results

Latencies of N1, P2, N2, P3, and MMN before and after the mental tasks were researched, all of which were collected at three different recording sites: Cz, Fz, and Pz. Tab. 1 showed the basic descriptive statistics for the Latencies of each AEP component before and after the mental tasks.

Tab. 1 The basic descriptive statistics for the Latencies of each AEP component before and after the mental tasks (expressed in the form of "Mean±SD", ms)

		N1	P2	N2	P3	MMN
Fz	Before	92±16	166±14	223±23	322±16	182±18
	After	109±15*	182±26*	255±20*	351±17*	203±23*
Pz	Before	94±16	165±16	221±25	320±16	180±18
	After	109±17*	181±28*	258±21*	350±19*	201±24*
Cz	Before	92±16	164±17	219±26	319±17	180±21
	After	110±17*	181±29*	257±21*	352±20*	201±25*

* $p < 0.01$ vs the value of baseline (the datum of "before").

From Table 1, it could be seen clearly that for all of the three recording sites:

Cz, Fz, and Pz, the Latencies of N1, P2, N2, P3, and MMN all increased after the mental tasks were adopted. The outcomes of the Paired Samples t‐test for the Latencies of each AEP component were also shown in Tab. 1. It was demonstrated clearly that, for all of the three recording sites: Cz, Fz, and Pz, the prolongation in the Latencies of N1, P2, N2, P3, and MMN after the mental tasks were all statistically significant ($p<0.05$). Repeated Measures ANOVA showed that there were no significant differences for the prolongation in the AEP Latencies among the three different recording sites: Cz, Fz, and Pz. ($p>0.05$).

3.2   Salivary Cortisol and s‐IgA Results

Concentrations of salivary cortisol and s‐IgA before and after the mental tasks were seen in Tab. 2. The test data were all normally distributed ($p>0.05$, Shapiro-Wilk Test).

Tab. 2   The basic descriptive statistics for concentrations of salivary cortisol and s‐IgA before and after the mental tasks (expressed in the form of "Mean±SD")

	Cortisol/(nmol/l)	s‐IgA/(μg/ml)
Before	21.2±10.7	280.3±107.6
After	23.6±11.2	274.3±110.2

Paired Samples t‐test showed that there were no significant statistical differences for the changes of salivary cortisol and s‐IgA concentrations after the mental tasks were adopted (For cortisol: $t=-0.824$, $p=0.420$; For s‐IgA: $t=0.181$, $p=0.858$). So, in this experiment, changes of salivary cortisol and s‐IgA levels due to mental tasks were not significant.

4   Conclusions

Long-latency AEP components are used to evaluate mental workload with task-irrelevant auditory probes. With the introduction of mental tasks and increases in the difficulty of the primary task (mental task), more processing resources are allocated to the primary task. However, the brain capacity is not unlimited. Increases in the difficulty of the primary task will result in decreased processing resources available for the secondary task (auditory task), which is reflected on the increases in the Latencies of probe-evoked AEP components. Attention would be allocated to task-irrelevant probes. Stimulus which suddenly appears in the environment will capture subjects attention without their intentions. The deviant irrelevant probes capture attention automatically. Attention which is not focused on the mental task is captured by the deviant tones and is reflected on the Latencies of probe-evoked AEP components. As additional resources are required for the mental task, fewer resources are available for the MMN and other AEP components [14].

Blood sampling would cause subjects to be afraid, nervous and painful. This additional nervous emotion and pain would become an extra "Confusion Factor" to this kind of neuropsychological experiments and resulted in misleading or even wrong results. So, salivary sampling is widely used instead of blood sampling and the results of salivary cortisol studies are similar with that of plasma cortisol studies [3, 4, 5, 16, 26].

Effects of mental workload on cortisol levels are different: While some studies demonstrated increased cortisol levels resulted from mental workload [3, 5, 16, 26], some other studies concluded that changes of cortisol levels due to mental workload were not significant [4, 28]. Besides, some psychological factors might even decrease the cortisol response [3]. It is much likely that some subjects participating in the experiment were just in the state of fatigue, burnout, exhaustion, posttraumatic stress syndrome, or had a feeling of happiness and optimism. All these factors would led to the increases in their salivary cortisol responses due to mental workload not so significant and even decrease.

And also, effects of mental workload on s - IgA levels are different: While some studies demonstrated increased s - IgA levels resulted from academic stress, mental arithmetic task, or memory search [19, 22, 23, 32, 33], some other studies showed that mental workload resulted in decreased s - IgA levels [9, 32]. In the study of Wetherell MA et al, it was found that subjects who perceived the task as too difficult and overwhelming, perhaps due to the lack of skills or abilities, tended to show a decrease in salivary s - IgA secretion. Decreased salivary s - IgA level was associated with increases in the perception of mental workload [32]. It was also suggested that changes of s - IgA were relevant to alpha-adrenergic mechanisms [22], the number of CD4 + T cells, the CD4/8 ratio, total lymphocytes, and NK cells [34].

The cerebral cortex consists of a large number of neurons and is associated with different types of cortical cells, molecules, gene expression patterns, etc. Mental workload and stress often activate the hypothalamic-pituitary-adrenal (HPA) axis and sympathetic adrenomedullary system. Cortisol plays an important role in the responses to mental workload. Corticotrophin-releasing hormone - 1 (CRH - 1) receptors regulate the behavioral, sympathetic, and HPA responses. In the brain, there are two kinds of corticoid receptors exerted by the corticosteroid

hormones: mineralocorticoid receptors (MR) and glucocorticoid receptors (GR).
While MR function as the maintaining of homeostasis induced by stress and mental
workload, GR function as the recovery and restorage. Cortisol is mediated by MR
and GR, which regulate emotion and cognition [11, 12].

Long-latency AEP could also be evoked in animals, such as rats and mice [15,
25, 30]. Changes of cortisol levels induced by mental workload are not only
available in animal saliva samples. Some neurosteroids glucuronides, such as
cortisol and corticosterone, were discovered in mouse cortex, hippocampus,
hypothalamus, and mid-brain. Acute stress might increase the mRNA levels of c-
fos and corticotrophin-releasing factor in rat brains. Acute stress might activate
Relaxin – 3 expression and alter Relaxin – 3 gene transcription by activation of
corticotrophin-releasing factor – 1 receptors on nucleus incertus neurons [2]. In
our future studies, effects of mental workload on long-latency AEP, salivary
cortisol, and s – IgA would be researched at the molecular, cellular, and genetic
levels. Besides, some adequate animal models, such as foot shock, maze, repeated
forced swimming, would be used for our future studies [2, 7, 13, 20].

Acknowledgement

This research is supported by 'Chinese National Key Basic Research Program
(973 Program)' with the number of : 2010CB734103.

References

[1] Allison B Z, Polich J. Workload assessment of computer gaming using a single-stimulus
event-related potential paradigm [J], Biol. Psychol. 2008,77:277 – 283.

[2] Banerjee A, Shen P J, Ma S, et al. Swim stress excitation of nucleus incertus and rapid
induction of relaxin – 3 expression via CRF1 activation [J], Neuropharmacology 2010,58:
145 – 155.

[3] Chida Y, Steptoe A, Cortisol awakening response and psychosocial factors: a systematic
review and meta-analysis [J], Biol. Psychol. 2009,80:265 – 278.

[4] Elmenhorst E M, Vejvoda M, Maass H, et al. Pilot workload during approaches:
comparison of simulated standard and noise-abatement profiles [J], Aviat Space Environ.
2009,80:364 – 370.

[5] Fibiger W, Evans O, Singer G, Hormonal responses to a graded mental workload [J], Eur.
J. Appl. Physiol. Occup. Physiol. 1986,55:339 – 343.

[6] B. Fowler, P300 as a measure of workload during a simulated aircraft landing task [J],
Hum. Factors. 1994,36:670 – 683.

[7] Funk D, Li Z, Lê A D, Effects of environmental and pharmacological stressors on c-fos and
corticotropin-releasing factor mRNA in rat brain: Relationship to the reinstatement of alcohol

seeking [J], Neuroscience 2006,138:235 – 243.

[8]　Hyde M, The N1 response and its applications [J], Audiol. Neurootol. 1997,2:281 – 307.

[9]　Jemmott J B, Borysenko J Z, Borysenko M, et al. Academic stress, power motivation, and decrease in secretion rate of salivary secretory immunoglobulin A [J], Lancet 1983,1: 1400 – 1402.

[10]　Kelly A S, Purdy S C, Thorne P R, Electrophysiological and speech perception measures of auditory processing in experienced adult cochlear implant users [J], Clin. Neurophysiol. 2005,116:1235 – 1246.

[11]　Kloet E R, Derijk R, Signaling pathways in brain involved in predisposition and pathogenesis of stress-related disease: genetic and kinetic factors affecting the MR/GR balance [J], Ann. N. Y. Acad. Sci. 2004,1032:14 – 34.

[12]　Kloet E R, Stress: a neurobiological perspective [J], Tijdschr Psychiatr. 2009,51:541 – 550.

[13]　Korte S M, De Boer S F, A robust animal model of state anxiety: fear-potentiated behaviour in the elevated plus-maze [J], Eur. J. Pharmacol. 2003,463:163 – 175.

[14]　Kramer A F, Trejo L J, D. Humphrey, Assessment of mental workload with task-irrelevant auditory probes [J], Biol. Psychol. 1995,40:83 – 100.

[15]　Lazar R, R. Metherate, Spectral interactions, but no mismatch negativity, in auditory cortex of anesthetized rat [J], Hear. Res. 2003,181:51 – 56.

[16]　Leino T K, Leppäluoto J, Ruokonen A, et al. Neuroendocrine responses to psychological workload of instrument flying in student pilots [J], Aviat Space Environ. 1999,70:565 – 570.

[17]　Näätänen R, Mismatch negativity (MMN): perspectives for application [J], Int. J. Psychophysiol. 2000,37:3 – 10.

[18]　Nikjeh D A, Lister J J, Frisch S A, Preattentive cortical-evoked responses to pure tones [J], harmonic tones, and speech: influence of music training, Ear. Hear. 2009,30:432 – 446.

[19]　Otsuki T, Sakaguchi H, Hatayama T, et al. Secretory IgA in saliva and academic stress [J], Int. J. Immunopathol. Pharmacol. 2004,17:45 – 48.

[20]　Passerin A M; Cano G, Rabin B S, et al. Role of locus coeruleus in foot shock-evoked Fos expression in rat brain [J], Neuroscience 2000,101:1071 – 1082.

[21]　Prinzel L J, Freeman F G, Scerbo M W, et al. Effects of a psychophysiological system for adaptive automation on performance, workload, and the event-related potential P300 component [J], Hum. Factors. 2003,45:601 – 613.

[22]　Ring C, Drayson M, Walkey D G, et al. Secretory immunoglobulin A reactions to prolonged mental arithmetic stress: inter-session and intra-session reliability [J], Biol. Psychol. 2002, 59:1 – 13.

[23]　Ring C, Carroll D, Hoving J, et al. Effects of competition, exercise, and mental stress on secretory immunity [J], J. Sports. Sci. 2005,23:501 – 508.

[24]　Roman S, Canévet G, Marquis P, et al. Relationship between auditory perception skills and mismatch negativity recorded in free field in cochlear-implant users [J], Hear. Res. 2005, 201:10 – 20.

[25]　Ruusuvirta T, Penttonen M, Korhonen T, Auditory cortical event-related potentials to pitch

deviances in rats [J], Neurosci. Lett. 1998,248:45 - 48.

[26] Simoens V L, Istók E, Hyttinen S, et al. Psychosocial stress attenuates general sound processing and duration change detection [J], Psychophysiology 2007,44:30 - 38.

[27] Sirevaag E J, Kramer A F, Wickens C D, et al. Assessment of pilot performance and mental workload in rotary wing aircraft [J], Ergonomics 1993,36:1121 - 1140.

[28] Sudo A, Evaluation of workload in middle-aged steel workers by measuring urinary excretion of catecholamines and cortisol [J], Sangyo Igaku. 1991,33:475 - 484.

[29] Ullsperger P, Freude G, Erdmann U, Auditory probe sensitivity to mental workload changes-an event-related potential study [J], Int. J. Psychophysiol. 2001,40:201 - 209.

[30] Umbricht D, Vyssotki D, Latanov A, et al. Deviance-related electrophysiological activity in mice: is there mismatch negativity in mice? [J] Clin. Neurophysiol. 2005,116:353 - 363.

[31] Wable J, Van Den Abbeele T, Gallégo S, et al. Mismatch negativity: a tool for the assessment of stimuli discrimination in cochlear implant subjects [J], Clin. Neurophysiol. 2000,111:743 - 751.

[32] Wetherell M A, Hyland M E, Harris J E, Secretory immunoglobulin A reactivity to acute and cumulative acute multi-tasking stress: relationships between reactivity and perceived workload [J], Biol. Psychol. 2004,66:257 - 270.

[33] Willemsen G, Ring C, McKeever S, et al. Secretory immunoglobulin A and cardiovascular activity during mental arithmetic: effects of task difficulty and task order [J], Biol. Psychol. 2000,52:127 - 141.

[34] Willemsen G, Carroll D, Ring C, et al. Cellular and mucosal immune reactions to mental and cold stress: associations with gender and cardiovascular reactivity [J], Psychophysiology 2002,39:222 - 228.

[35] Zhang F, Eliassen J, Anderson J, et al. The time course of the amplitude and latency in the auditory late response evoked by repeated tone bursts [J], J. Am. Acad. Audiol. 2009,20: 239 - 250.

The Effects of Perceptual Load Related to Flight Task on Auditory ERPs

Wanyan Xiaoru[1], Zhuang Damin[1], Lu Shasha[1], Liu Wei[2]

1. School of Aeronautic Science and Engineering, Beijing University
of Aeronautics and Astronautics, Beijing, 100191

2. School of Automation, Beijing University of Posts and
Telecommunications, Beijing, 100876

Abstract　To investigate the effects of perceptual load related to flight task on the auditory ERPs, 14 subjects participated in a flight simulation experiment, and were asked to perform the indicator-monitoring task based on different perceptual loads. Auditory probes were simultaneously sent binaurally through headphones under the oddball paradigm. Subjects were instructed to pay attention to the flight simulation task and ignore auditory probes. The auditory ERPs of mismatch negativity (MMN), N1 and P2 were recorded as the evaluation indexes of perceptual load. The results revealed that the frontal-central MMN was affected by perceptual load significantly. The MMN average amplitude was enhanced under the high perceptual load, reflecting an improvement of automation processing ability of task-irrelevant information. Furthermore, the MMN average amplitude was positively correlated with the accuracy rate of detecting abnormal information. The present study indicated that the frontal-central MMN is well sensitive to perceptual load, and it may offers effective electrophysiological reference for the perceptual load assessment in complex flight tasks. However, the results also showed that the perceptual load had little effect on the auditory early components N1 and P2 as well as the temporal MMN, which manifested the complicated effects of auditory ERPs on measuring the perceptual load related to flight task to some extent.

Keywords　ERPs, perceptual load, mismatch negativity, N1, P2, flight simulation

1　Introduction

When operating an aircraft, the pilot often needs to keep monitoring multiple indicators simultaneously, and getting information effectively highly depends on reasonable selected attention according to the information priority [1]. Relative researches have concluded that the pilot's attention relates closely to his perceptual load. Low level arousal of perceptual resources (e. g. transoceanic flight) or perceptual overload (e. g. rescue operation of helicopter or single flight in bad weather) will all probably make the pilot hard to keep an appropriate attention level, which subsequently results in human errors that threat flight safety [2]. As the perceptual workload can be observed as the attention paid for meeting the requirements of performance [3], researching the differences of operator's attention to information under different perceptual loads can offer effective references for the further study of selected attention mechanism; moreover, it may be applied to situation awareness (SA) analysis, vigilance evaluation and perceptual load assessment of pilot in complex tasks.

Up to now, the existing studies on pilot attention activities have been confined to the attention behavior itself and the latter decision-making [4, 5]; however, the stage of pilot preattentive processing has been more or less ignored. Preattention which happens before human consciousness belongs to unconscious attention, reflecting the automatic information processing of cerebral cortex, which is considered as the beginning or preparing stage of selective attention. Therefore, the automatic processing level of information in the preattention stage may significantly affect the later selective attention behavior as well as the cognitive processing of information. Converging evidence showed that the MMN is the main electrophysiological index of preattentive processing which effectively mirrors the brain's automatic detection ability of information varieties [6]. Recently, similar studies that have been carried out mainly investigated the effect of task difficulty on auditory MMN under the single or dual task conditions [7 - 10]. Most of these researches are based on abstract psychological experiments, and show a lack of aviation background. However, due to the complexity and risk of fight operations, the actual utility of such research conclusions needs to be verified in an in-flight environment.

In the previous study, multitasks which including monitor, calculation, memory, attack and detection were researched associated with different perceptual loads during simulation flight, and the perceptual loads were manipulated by changing the working speed rhythm. The experiment manifested that the task difficulty was directly proportional to perceptual load. Besides, the frequency and high oscillation amplitude waves of electroencephalogram (EEG) can reflect the perceptual load intensity to some extent [11]. According to measuring and comparing the auditory MMN under different perceptual loads, the effect of different perceptual loads related to flight task on preattentive processing has been discussed in the present study. Besides, as the exogenous components such as N1 and P2 are also affected by psychological and physiological activities, and their perceptual load effects have not been systematically investigated [12]. Therefore, the perceptual load effects of auditory early components N1 and P2 are studied simultaneously, hoping to propose a kind of pilot perceptual load assessment method based on auditory ERP technology.

2 Methods

2.1 Subjects

As 10 to 15 participants are suggested for obtaining reliable data [13], 14 male flying cadets from Beijing University of Aeronautics and Astronautics participated in the present study. All subjects (ranging from 22 to 28 years, mean age 25. 6 years) were right-handed with normal or corrected to normal vision. Written informed consent was obtained prior to the study.

2.2 Flight Simulation Task

The subjects were asked to perform the whole dynamic process of flight simulation in a flight simulator, including take-off, cruise and landing. It took approximately 13 minutes for once flight simulation task. During the simulation flight, each subject was instructed to keep monitoring the flight indicators presented on the head-up display. When abnormal information (i. e. , visual target) was detected, subject pressed a certain key to recover the information state as quickly and accurately as possible. The perceptual load of the subject was manipulated by setting the quantity of flight indicators and the refresh frequency of information [14]. The parameters of the high (quantity of flight indicators: 9; duration of abnormal information: 1s; inter-stimulus interval between abnormal information: 0. 5 s) and low (quantity of flight indicators: 3; duration of abnormal information: 2 s; inter-stimulus interval between abnormal information: 2 s) perceptual loads were set before experiment.

2.3 Oddball Task

During visual tasks, auditory stimuli were presented binaurally through headphones with an oddball paradigm, in which 1,200 standards (frequency: 1,000 Hz; 65 dB SPL; probability: 80%) and 300 deviants (frequency: 1,100 Hz; 65 dB SPL;

probability: 20%) were included. All the auditory stimuli were presented for 50 ms with the stimulus onset asynchrony (SOA) of 500 ms. The MMN can be elicited in both attended and unattended conditions [15]. As the preattentive processing of auditory information was researched in the present study, elicitation method under involuntary attention condition was adopted. Therefore, the subjects were instructed to ignore these auditory stimuli and pay attention to the flight simulation task.

2.4 Experimental Design

All the subjects conducted two flight simulation tasks under the high and low perceptual load conditions respectively, and the task order was counterbalanced across participants. Due to the reversed polarities, in addition, repeated measures ANOVAs were employed for the analysis of the fronto-central MMN and the temporal MMN, respectively. For the fronto-central region, perceptual load (high; low), brain region (left: F3, FC3, C3; middle: FZ, FCZ, CZ; right: F4, FC4, C4) and site (frontal: F3, FZ, F4; fronto-central: FC3, FCZ, FC4; central: C3, CZ, C4) were as the within-subject factors. For the temporal region, the MMN was analyzed by the perceptual load (high; low) and site (M1; M2) as the within-subject factors. For the N1 and P2 components in the fronto-central region, the same statistical analysis method which has been used for fronto-central MMN was adopted.

2.5 EEG Recording and Analysis

EEG signals were recorded from 30 electrode sites (F7, FT7, T3, TP7, T5; FP1, F3, FC3, C3, P3, O1; FZ, FCZ, CZ, CPZ, PZ, OZ; FP2, F4, FC4, C4, P4, O2; F8, FT8, T4, TP8, T6; M1, M2) using Neuroscan Nuamps amplifier. The tip of nose was as the reference. The vertical and horizontal electro-encephalogram (EOG) signals were also recorded. Both the EEG and EOG signals were continuously recorded with an online band-pass filter of 0.1—100 Hz and the sample rate of 500 Hz. After the eye movements were corrected, the EEG data were segmented by the epoch of 450 ms, including a 50 ms pre-stimulus baseline. Any epoch with artifact voltages exceeding $\pm 70\ \mu V$ was rejected. The MMN waveforms were obtained by subtracting the ERPs elicited by the standard tones from the ERPs elicited by the deviant tones.

3 Results

3.1 Behavioral Assessments

Comparisons of the behavioral performances that included the accuracy rates of detecting abnormal information and reaction times between different perceptual loads were conducted by one-way repeated measures ANOVA. As expected, compared with the low perceptual load (96.1%; 696.8 ms) the accuracy rate decreased significantly (65.4%; $p < 0.001$) and the reaction time was longer (736.6 ms; $p = 0.05$) under the high perceptual load.

3.2 ERP Results

According to the grand average MMN waveforms presented in Fig. 1, the selected time window of MMN was set as 150 – 300 ms post-stimulus. As the peak amplitude of MMN was not clear across subjects, the average amplitude of MMN was selected as the evaluation index. The whole time window were divided into three segments (150 – 200 ms, 200 – 250 ms, 250 – 300 ms) and the MMN average amplitudes of each segment were analyzed statistically.

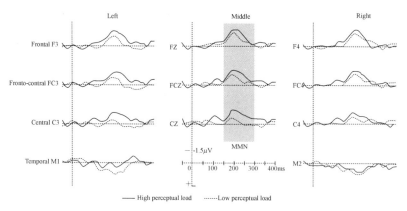

Fig. 1　The grand average MMN waveforms under the high (thick line) and low (dashed line) perceptual loads at the fronto-central and temporal electrode sites

For the fronto-central region, three-way repeated measures ANOVAs manifested that the main effect of perceptual load was significant for the MMN average amplitude ($p = 0.025$) in 200 – 250 ms time window, showing that the MMN was enhanced under the high ($-0.89 \mu V$) than low ($-0.43 \mu V$) perceptual load. The main effect of brain region was also significant, especially in 150 – 200 ms ($p = 0.021$) and 200 – 250 ms ($p = 0.028$) time window, reflecting a right hemisphere distribution for the right (150 – 200 ms: $-0.59 \mu V$; 200 – 250 ms: $-0.71 \mu V$), middle (150 – 200 ms: $-0.57 \mu V$; 200 – 250 ms: $-0.68 \mu V$) and left (150 – 200 ms: $-0.41 \mu V$; 200 – 250 ms: $-0.59 \mu V$) brain regions respectively, as shown in Fig. 2. In addition, the main effect of site was significant in 150 – 200 ms ($p = 0.001$) and 200 – 250 ms ($p < 0.001$) time window, showing a fronto-central distribution of MMN for the frontal (150 – 200 ms: $-0.65 \mu V$; 200 – 250 ms:

$-0.84\,\mu\text{V}$), fronto-central ($150-200\,\text{ms}$: $-0.53\,\mu\text{V}$; $200-250\,\text{ms}$: $-0.68\,\mu\text{V}$), and central ($150-200\,\text{ms}$: $-0.38\,\mu\text{V}$; $200-250\,\text{ms}$: $-0.46\,\mu\text{V}$) sites, as shown in Fig. 2.

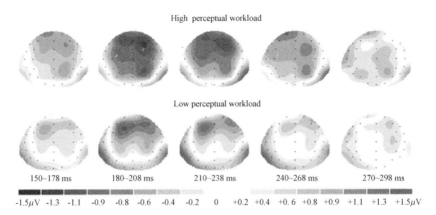

High perceptual workload

Low perceptual workload

150~178 ms   180~208 ms   210~238 ms   240~268 ms   270~298 ms

-1.5µV   -1.3   -1.1   -0.9   -0.8   -0.6   -0.4   -0.2   0   +0.2   +0.4   +0.6   +0.8   +0.9   +1.1   +1.3   +1.5µV

Fig. 2   The 2D-mapping distribution of MMN amplitudes between 150 and 300 ms under the high and low perceptual loads

In $250-300\,\text{ms}$ time window, the perceptual load × brain region interaction was significant ($p=0.003$). Under the low perceptual load, the MMN average amplitude was larger under the right ($-0.34\,\mu\text{V}$) than middle ($-0.12\,\mu\text{V}$) brain region, and the MMN average amplitude of the middle brain region was larger than that of the left ($0.11\,\mu\text{V}$) brain region. In the left brain region, the MMN average amplitude was larger under the high ($-0.49\,\mu\text{V}$) than low ($0.11\,\mu\text{V}$) perceptual load. In $250-300\,\text{ms}$ time window, the perceptual load × site interaction was significant ($p=0.002$), which showed that the MMN average amplitude of the frontal ($-0.23\,\mu\text{V}$) sites was larger than that of the central ($-0.01\,\mu\text{V}$) sites under the low perceptual load.

For the temporal region, two-way repeated measures ANOVAs showed that the MMN average amplitude did not differ significantly between different perceptual loads ($150-200\,\text{ms}$: $p=0.320$; $200-250\,\text{ms}$: $p=0.128$; $250-300\,\text{ms}$: $p=0.229$) and sites ($150-200\,\text{ms}$: $p=0.874$; $200-250\,\text{ms}$: $p=0.964$; $250-300\,\text{ms}$: $p=0.502$) in $150-300\,\text{ms}$ time window. The interaction effects between perceptual load and the site ($150-200\,\text{ms}$: $p=0.272$; $200-250\,\text{ms}$: $p=0.199$; $250-300\,\text{ms}$: $p=0.179$) also failed to reach significance in the selected time window.

For the auditory early components N1 (latency range of $60-140\,\text{ms}$ post-stimulus) and P2 (latency range of $160-240\,\text{ms}$ post-stimulus) in fronto-central region, the peak amplitude and peak latency were chosen as the evaluation indexes. The grand average N1 and P2 waveforms under different perceptual loads were shown as Fig. 3. For the peak amplitude and peak latency of N1 elicited by standards, three-way repeated measures ANOVAs manifested that the main effect of perceptual load was not significant ($p = 0.242$ and $p = 0.339$, respectively). For the peak amplitude and peak latency of N1 elicited by deviants, the main effect of perceptual load was not significant ($p = 0.430$ and $p = 0.214$, respectively). For the peak amplitude and peak latency of P2 elicited by standards, the main effect of perceptual load was not significant ($p = 0.258$ and $p = 0.325$, respectively). For the peak amplitude and peak latency of P2 elicited by deviants, the main effect of perceptual load was not significant ($p = 0.625$ and $p = 0.214$, respectively).

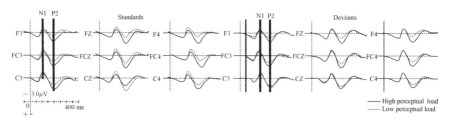

Fig. 3　The grand average N1 and P2 waveforms under the high (thick line) and low (dashed line) perceptual loads at the fronto-central electrode sites

3.3　Correlation Test

In $200 - 250\,\text{ms}$ time window, the site FZ, FCZ and CZ were selected to test the correlation between the MMN average amplitude and task performance. At FZ, FCZ and CZ sites, there were positive correlations between the MMN average amplitudes and the accuracy rates ($r = 0.39$, $p < 0.05$; $r = 0.39$, $p < 0.05$; $r = 0.46$, $p < 0.05$, respectively), but not between the MMN average amplitudes and the reaction times ($r = -0.14$, $p > 0.05$; $r = -0.22$, $p > 0.05$; $r = -0.25$, $p > 0.05$, respectively). As the main effect of perceptual load was not remarkable for the temporal MMN, N1 and P2, the correlation tests between the ERPs and behavioral data were not performed.

4　Discussions

The behavioral data showed that the difference of performances under the high and low perceptual loads was obvious. Increasing perceptual load was associated with lower accuracy rate of detecting abnormal information and longer reaction time. As the quantity of flight indicators which needed to be monitored simultaneously was increased under the high perceptual load, the average attention resource that allocated to a certain flight indicator was decreased.

The ERP results demonstrated that the MMN at the mastoid sites exhibited an inverted component compared to the MMN at the fronto-central region. Besides, the MMN average amplitudes of the right brain hemisphere were larger than the left one, and the MMN was significantly elicited in the frontal region, which were consistent with the well-know conclusions [16]. Furthermore, the ERP results indicated that the fronto-central MMN was affected by perceptual load obviously. Compared with the low perceptual load, the MMN average amplitude under the high perceptual load increased significantly, and there was a positive correlation between the MMN average amplitude and the accuracy rate during simulation flight. However, the temporal MMN, N1 and P2 were not significantly modulated by the perceptual load.

It is accepted widely that there are at least two intracranial resources for generating MMN. First, a bilateral supratemporal automatic process of information varieties generates the supratemporal MMN subcomponent. Then, a predominantly right-hemispheric frontal process of information varieties generates the frontal MMN subcomponent and the subsequent P3a which associates with the involuntary attention orientation [6]. In contrast with the perceptual load hypothesis based on the passive resource competition, the cognitive control theory proposed that some advanced cognitive control functions can keep the priority of information processing order, which enables the task-relevant information be processed at first. At present, the cognitive control theory has been proved in several behavioral experiments [17]. Moreover, the frontal cognitive control function is considered to play an important role in the active cognitive control mechanism, which can select the useful information and inhibit task-irrelevant information effectively [18]. In the present study, the average amplitude of fronto-central MMN increased significantly under the high perceptual load, showing that the task-irrelevant information (i. e., auditory stimuli) got more processing. The probably reason is that critical task demands and higher time pressure resulted in the increase of attention load, which consumed the resources used for actively suppressing task-irrelevant information processing. Therefore, a decrease of subject's gating ability to the task-irrelevant information led to the enhancement of MMN component. Clearly, this result offers strong neurophysiological support for the prediction by the cognitive control theory and is in line with reference [9]. Compared with the current study, the absence of obvious discrimination between different perceptual loads may led to the failure of finding the perceptual load effect on MMN in some previous studies.

In our experiment, the perceptual load did not affect the temporal MMN. Generally, the temporal region is important for distinguishing and storing auditory information but cannot reject the task-irrelevant information. Therefore, the temporal MMN was not sensitive to perceptual load.

Relative research proposed that the N1 and P2 components were sensitive to the perceptual loads with different types and difficulties, however, whether the N1 and P2 were sensitive to the perceptual loads with the same type and different difficulties need to be further tested [12]. According to the results in our study, there were no obvious relations between the N1/P2 components and perceptual loads with the same type and different difficulties, which manifested the complexity and difficulty of measuring the perceptual load related to flight task by ERP method to some extent.

Presently, there are still some drawbacks existing in the study. Firstly, It should be noticed that the task performance adopted in the present study is just one of the main aspects of perceptual load, and the correlation between the fronto-central MMN and the perceptual load should be further tested combining with the comprehensive workload gauges such as the SWAT (subjective workload assessment technique) or NASA - TLX (national aeronautics and space administration-task load index). Secondly, although the experiment performed in the present study was based on the flight simulator and dynamic flight tasks, there exists a gap compared with the actual flight mission and environment. Therefore, the obtained conclusions need to be verified in the practical applications. Despite this, the fronto-central MMN shows good sensitivity to the perceptual load related to flight task. It may provide effective electrophysiological evidence for flight perceptual load assessment in complex tasks.

5 Conclusions

In conclusion, the current study investigated the auditory information processing under different perceptual loads related to flight task as indexed by MMN, N1 and P2. The fronto-central MMN was modulated by perceptual load significantly, and MMN was enhanced under the high than low perceptual load, reflecting an improvement of information automation processing. Positive correlation was found between the MMN average amplitude and the performance, suggesting that the fronto-central MMN may offer an effective index for the assessment of flight perceptual load. Perceptual load had little effect on the temporal MMN, N1 and P2. As the MMN was elicited in involuntary attention conditions and there was no need for subject to identify the probes initiatively, the intrusion to the primary task falls to the lowest so that the practical application can be assured.

Acknowledgments

This study is supported by National Basic Research Program of China (Program Grant No. 2010CB734104). The authors wish to thank the Visual Art & Brain Cognition Lab and the Beijing SKYL Technology Co. Ltd. for their valuable suggestions. The authors also thank the anonymous reviewers for their helpful comments to improve the manuscript.

References

[1] Guo X C, Liu B S, Ma X S, et al. General information of display and its priority for advanced fighter cockpit [J]. Journal of Beijing University of Aeronautics and Astronautics, 2006,17(4):260 – 263.

[2] Wickens C D, Lee J D, Liu Y, et al. An introduction to human factors engineering [M]. New Jersey: Prentice Hall, 2004.

[3] Young M S, Stanton N A. Mental wokeload: theory, measurement, and application [M]. Karwowski W (ed.), International Encyclopedia of Ergonomics and Human Factors. London: Taylor & Francis, 2001:507 – 509.

[4] Miller S M, Kirlik A, Kosorukoff A, et al. Ecological Validity as a Mediator of Visual Attention Allocation in Human-Machine Systems [S]. AHFD– 04 – 17/NASA– 04 – 6, 2004.

[5] Liu Z Q, Yuan X G, Liu W, et al. Eye-movementanalysis of simulating airplane landing course [J]. Journal of Beijing University of Aeronautics and Astronautics, 2002,28(6): 703 – 706.

[6] Naatanen R, Paavilainen P, Rinne T, et al. The mismatch negativity (MMN) in basic research of central auditory processing: A review [J]. Clinic Neurophysiology, 2007,118 (12):2544 – 2590.

[7] Yucel G, Petty C, McCarthy G, et al. Visual task complexity modulates the brain's response to unattended auditory novelty [J]. NeuroReport, 2005,16:1031 – 1036.

[8] Kramer A F, Trejo L J, Humphrey D. Assessment of perceptual load with task-irrelevant auditory probes [J]. Biological Psychology, 1995,40(1 – 2):83 – 100.

[9] Zhang P, Chen X C, Yuan P, et al. The effect of visuospatial attentional load on the processing of irrelevant ascoustic distractors [J]. Neuroimage, 2006,33(2):715 – 724.

[10] Muller-Gass A, Stelmack R M, Campbel K B. The effect of visual task difficulty and attentional direction on the detection of acoustic change as indexed by the mismatch negativity [J]. Brain Research, 2006,1078:112 – 130.

[11] Zeng Q G, Zhuang D M, Ma Y X. Perceptual load and Target Identification [J]. Acta

Aeronautica et Astronautica Sinica，2007，28(Sup.)：76 - 80.

[12]　Zhao L，Wei J H，Jiang Y，et al. The effects of task load on early and late components of auditory event related potentioals [J]. Space Medicine & Medical Engineering，2002，15(1)：12 - 16.

[13]　Luck S J. An introduction to the event-related potential technique [M]. Cambridge：MIT Press，2005.

[14]　Zhang L，Zhuang D M，Wanyan X R. The color coding of information based on different perceptual load and task type [J]. Acta Armamentarii，2009，30(11)：1522 - 1526.

[15]　Xiao J，Huang W，Wei J H. The Effects of Selective Attention on ERPs Elicited by Visual & Auditory Deviant Stimuli [J]. Acta Scientiarum Naturalium Universitatis Pekinensis，1999，35(3)：421 - 428.

[16]　Marco-Pallares J，Grau C，Ruffini G. Combined ICA-LORETA analysis of mismatch negativity [J]. Neuroimage，2005，25：471 - 477.

[17]　Lavie N. Distracted and confused?：Selective attention under load [J]. Trends in Cognitive Neuroscience，2005，9(2)：75 - 82.

[18]　Lavie N，Hirst A，de Fockert J W，et al. Load theory of selective attention and cognitive control [J]. Journal of Experimental Psychology：General，2004，133(3)：339 - 354.

Study of Fatigue Measurement Based on Eye Tracking Technique

Wang Lei

Research Institute of Civil Aviation Safety, Civil Aviation
University of China, Tianjin, China

Abstract   This paper aimed to introduce eye tracking concepts and technique to study fatigue measurement issues in aviation field. A fatigue measurement method based on PERCLOS value calculation was proposed out in this paper and its frame, process and algorithms were discussed meanwhile. Then the feasibility and reliability of using PERCLOS value as the fatigue judgment index were studied by eye tracking experiment. The experiment results showed that PERCLOS value is a perfect index for measuring fatigue and the threshold was suggested to be set as 0.5 when eye close threshold is set as 70%. Finally it concluded that the fatigue measurement method based on eye tracking is effective and reliable, and the applicable Fatigue Measurement System was expected to be developed and used in future.

Keywords   Fatigue measurement, eye tracking, PERCLOS, aviation

1   Introduction to Fatigue Issues in Aviation

Fatigue refers to a degradation of mental and physical abilities and a demotion of emotional status due to the isolated or combined effects of insufficient sleep, working/resting against the body's natural circadian rhythms, and certain aspects of the work demands and workload such as time on task, ergonomic considerations and so on [1, 2].

Fatigue is seemed as a threat to aviation safety because of the impairments in alertness and performance it creates. Fatigue risk exists widely among flight and cabin crews, air traffic controllers, technicians, mechanics, dispatchers and ramp workers. Especially for pilots and controllers who should suffer the tire brought by sleep loss, night and shift work, and long duty cycles, and also the pressure of remaining alert by their actions, observations and communications. As 70% of fatal accidents in commercial aviation relate to human error the risk of fatigue contribute 15%-20% to the overall accident rate [3]. NASA is cited as confirming that "fatigue plays a part in anywhere between 3.8% and 21% of aviation accidents" [4]. For example, fatigue played an important role in the crash of Comair Flight 5191, according to US NTSB's report where pointed out that lack of sleep hindered the performance of three important players in the crash — the captain, the co-pilot and the air traffic controller [5]. The US FAA has now recognized that "incorporating fatigue risk management systems into everyday operations is the ultimate goal, but doing so will take innovation in addressing a myriad of regulatory issues" [6]. The latest recommendation provides that the FAA oversee the implementation of a "fatigue management system" that would address "the problems associated with fatigue in an operational environment" and "take a comprehensive, tailored approach to the problem of fatigue" within the industry [7]. Not only in US, but also in other countries or organizations the fatigue issue attracted more and more attentions in recent years. For example EASA implemented new regulation on Flight Time Limitations (FTL) for controlling flight fatigue in 2008, IATA established its own FTL Task Force under its Operations Committee and ICAO proposed to mandate pilot fatigue risk management and addressed crew member fatigue in ANNEX 6 [8, 9].

Generally to say, fatigue risks exist in aviation industry universally and chronically. Especially in the countries or regions where there is large air traffic volume or air transport volume is increasing rapidly, the fatigue problem is becoming more and more visible and serious. Monitoring and managing fatigue risk will be a new and effective way of improving aviation safety.

2 Summary on Methods of Measuring Fatigue

Fatigue measurement is the first and key step of managing fatigue risks. Currently most of fatigue measurement research focused on road driving fatigue and there are few research products referred with aviation operations.

Among those previous researches, the methods of measuring fatigue generally were divided as subjective measurement and objective measurement. The subjective measurement of fatigue is based on multi-dimensional scaling such as Pearson Fatigue Scale, Fatigue Severity Scale, and Fatigue Impact Scale and so on [10,

11]. Though subjective measurement method is easy to use, it is difficult to quantify fatigue degree accurately. Meanwhile the measurement result is not real time display and its reliability could be influenced by personnel difference easily.

The objective measurement is based on the physiological phenomena and which can be accomplished by two ways. One way is to measure the changes of physiological signals, such as brain waves, heart rate, pulse rate, and skin electric potential, as means of detecting a drowsy situation [12]. The approach is suitable for making accurate and quantitative judgments of alertness levels; however, it must annoy testee to attach the sensing electrodes on the body directly. Thus, it would be difficult to use based on the sensors under real-world condition. The approach has also the disadvantage of being ill-suited to measure over a long period of time owing to the large effect of perspiration to the sensors. The other way focuses on the physical changes, such as the inclination of the testee's head, sagging posture, or the open/closed state of the eyes. The measurements of these physical changes are classified into the contact and the non-contact types. The contact type involves the detection of movement by direct means, such as using a hat or eye glasses or attaching sensors to the testee's body. The non-contact type uses optical sensors or video cameras to detect the changes. The non-contact or non-intrusive method of measuring fatigue mainly depends on the face recognition and eye recognition technique. Face recognition is one of the few biometric methods that possess the merits of both high accuracy and low intrusiveness [13]. It has the accuracy of a physiological approach without being intrusive. For this reason, the methods of monitoring fatigue based on face and eye recognition techniques were researched continuously in the past decades. There were many results in road driving field but few of them were used in aviation industry. The comparison of above fatigue measurement methods is as the Tab. 1.

Tab. 1 Comparison of fatigue measurement methods

Method	Is it real time?	Is it intrusive?	Accuracy	Operation Convenience
Questionnaire Survey	No	No	Low	Good
Reaction Time Measurement	No	No	High	Not good
Physiological Signal Measurement	Yes	Yes	High	Not good
Eye-moving Feature Measurement	Yes	No	High	Good

Comprehensively seeing from above table, it is easily to find that the method of measuring eye-moving feature which has the characteristics of real time, non-intrusive measuring, high accuracy and good operation convenience is an applicable and reliable way to detect fatigue.

3 Fatigue Measurement System Based on Eye Tracking

3.1 Framework of Fatigue Measurement System

Combining with face recognition and eye tracking technique, a framework of real time Fatigue Measurement System (FMS) was proposed out as follow Fig. 1.

The system consists of four components: face detection, eye detection, eye tracking, and fatigue detection. Firstly a frame of image will be captured from an ordinary camera or video files. The first frame will be processed by using image process methods and then used for initial face detection and eye location. If any one of these detection procedures fails, then go to the next frame and restart the above detection processes. Otherwise, the current eye images are used as the dynamic templates for eye tracking on subsequent frames, and then the fatigue detection process is performed. If eye tracking fails, the face detection and eye location restart on the current frame. These procedures continue until there are no more frames. When all frames in a time limit were detected over and a statistic regarded with eye-closed frames will be made later. The statistic value is called PERCLOS which is an accredited threshold of judging fatigue.

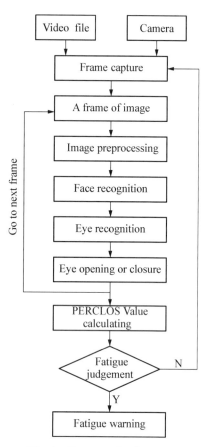

Fig. 1　Framework of Fatigue Measurement System

3.2 Hardware Requirements

The Fatigue Measurement System is mainly made up of a camera and a processor. The camera could be a common CMOS or CCD lens and the processor could be the PC or Digital Signal Processor. Surely a video capture device is needed to record video if the DSP is chose. In addition, the IR light could also be select hardware for better detecting and tracking.

3.3   Recognition Algorithms

Finding an accurate and speedy algorithm is an important task in the process of developing this system. A lot of algorithm has been developed in the area of face and eye recognition. Among them the algorithms based on skin color segmentation are generally accepted and well-used. Here we won't explain any algorithm in detail but give out the common calculation principles of detecting face and eye. The face recognition is performed in three steps. The first step is to classify each pixel in the given image as a skin pixel or a non-skin pixel. The second step is to identify different skin regions in the skin-detected image by using connectivity analysis. The last step is to decide whether each of the skin regions identified is a face or not. After the probable location of the face is found the left and the right edges of the face are determined.

The Template Matching Method and Gray Projection Method are two common methods of positioning the eyes and they are often combined to use together [14]. Generally the first step is to locate two rough regions of eyes by using an efficient feature based on Gray Projection Method. Then on the basis of these two regions, the sizes of two eyes will be evaluated, and the templates of eyes will be created according to the estimated sizes. Finally, the precise locations of the two centers of iris will be found out after template matching is applied in these two rough regions.

4   Perclos

PERCLOS means the percentage of the intervals with closed eyes in a fixed time window, disregarding regular blinks. Walt Wierwille first studied PERCLOS in the 1980s and 1990s, studies have shown that fatigue related with lacking sleep, pupil diameter, staring eyes, the rapid rotation of eyeball, eyes sweeping as well as other factors [15]. PERCLOS is the most potential and best way of fatigue menstruation and its data can really represent fatigue. In April 1999 the United States Federal Highway Administration convenes experts and scholars to contrast the validity of PERCLOS and other measure methods of eyes activities. The study considers that priority should be given to the measurement of PERCLOS of motor vehicles drivers, and this can be real-time, non-contact method for the fatigue evaluation [16].

The definition of PERCLOS has shown that the PERCLOS value is to calculate the percentage of eye closure time in a unit time. As far as the Fatigue Measurement System is concerned, the frequency of frame capture is over 20 fps which is higher than regular eye blink. So the simple way of calculating PERCLOS is to calculate the proportion of eye closure frames to the total frame number in a time limit.

$$ p = \frac{\sum f_c}{f_t} $$

(f_c means eye closure frame, f_t means total number frames in a time limit.)

But whichever algorithm we choose, the final calculation result will be affected by a parameter which is the threshold c for judging eye closure. Typical values for the threshold c are 70%, 80% and 100% which respectively means the degree of eyelid dropping down and covering iris is 70%, 80% and 100%. Obviously the PERCLOS value will change along with the change of c, and then the threshold of judging fatigue will be different at the same time. In next section we will take experiments to study PERCLOS changes under different c and try to make a more precise PERCLOS threshold to judge fatigue.

5   Experiment and Analysis on Perclos

5.1   Objective

The objective of the experiment is to study the PERCLOS change in case of fatigue & non-fatigue state and also different eye closure threshold. It also aimed to find an appropriate PERCLOS value for judging fatigue.

5.2   Instruments

The Instrument used in the experiment is FaceLAB Eye Tracker which is made up of 1 portable computer and 2 advanced cameras. FaceLAB eye tracking system could produce accurate head and eye-movement data on-time, in the toughest of tracking environments. It also delivers data with the addition of a real-time PERCLOS fatigue assessment including raw data on the minutia of eyelid behavior. The data rate or sampling frequency of the system is 60 Hz.

5.3   Steps

2 volunteers were involved in the experiments to receive testing. 4 sets of eye movement data were collected in the condition of 2 volunteers' fatigue and non-fatigue state. Then 4 sets of PERCLOS value were selected out and analyzed for finding their change rules and confirming the threshold of judging fatigue. The steps are as following.

- Firstly to collect data in non-fatigue condition and ensure the volunteers' physical and mental status is normal instead of tired or sleepy.

- To brief experiment objective, process and attention tips to the 2 testers.
- To debug experiment instruments and create Stereo Head Model of eye tracker.
- To let 2 testers take the seat in sequence and keep right position in the front of 2 cameras, set up their Face Model and adjust the tracking quality.
- To set the Eye Closure Threshold (ECT) as 70%, 80% and 90% respectively, and then log 1 minute of data when tracking became steady.
- Then to collect data in fatigue condition and repeat Step 4 and 5.

5.4 Results and Discussions

The average PERCLOS value was calculated by using the recorded data and the final result is as Tab. 2.

Tab. 2   Calculation result of PERCLOS value

	Eye Closure Threshold/%	PERCLOS (non-fatigue)	PERCLOS (fatigue)
Testee 1	70	0.003	0.564
	80	0.002	0.189
	90	0.001	0.089
Testee 2	70	0.002	0.564
	80	0.002	0.148
	90	0.001	0.137

As seen from the above table, we can find that:

- Whatever the Eye Closure Threshold we chose, the PERCLOS value changed greatly from non-fatigue to fatigue condition and the discrepancy was almost over 100 times.
- The Eye Closure Threshold would also affect PERCLOS value. Especially in fatigue case, there is remarkable difference on PERCLOS value between 70% ECT and 80% ECT. It is as showed in Fig. 2 and Fig. 3.
- Combining with 2 points above, we can conclude that it is reliable to use PERCLOS value as the index of judging fatigue. The Eye Closure Threshold is suggested to set as 70%, and then the PERCLOS threshold of judging fatigue would be set as 0.5 accordingly.

6   Conclusions

Fatigue risks exist in aviation industry universally, especially for flight

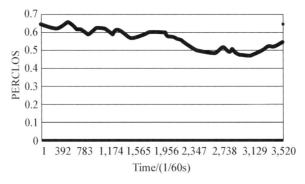

Fig. 2　PERCLOS change of fatigue when ECT was set as 70%

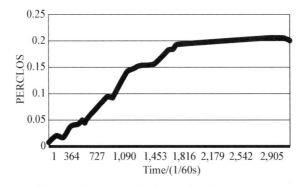

Fig. 3　PERCLOS change of fatigue when ECT was set as 80%

operation and ATC operation where is confronted with ruleless working hours and high working pressure. The fatigue problem becomes more visible and serious along with rapid increasing of air transport volume. Most of civil aviation authorities have begun to pay more attention on fatigue and some have proposed to develop Fatigue Risk Management System and integrate it into Safety Management System.

Currently few methods and tools have been developed and applied into aviation operation practice, but lots of research has proved that it is a reliable and applicable method to measure fatigue by using face recognition and eye tracking technique.

A fatigue measurement method based on PERCLOS value calculation was proposed out in this paper and its frame, process and algorithms were discussed meanwhile. The feasibility and reliability of using PERCLOS value as the fatigue judgment index was studied. The experiment results showed that PERCLOS value is a perfect index for detecting fatigue and the threshold was suggested to be set as 0.5 when eye close threshold is set as 70%.

The fatigue measurement method put forward in this paper is expected to be a foundation of developing applicable Fatigue Monitoring and Warning System which could be used in aviation practice. The work of next step will mainly focus on developing face and eye recognition algorithm to realize real time measurement on eye movement and fatigue state. The final research product would be an effective tool to manage fatigue risks in aviation.

Acknowledgements

This research was supported by the National Basic Research Program of China (No. 2010CB734105) and Research Fund (No. 09CAUC_E08) of Civil Aviation University of China.

References

[1]　Gander P H, Rosekind M R, Gregory K B. Flight crew fatigue Ⅵ: an integrated overview [J]. Aviation, Space, and Environmental Medicine 1998,69:B49 – B60.

[2]　Dawson D, McCullough K. (2004). Managing Fatigue as an Integral part of a Fatigue Risk Management System [C]. ISASI, 2004, Australia.

[3]　Bennett S A. Flight Crew Stress and Fatigue in Low-Cost Commercial Air Operations — An Appraisal [J], International Journal of Risk Assessment and Management, 4:207 – 231.

[4]　Bennett S A. Pilot Stress and Fatigue in Low-Cost Operations, Journal of the Institute of Civil Defense and Disaster Studies, Summer 2004, London: Institute of Civil Defenses and Disaster Studies, http://www. icdds. org/downloads/aler%20june%202004. pdf.

[5]　NTSB: Air controller fatigue contributed to 4 mishaps. http://www. cnn. com/2007/US/04/10/controller. fatigue/index. html.

[6]　NTSB Most Wanted Transportation Safety Improvements Federal Issues http://www. ntsb. gov/recs/mostwanted/fatigue. htm.

[7]　The NTSB and FAA Address Fatigue in Aviation Operations. http://www. faa. gov/news/press_releases/news_story. cfm? newsId=10246.

[8]　Cabon P, Mollard R, Debouck F, et al. Toward a fatigue Risk Management System: application for the regional French airlines [C]. In : Proceedings of the 79th Aerospace Medical Association Annual Scientific Meeting, Boston, MA, 4 – 7 May 2008.

[9]　Signal T L, Ratieta D, Gander P H. Flight crew fatigue management in a more flexible regulatory environment: an overview of the New Zealand aviation indus-try. Chronobiology International, 2008,25(2 – 3):373 – 388.

[10]　Guo B Y, Fang W N. Physiological Fatigue Determination Based on Percentage of Eyelid Closure with Eye Tracking System [J]. Chinese Journal of Clinical Rehabilitation. 2005. 7,9 (26):246 – 248.

[11]　AWAKE-System for Effective Assessment of Driver Vigilance and Warning According to Traffic Risk Estimation, http://www. awake-eu. org/system. html.

[12] Liu Y L. Driver Fatigue Monitoring Method based on Eye State Classification [C]. 2008 Chinese Control and Decision Conference，2008，6：2257 - 2260.

[13] Liu G. Research on Driver Fatigue Detection based on Video [D]. Xi'an University of Technology. Xi'an，2006，26 - 28.

[14] Jones，Computer M J，Viola P. Fast Multi-view Face Detection [C]，IEEE Conference on Vision and Pattern Recognition (CVPR)，June，2003.

[15] Carlos H M，Marcio R M M. Eye gaze tracking techniques for interactive applications [J]. Computer Vision and Image Understanding，2005，98：4 - 24.

[16] Zheng P. Recognition Algorithm of Driver Fatigue Detection based on PERCLOS [J]. Agriculture University of China，2002，7(2)：104 - 109.

中航工业金城南京机电液压工程研究中心

Nanjing Engineering Institute of Aircraft Systems, Jincheng, AVI

实现客户的期望
追求航空人的理想

Realize customers' expectation
Pursue aviation people's ideal

传承优势，创新发展，成为国内领先、国际一流的机电研究中心

Carrying on advantages, development on innovation, building a world-class
electromechanical R&D center.

地址：中国.江苏省南京市水阁路3
电话：025-51819021
传真：025-51819900
邮箱：neias@avic.com

中航工业金城南京机电液压工程研究中心

NANJING ENGINEERING INSTITUTE OF AIRCRAFT SYSTEMS，JINCHENG，AVIC

中航工业金城南京机电液压工程研究中心是由原中国航空附件研究所整体与原金城集团有限公司航空业务部分组成，是我国航空机载机电系统的研发中心和生产基地，具有完整的预先研究、型号研制、设计制造、试验交付和维修服务的手段和能力。中心现有职工 3000 余人，其中研究员 40 余人，高级专业技术人员 300 余人，中级专业人员及工人技师 600 余人。中心设有 7 个研究部、14 个分厂、26 个部(处)机关和 6 个专业子公司。

南京机电液压工程研究中心现有的专业研究领域包括：飞机机电控制系统、飞机液压操纵系统、飞机燃油系统、飞机环境控制系统、飞机第二动力系统、飞机电源传动系统等。

南京机电液压工程研究中心拥有完善的军工产品质量保证体系和计量检测体系，集综合性多专业的技术优势和综合实力，先后为 60 多个机型 37 大机电系统提供产品和配套附件。目前拥有的空中加油装备、恒速传动装置、应急动力装置、空气涡轮起动机、燃气涡轮起动机、高压除水环境控制系统、三轮涡轮冷却器、高性能电液伺服阀、燃油泵、飞机地面操纵系统等核心技术和产品，在国内具有不可替代性；先后获得 4 项国家级科技进步特等奖。

南京机电液压工程研究中心拥有 BOEING 系列、AIRBUS 系列、TY-154、BAE-146、运七等多种机型一千多项附件的民航产品维修许可权，民航机电产品维修居于国内领先地位。

南京机电液压工程研究中心共获省部级以上科技奖 300 多项。其中，国家级科技进步特等奖四项；国家级科技进步奖 20 项。

南京机电液压工程研究中心先后与汉胜、派克、霍尼韦尔、利勃海尔等国际知名公司开展深入合作，组建合资公司，并为波音、空客系列提供燃油、液压、环控等系统及产品。

"传承优势，创新发展，成为国内领先、国际一流的机电研究中心"，是南京机电液压工程研究中心的宏伟愿景，也是全体员工矢志不渝的奋斗目标。

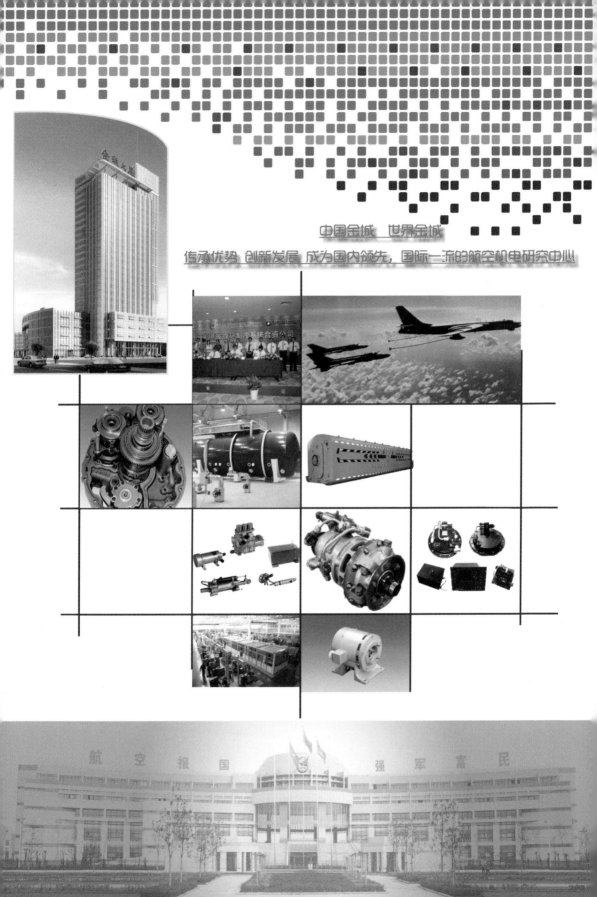

中国金城　世界金城

传承优势 创新发展 成为国内领先，国际一流的航空机电研究中心

航 空 报 国　　　强 军 富 民

地址：南京市江宁区水阁路 33 号　　　邮编：211106　　　传真：025-51819900

大飞机出版工程
书 目

一期书目（已出版）

　　《超声速飞机空气动力学和飞行力学》(俄译中)

　　《大型客机计算流体力学应用与发展》

　　《民用飞机总体设计》

　　《飞机飞行手册》(英译中)

　　《运输类飞机的空气动力设计》(英译中)

　　《雅克-42M 和雅克-242 飞机草图设计》(俄译中)

　　《飞机气动弹性力学及载荷导论》(英译中)

　　《飞机推进》(英译中)

　　《飞机燃油系统》(英译中)

　　《全球航空业》(英译中)

　　《航空发展的历程与真相》(英译中)

二期书目（已出版）

　　《大型客机设计制造与使用经济性研究》

　　《飞机电气和电子系统——原理、维护和使用》(英译中)

　　《民用飞机航空电子系统》

　　《非线性有限元及其在飞机结构设计中的应用》

　　《民用飞机复合材料结构设计与验证》

　　《飞机复合材料结构设计与分析》(英译中)

　　《飞机复合材料结构强度分析》

　　《复合材料飞机结构强度设计与验证概论》

　　《复合材料连接》

　　《飞机结构设计与强度计算》

　　《飞机材料与结构的疲劳与断裂》(英文版)

三期书目

　　《适航理念与原则》

　　《适航性：航空器合格审定导论》(译著)

《民用飞机系统安全性设计与评估技术概论》

《民用航空器噪声合格审定概论》

《机载软件研制流程最佳实践》

《民用飞机金属结构耐久性与损伤容限设计》

《机载软件适航标准 DO－178B/C 研究》

《运输类飞机合格审定飞行试验指南》(编译)

《民用飞机复合材料结构适航验证概论》

《民用运输类飞机驾驶舱人为因素设计原则》

四期书目

《航空燃气涡轮发动机工作原理及性能》

《航空发动机结构》

《航空发动机结构强度设计》

《风扇压气机气动弹性力学》(英文版)

《燃气轮机涡轮内部复杂流动机理及设计技术》

《先进燃气轮机燃烧室设计研发》

《燃气涡轮发动机的传热和空气系统》

《航空发动机适航性设计技术导论》

《航空发动机控制》

《气动声学基础及其在航空推进系统中的应用》(英文版)

《叶轮机内部流动试验和测量技术》

《航空涡轮风扇发动机试验技术与方法》

《航空轴流风扇压气机气动设计》

《燃气涡轮发动机性能》(译著)

其他书目

《民用飞机环境监视系统》

《民用飞机飞行管理系统》

《飞机内部舒适性设计》(译著)

《航空航天导论》

《航空计算工程》

《涡动力学》(英文版)

《尾涡流控制》(英文版)

《动态工程系统的可靠性分析:快速分析方法和航空航天应用》(英文版)

《国际航空法导论》(译著)